THE ECONOMIC LAWS OF SCIENTIFIC RESEARCH

The Economic Laws of Scientific Research

Terence Kealey
Department of Clinical Biochemistry
University of Cambridge

First published in Great Britain 1996 by
MACMILLAN PRESS LTD
Houndmills, Basingstoke, Hampshire RG21 6XS
and London
Companies and representatives
throughout the world

A catalogue record for this book is available
from the British Library.

ISBN 0–333–56045–0 hardcover
ISBN 0-333-65755-1 paperback

 First published in the United States of America 1996 by
ST. MARTIN'S PRESS, INC.,
Scholarly and Reference Division,
175 Fifth Avenue,
New York, N.Y. 10010

ISBN 0–312–12847–9 hardcover
ISBN 0–312–17306–7 paperback

Library of Congress Cataloging-in-Publication Data
Kealey, Terence.
The economic laws of scientific research / Terence Kealey.
p. cm.
Includes bibliographical references and index.
ISBN 0–312–12847–9 (cloth)
1. Research, industrial—Economic aspects—History. 2. Science–
–History. 3. Economic history. I. Title.
HC79.R4K43 1996
330'.072—dc20   95–14914
                                                                              CIP

11   10   9   8   7   6
05   04   03   02   01   00   99   98

*This book is dedicated, alphabetically, to
Carol Clark of Balliol College, Oxford
and to
the McGowan family of Northumberland*

Contents

List of Figures

viii

List of Tables

ix

Preface and Acknowledgements

I have often felt lonely ploughing my *laissez faire* science furrow but I have been supported by countless good friends and I am thrilled to thank them here.

I thank Charles Moore, Andrew Gimson and Dominic Lawson of the *Spectator* for publishing my first articles on science funding, and I thank Oliver Knox of the Centre for Policy Studies for publishing my pamphlet *Science Fiction*. I thank Graham Brady and Nicholas Finney of the Waterfront Partnership for proposing an 'Aims of Industry' award for my studies on the economics of science, and I thank the Prime Minister, Mr John Major, for presenting it (although he spoilt the day for me by launching into an unreasoned defence of ERM and by promising never to leave it – six weeks later, fortunately for Britain, he was forced to). I thank Norman Stone for his Save British Science-bashing pieces which deflected some of the flak from me (scientists who believe in the public funding of their own work can turn nasty). And I thank Jonathan Clark for including an essay of mine in the book he edited, and Macmillan published in 1990, *Ideas and Politics in Modern Britain* (Sally, my wife, does not thank Jonathan Clark as I wrote the essay on our honeymoon).

A number of very distinguished scholars of science policy have poured help at me, an amateur, and I really thank them. Nathan Rosenberg of the Department of Economics, Stanford University, and Tibor Braun of the Information Science and Scientometrics Research Unit, Budapest, were stunningly generous of their time and expertise. Nick Segal of Segal, Quince, Wicksteed, and K. K. Cheung of the Department of Community Medicine, Cambridge University, were both also enormously helpful. Alan Gomersall and Lesley Grayson of the British Library, the editors of *Science and Technology Policy,* have been staunch. Professor Tom Sherwood, the Dean of the Faculty of Clinical Medicine, Cambridge, educated me in Francis Bacon. David Edgerton of the Department of the History of Science, Imperial College, and Robin Matthews of the Department of Economics and of Clare College, Cambridge, read early drafts of the manuscript and provided incisive, kind and shrewd advice. Sadly, I have sometimes ignored the comments of

Rosenberg, Braun, Segal, Edgerton and Matthews. This is my book, faults and all, and only I can be blamed for it.

In my field of clinical biochemistry, I thank Paul Luzio of my department for vast help and I thank Nick Hales, our head of department, for championing my freedom to articulate views with which he disagrees. If more academics were like him and Paul, our universities would be healthier. I thank my colleagues in the group in the lab, Mike Philpott, Robert Guy, Debs Sanders, Darren Thompson, Nick Dove, Sarah Rudman, Cheryl Smythe, Alison Shanks, Sally Darracott-Cankovic, Wyn Rees and Michaela Tillmanns for the beautiful research they have continued to produce while I have been distracted by this book. And I thank Martin Green and our other colleagues in Unilever, Jon Beck, Paul Bowser, Alan Evenson, Walter Gibson, Jim Leslie, Bill Parrish, David Pawson, Ian Scott, Frans Van Der Ouderaa, and Gill Westgate for their collaborations and for their continued, wise, financial support of our biochemical research.

Let me commit an entire paragraph to restating this. Without the support of my colleagues in the lab in Cambridge, and also in Unilever, it would have been impossible to have pursued both my science and this book. I thank them all.

Now for my enemies. It is obviously invidious to name too many, but I must single out two whose protracted campaigns of deviousness, disinformation and frank intellectual sabotage have been disgraceful and deserving of public condemnation. I refer to Helena and Teddy, our children, whose teething, nappy-changing and demands for entertainment have set this book back by at least three years. Another troublemaker is Barbara Langlois, who complained about my incompetence, inconsistency and pedantry throughout her typing and re-typing of this book (she has a horror of references, and maintains that no one ever reads them). This book would never have been completed without her constant, unfailing, humorous kindness. I thank her.

Now for my opponents. I must acknowledge them because they prompted this study. This book was conceived the day Dennis Noble *et alia* helped orchestrate Oxford University's rejection of Mrs Thatcher's honorary degree in 1984 (an episode to which, intriguingly, she made no reference in her memoirs). The Oxford academics claimed that Mrs Thatcher was destroying British science, but as I had been asked to leave Oxford in 1982 because of a shortage of space, I knew their claims were false (our department's labs were only three years old, yet money was so prolific they were already crowded). Newcastle University, my next home, had even newer labs, even more crowded, and

Oxford, when I returned in 1986, pullulated with scientists. Cambridge today is like the Tokyo subway. It is also one big building site. There is no decline of British science.

Now for an apology. Kind friends have complained that the book, perfect in all other ways, betrays an occasional peevishness. Well, it does. So does *The Wealth of Nations* – and for the same reasons.

The notes and references to each chapter include the reading that under-pins this study with the exception of two books that could not be easily referenced: Thomas Africa's *Science and the State in Greece and Rome* (New York and London: Wiley, 1968) which informed Chapter 2, and Paul Johnson's *Intellectuals* (London: Weidenfeld and Nicolson, 1988) which provided some disreputable *aperçus* for Chapter 12.

Finally, I acknowledge Carol Clark and the McGowans, to whom this book is dedicated. Carol, Jim, Mary, Charles, Hugo and Lorna each deserve a library of dedications, but as I hope never to write another book, I shall distribute this one between them. My years in Newcastle, 1982–6, were bitter. It was the wrong university for my science at that time, and although I worked with Barry Argent and some other super colleagues – let me acknowledge Lorrane Agius, George Alberti, Kim Bartlett, Mike Laker and Andrew Skillen for all their companionship – my research suffered. It was no fault of Newcastle University: I should not have been prosecuting that particular project at that moment in that place. Moreover I was lonely, far from my mother, my brother Gavin and Karen his wife, and although I made good friends, and old ones visited – let me say hello to Marcus, Mimi, Stafford and Jo-Jo Everard, Oliver and Carol Griffiths, Bill and Ghillie Gutteridge, Mona and Don Hughes, Jonathan and Jane Kay, Alex and Ruth Macleod, Pete and Jen Parker, Dick and Kate Philbrick, Peter and Carol Saunders, Lawrence and Camilla Weaver – I was not happy. Into my life, like fairy-godparents, stepped the McGowans, who welcomed me into their beautiful house, fed me delicious food, laughed with and at me, and made life bearable. I will never be able to thank them enough.

I returned to Oxford in 1986, discouraged and distracted, and Carol Clark lent me a room (for no rent) while I sorted myself out. She taught me to write, guided my reading, re-integrated me into intellectual society, restored my self-confidence, and supported me through my fellow scientists' denunciations. I owe her a debt of absolute and utter gratitude.

And then Andrew introduced me to Sally, and we have lived happily ever after (Sally and I that is).

1 Francis Bacon and Adam Smith

On the night of 10 April 1988, at the height of one of the recurring national debates over the health of British science, Professor Sir George Porter, the President of the Royal Society, addressed the British nation. Delivering the Dimbleby Lecture[1] on prime-time BBC1 television, Sir George demonstrated a small piece of machinery and gave a little history: 'On 29 August 1831, Faraday discovered electromagnetic induction. . . . This is the actual coil that he used to make that historical discovery. . . . The next year, in France and, a little later, in England, small dynamos like this were made [to] generate a continuous electric current. With some further development, here is Battersea Power Station!' After a pause for dramatic effect, Sir George continued: 'When Faraday showed the [electromagnetic] effect to the Chancellor of the Exchequer, Mr Gladstone asked "But what is the use of your discovery?", to which Faraday replied, "I know not, Sir, but I'll warrant one day you'll tax it."'

In Sir George's words 'this lesson hardly needs any elaboration'. Science is a crucial investment, but politicians consistently underfund it from ignorance, short-sightedness or just plain folly. How rich, how informed, how happy we would be, if only politicians were not such beasts over their funding of science!

Inspection, however, casts a certain doubt on Sir George's story (to be fair to Sir George, it is often retold by science lobbyists the world over – he did not coin it himself). First, both Faraday and Gladstone left voluminous diaries – neither mentions the episode. Secondly, both men have had their biographies written many times – no standard biography mentions it. Thirdly, Faraday described electromagnetic induction in 1831, but Gladstone did not become Chancellor of the Exchequer until 1852, twenty-one years later – by which time the significance of electricity was well understood. But fourthly, and most importantly, neither Gladstone nor any other British politician did support science to any significant extent during the nineteenth century – yet that did not prevent Britain from growing into the richest and most industrialised country in the world, nor from producing scientists such

1

as Davy, Kelvin, Maxwell, Lyell and Darwin. Curiously, nineteenth-century France and Germany, whose governments did fund science expansively, trailed behind. Can the government funding of science be so important?

During the twentieth century the governments of all advanced countries started to fund research. The extent of that funding, however, has varied widely. The Soviet and Chinese governments, for example, funded all of their countries' research, generously, in the belief that it would fuel economic and social growth. The governments of Switzerland or Japan, however, spend relatively little, yet those countries thrive, both economically and scientifically. This book was written to make an audit of these different policies. Should governments fund science? If so, by how much? If not, why not? The debate is an old one, and it goes back to two Britons, Francis Bacon and Adam Smith.

FRANCIS BACON (1561–1626)

Francis Bacon was a lawyer, politician, embezzler and torturer (see the postscript to this chapter) but he has long been revered. Robert Boyle hailed him as the presiding spirit of the new learning. Voltaire called him the father of experimental philosophy. Immanuel Kant dedicated his *Critique of Pure Reason* to him. Bacon earned his veneration, albeit posthumously, through his books, of which the most important was *The Advancement of Learning* (1605).

In Bacon's day, scholars suffered from a peculiar form of intellectual distortion – they were doing one thing while professing another. Europe was formally Aristotelian, and the universities taught that Ptolemy, Galen, Vitruvius and the other ancients had discovered all there was to discover about the laws of nature. This certainty was imposed by the Church, which claimed a monopoly on learning. The universities, therefore, taught that truth could only be elucidated through deduction. An investigator wishing to unravel a new, particular truth could only do so by deducing it from the established laws of nature. But, increasingly, late medieval scientists were disproving the ancients. As early as 1530 Copernicus wrote *De Revolutionibus*, where he showed the sun to be the centre of the universe (Aristotle, of course, had placed the earth there). In *De Humani Corporis Fabricia* (1543) and *Opuscula Anatomica* (1654) Vesalius and Eustachio respectively had disproved many of Galen's anatomical tenets. By Bacon's day, science was in ferment. Kepler and Galileo were creating a new cosmology, Fallopius was producing pupils such as William Harvey who was to forge a new physiology, and even humble little Colchester, in Essex, was to spawn William Gilbert, whose *De Magnete* (1600) conferred upon him the title of Father of Magnetism. Sixteenth-century scientists, clearly, were not restricted to deduction, but they had been brought up to believe they were. How did they, and their non-scientific contemporaries, relieve the tension?

Some innovative scientists, of course, were simply victimised. Vesalius, the anatomist, was driven out of his professorship at Padua by the Galenists, and in 1563 he was condemned to death by the Inquisition (the sentence was eventually commuted to a pilgrimage to Jerusalem, but a shipwreck drowned him *en route*, to the delight of his enemies who attributed it to divine intervention). In 1633 Galileo was shown the instruments of torture by Pope Urban VIII, which rapidly converted him back to intellectual orthodoxy (even if he did whisper 'but it still moves' during his public retraction). And the first great Bacon, Roger Bacon of Oxford, the experimental philosopher who invented, amongst other things, the magnifying glass, was imprisoned between 1277 and 1291 for 'suspected novelties'.

Sometimes the orthodox thinkers battled it out honorably with the new. Tycho Brahe, for example, uncovered so many errors in the astronomical tables that he showed that neither Copernicus nor Kepler actually had enough evidence for heliocentricity. More usually, however, medieval society simply resorted to euphemism. The doctrine was promulgated that the Renaissance, for all its ferment, was just that – a rebirth of the old learning. This fiction survives in the word 'research' which incorporates the belief that 'discovery' is only 'recovery'. Bacon denied all that. He believed that facts could only be reliably determined by direct observation: 'Those who determine not to conjecture and guess but to find out and know . . . must consult only things themselves' (*The Advancement of Learning*, 1605). Bacon went further – he invented a new intellectual model. Where medieval philosophers had advocated *deduction*, Bacon honoured *induction* as the finest of scientific achievements: Bacon believed that a researcher, on discovering new facts, could thus induce new laws of nature. Bacon went further still – he invented the concept of progress, or progression as he called it. Remarkably, that concept is less than five hundred years old. J. B. Bury, in his *Idea of Progress* (1920)[2] showed that Democritus, in the fifth century BC, hinted at it, as did the other Bacon, Roger, during the fourteenth century, but an overt definition had to await Francis Bacon's *Advancement of Learning*. Progress he defined as the addition of new knowledge to old, to promote the creation of yet more.

Bacon was to have a huge impact. Within thirty years of his death, the scientists had grown to love him. Partly this was because he was so flattering. He saw scientists as heroes: 'the utility of some one particular invention so impresses men that they exalt to superhuman rank the man who is responsible for it' (*Cogitata et Visa*, 1607). Bacon elevated science itself: 'And indeed it is this glory of discovery that is the true ornament of mankind'.[3] Bacon even thought that science would elevate people, 'the improvement of man's mind and the improvement of his lot are one and the same thing'.[4] No wonder men like Robert Boyle or Robert Hooke, the authors of Boyle's and Hooke's Laws and two of the founders, in 1662, of the Royal Society, were to devote years in writing essays in praise of Bacon. It was the least they could do. But Bacon did more than legitimise the scientific method – he was to argue for the government funding of science. Much of the second part of *The Advancement of Learning* consists of a plea to King James I for the state support of academic science.

Much of Bacon's philosophy has been superseded. A quick tour of a modern scientific laboratory, for example, would disabuse most visitors of the suggestion that the practice of science improved people's character, and Karl Popper and other philosophers have considerably refined Bacon's concept of a scientific hypothesis. But the arguments with which Bacon tried to persuade King James that governments should support science have survived unchanged and unchallenged. Freshly minted in 1605, his arguments are repeated verbatim in the common-rooms of universities to this day across the globe. First, Bacon justified the government support of science for its own sake: 'To the King . . . there is not any part of good government more worthy than the further endowment of the world with sound and fruitful knowledge.'[5] Bacon wanted more pure science: 'Amongst so many great foundations of colleges in Europe, I find strange that they are all dedicated to professions, and none left free to arts and sciences at large.'[6] Bacon then developed an economic argument. He believed that pure science underpinned applied science or technology: 'If any man think philosophy and universality to be idle studies, he doth not consider that all professions are from thence served and supplied.'[7] Technology, Bacon believed, created wealth: 'The benefits inventors confer extend to the whole human race' (*Novum Organum*, 1620).

It was Bacon, therefore, who first proposed the 'linear' model of technological advance:

government
funded   → pure science → applied science → economic growth.
academic   or
research   technology

To forge the chain, Bacon proposed a series of government initiatives. He wanted the State to fund the building of university laboratories: 'there will hardly be any main proficience in the disclosing of nature, except that there will be some experiments appertaining to Vulcanus or Daedalus, furnace or engine or any other kind'.[8] Bacon wanted research-fellows: 'inquirors concerning such parts of knowledge as may appear not to have been already sufficiently laboured or undertaken'.[9] In *Sylva Sylvarum* (1627) Bacon suggested a thousand different experiments that his research-fellows might undertake. These included investigations into refrigeration, perfumery, microscopy, perpetual motion, and the development of telescopes, submarines and aeroplanes. Bacon put no limits on research: 'The knowledge that we now possess will not teach a man even what to wish' (*De Interpretatione Naturae*, 1603). Bacon wanted travelling fellowships: 'more intelligence mutual between the universities of Europe . . . to take themselves to have a kind of contract, fraternity and correspondence one with the other'.[10] Finally, on a very contemporary note, Bacon wanted higher academic salaries: 'And because founders of colleges do plant, and founders of lectures do water, it followeth well in order to speak of the defect which is in public lectures; namely, in the smallness and meanness of the salary or reward which in most places is assigned unto them. For it is necessary to the progression of sciences that readers be of the most able and sufficient men . . . this cannot be, except their condition and endowment be such as may content the ablest man to appropriate his whole labour'.[11]

Bacon was an unexpected man. Few law officers or professional politicians set out, in their spare time, to catalogue all existing knowledge or to suggest new ways of acquiring it. Bacon believed himself to be unique: 'As I look round the world of learning, the last thing anyone would be likely to entertain is an unfamiliar thought' (*Cogita et Visa*, 1607). Actually, Bacon was less original than he claimed. Medieval Britain had produced a remarkable series of anti-Aristotelians. In his *Opus Maius* (1267) the first great Bacon, Roger, had written: 'it is necessary to check all things through experience' and: 'arguments are not enough, experience is necessary'. Duns Scotus (1265–1308) described theology as *sapienta* not *scientia*. In *Summa Logices* (1488)

Ockham defined knowledge as that which is known *per experientiam*, while the twelfth-century writer Adelard of Bath had denounced authority as a *capistrum* or halter. Nor was Bacon as original as he thought he was in his prescription of a new scientific method because, as we have seen, he was largely describing what his great contemporaries were doing anyway. In the words of Sir Richard Gregory, the editor of *Nature*, he was: 'not the founder but the apostle' of the scientific method.[12] But, original or not, Francis Bacon was hugely important because his work became immensely influential. The great scholars of the French Enlightenment, Diderot, d'Alembert and later Comte claimed him as their mentor. In consequence, France and much of continental Europe developed national policies for science that were explicitly Baconian. In particular, continental countries' governments funded science, technology and education. Conversely, Britain's governments did not. That neglect by successive British governments has infuriated scientists who, since the Royal Society was founded on the model of Salomon's House that Bacon outlined in his *New Atlantis* (1627), have generally been Baconian. Indeed, Professor Sir George Porter entitled the Dimbleby Lecture which introduced this chapter 'Knowledge itself is power' which was one of the aphorisms Bacon coined in his *Religious Meditations (of Heresies)*. Now, in this book, we shall review the different outcomes of these different countries' policies for science.

ADAM SMITH (1723–1790)

One can be too original. Francis Bacon may not have been totally innovative, but he was sufficiently in advance of his time to moan 'I have not even a person with whom I can converse without reserve on such subjects . . . why, I met not long ago a certain evil-eyed old fortune-telling woman, who . . . prophesied that my offspring should die in the desert' (*Redargutio Philosophiarum*). Adam Smith never made this mistake. By propagating ideas that all progressive people admired as *avant garde*, brave and correct, because they had already been pioneered by Mandeville, Petty, Steuart, Hume, Cantillon, Quesnay and Turgot, he ensured vast popularity and enormous respect. At one public dinner in Wimbledon, near London, William Pitt (the Prime Minister), asked Smith to preside at the head of the table because 'we are all your scholars'. Ever the charmer, Smith returned the compliment by exclaiming that Pitt understood his own (Smith's) ideas better than he did himself.

Adam Smith wrote two great books, *The Theory of Moral Sentiments* (1759) and *The Wealth of Nations* (1776). *The Wealth of Nations*, the book which primarily concerns us here, was a study of the factors that, indeed, promoted the wealth of nations. It has become famous, pre-eminently because it showed that free trade, not protectionism nor mercantilism, promoted the creation of wealth. But economics was not enough; Smith agreed with Francis Bacon that technology, also, was important to wealth creation: 'In manufactures [i.e. factories] the same number of hands, assisted with the best machinery, will work up a much greater quantity of goods than with more imperfect instruments of trade. The expense which is properly laid out upon a fixed capital of any kind, is always repaid with great profit . . . all improvements in mechanics are always regarded as advantageous to every society'.[13] Smith, however, disagreed with Bacon over the source of technology. Bacon supposed it would flow from academic science, Smith observed that it arose from within industry itself. In his *Lectures on Jurisprudence*, delivered at Glasgow University between 1763 and 1764, he stated: 'If we go into the work house of any manufacturer in the new works at Sheffield, Manchester or Birmingham, or even some towns in Scotland, and enquire concerning the machines, they will tell you that such or such an one was invented by some common workman.'[14]

Later, in his more definitive *Wealth of Nations*, Smith enumerated the three major sources of new technology. Pride of place, again, went to the factories: 'A great part of the machines made use of in manufactures . . . were originally the inventions of common workmen who . . . naturally turned their thoughts towards finding out easier and readier methods of performing it [their work]. Whoever has been much accustomed to visit such manufactures must frequently have been shown very pretty machines, which were the inventions of such workmen in order to facilitate and quicken their own particular part of the work.'[15] Smith then gave an example of how, in the earlier steam engines ('fire-engines'), the valve between the boiler and the piston had to be opened and closed with each cycle, and how one of the boys who had that tedious job had built a link between the handle of the valve and the shaft of the piston which then opened and shut the valve automatically.

Almost as important as a source of new technology, in Smith's judgement, were the factories that supplied other factories: 'All the improvements in machinery, however, have by no means been the inventions of those who had occasion to use the machines. Many improvements have been made by the ingenuity of the makers of the machines.'[16]

Only third, and least important, was the input that flowed from academic sciences: 'and some improvements in machinery have been made by those who are called philosophers, or men of speculation'.[17] (Smith claimed all scientists as philosophers; he listed them as 'mechanical, chymical, astronomical, physical, metaphysical, moral, political and critical [literary and aesthetic].[18])

Smith, therefore, proposed a very different model from Bacon for the origins of technology. Where Bacon believed that it would flow from academic science, Smith maintained that it largely derived from the industrial development of pre-existing technology. Indeed, Smith went even further – he actually reversed the intellectual flow – he believed that most advances in science were not made by academics at all, but by industrialists or by others working outside academia: 'The improvements which, in modern times, have been made in several different branches of philosophy, have not, the greater part of them, been made in universities'[19] i.e.: advances in science flow from the technological advances made by industrialists.

We can, therefore, compare Bacon's linear model:

academic science → applied science/technology → wealth

with Adam Smith's

pre-existing technology → new technology → wealth

$$\uparrow\downarrow$$

academic science.

Why did Smith disgree with Bacon? One answer was that Smith, of course, was born one and a half centuries after Bacon – into a very different world. Attorneys-General did not personally supervise torture in Smith's day. Economically, technologically and scientifically, the world had advanced. Bacon wrote at the very dawn of the scientific revolution, 150 years before the onset of the Industrial Revolution. He could only argue from abstract principles, but Smith could do so from personal experience of science and new technology. Let us examine those experiences, and let us examine the three areas where Smith disagreed with Bacon, namely that (i) pre-existing technology (not science) spawned new technology, (ii) technology also spawned new science, and (iii) government funding was not necessary.

Technology Spawns Technology

Eighteenth-century industrial growth was led by the enormous expansion in textiles. During the two decades that Adam Smith was most creative, 1760–80, the consumption of raw cotton by Britain increased 540 per cent from 1.2 million pounds weight to 6.5 million.[20] That growth could not have happened but for certain crucial inventions. Wool and cotton come in fibres that are, individually, very short, yet cloths need to be woven from filaments that are tens or hundreds of yards long. Medieval craftsmen partially solved that problem with the spinning wheel, which enabled a worker (traditionally an unmarried girl living at home, hence 'spinster') to spin short individual fibres into endlessly long filaments. But the spinning, and the subsequent weaving, remained essentially manual crafts. In 1733, however, John Kay invented the flying shuttle, which mechanised weaving, and in 1770 James Hargreaves invented the spinning jenny which, as the name implies, mechanised spinning. These major developments in textile technology, as well as those of Wyatt and Paul (spinning frame, 1758), Arkwright (water frame, 1769) and Crompton (mule, 1779), presaged the Industrial Revolution, yet they owed nothing to science; they were empirical developments based upon the trial, error and experimentation of skilled craftsmen who were trying to improve the productivity, and so the profits, of their factories. These men were technologists building on technology – academic science did not influence them.

Technology Spawns Science

Smith believed that technology could breed academic science. The sort of observation that led him to that conclusion was Joseph Black's unexpected discovery of carbon dioxide in 1757. Black was a doctor, known to Smith, working in Glasgow, engaging in a classic piece of applied science – he was trying to find drugs that would dissolve bladder stones. Such stones are rare now, but were common in the eighteenth century, and they caused a lot of pain. During the course of his research Black found that such stones, on being heated, gave off a gas, which Black called 'fixed air' and which was to be characterised as carbon dioxide. That discovery helped Lavoisier, a few years later, prove the existence of oxygen and disprove the existence of phlogiston (it used to be believed that fire was caused by the release of 'phlogiston', the material that burned). Thus did a piece of highly applied science help yield a great theoretical, academic advance.

Government Funding

Smith had one further fundamental disagreement with Bacon – he did not believe in the government funding of universities or of academic science; he thought that academic institutions should find their own support. He approved of the conditions under which he taught at Glasgow, namely that Scottish university teachers were reimbursed, directly, by each student who attended a course of lectures. When, in view of his fame, the university authorities proposed paying him a fixed stipend, he refused – whence would come his motivation to teach well? He believed that a teacher's 'diligence is likely to be proportioned to the motive which he has for exerting it'.

Smith had been startled at how inferior Oxford and Cambridge were, in his day, to the Scottish universities. Whereas Glasgow could boast of philosophers like Hutcheson, or Edinburgh of its medical school, the two English universities of Oxford and Cambridge were sunk in decay: 'In the University of Oxford, the greater part of the public professors have, for these many years, given up altogether even the pretence of teaching.'[21] Smith attributed the English professors' decay to their guaranteed salaries and tenured jobs. The Scottish universities were better because, unlike Oxford and Cambridge, they had few endowments: 'Improvements were more easily introduced into some of the poorer universities which . . . were obliged to pay more attention to the current opinions of the world.' Smith believed that 'Improvements in philosophy [i.e. science]' would only come if academics' pay depended on their producing them. He did not believe that governments should commission science because he mistrusted politicians so much: 'That insidious and crafty animal, vulgarly called a statesman or politician.' Smith quoted Hamlet's 'the avarice and injustice of princes . . . the insolence of office'. Nor did Smith believe that scientists could be entrusted to spend the public's money disinterestedly; he believed that no-one could be so trusted: 'I have never known much good done by those who have affected to trade for the public good'. He wrote: 'people of the same trade seldom meet together, even for merriment and diversions, but the conversation ends in a conspiracy against the public'.

Smith was very suspicious of interest groups who appealed for public money. Such appeals 'ought always to be listened to with great precaution, and ought never to be adopted till after having been long and carefully examined, not only with the most scrupulous, but with the most suspicious attention.'[22] Such appeals generally came from 'men whose interest is never exactly the same with that of the public, who have generally an interest to deceive the public'. Smith was no anarchist. He believed that the State had to enforce strong laws to protect the free market, and he also believed that State might supply certain public works to promote social or commercial benefits to a society if they were 'of such a nature, that the profit could never repay the expense to any individual or small number of individuals'. [23] But Smith did not believe that science fell into that category.

In summary, Smith did not believe that applied science flowed very much from pure science; indeed, he believed the opposite was as likely to be true. Moreover, he believed that applied science or technology sprang from the market place, spawned by individuals or companies competing for profits. In as much as advances in technology did emerge from pure science, Smith did not believe that that justified government spending. Smith feared the economic consequences of the increased taxation that would entail, he distrusted any measure that increased the power of politicians, and he distrusted the capacity of academics to work without immediate goads. He believed that private sources (students' fees, endowments and consultancies) would provide enough funding for university science to meet the needs of the economy.

Was Smith right? Or was Bacon right to suggest that government funding for pure science was necessary? The two men produced very different models, and their differences can be tested. The models, of course, agreed that technology underpins wealth (everybody agrees on that) but Bacon's model supposes that the government funding of pure science is crucial to wealth creation, Smith's model denies that. Fortunately, different countries' governments have, over the centuries, pursued different policies towards science funding, so we, at the end of the twentieth century, can do what those seventeenth- and eighteenth-century thinkers could not: we can test their models against the historical experience.

POSTSCRIPT ON FRANCIS BACON

Bacon's posthumous reputation would have startled his contemporaries, most of whom ignored his writings. Uniformly, moreover, they disliked him – with good reason. Francis Bacon was not a nice man. Born in 1561, in London, of good family, Bacon was educated at Trinity College, Cambridge University. As was then the custom, he entered

Trinity young, at twelve and half years' of age. He left two years later. Thereafter, he first entered the civil service, then qualified as a barrister, and then became an MP. For much of his life he was successful, becoming Solicitor-General in 1607, Attorney General in 1613 and Lord Chancellor in 1618. In that same year he was created Baron Verulam of Verulam, Viscount St Albans. But he was a crook, a cheat and a fraudster who clawed his way to the top through a series of betrayals.

His first patron was Robert, Earl of Essex, Elizabeth I's favourite; but Essex's own career did not prosper, and by 1601, in desperation, Essex was plotting against the Queen. Bacon urged him to rebel, planned the rebellion with him, then betrayed the plans to the authorities and personally led the prosecution at the trial that condemned the unfortunate Essex to death. In 1616 Bacon again led the prosecution at the trial of another old friend, Robert Carr, Earl of Somerset, for the murder of Thomas Overbury, the poet.

A cruel man, Bacon, when Attorney-General, personally supervised the torture of suspects. His wicked ways finally undid him, and in 1621 he pleaded guilty at his own trial for bribery and corruption. Lucky to escape with his life, his estates were attainted and he was exiled in perpetuity from Parliament, the Court and central London.

The last word on Bacon, the man, should be left to one of his earliest biographers, Aubrey. In his *Brief Lives*, Aubrey (1626–97) described how Bacon, towards the end of his life, decided to actually do some science instead of just writing about it. The consequences were unfortunate: 'He was taking the aire in a coache with Dr. Witherborne (a Scotchman, Physitian to the King) towards High-gate, snow lay on the ground, and it came into my lord's thoughts, why flesh might not be preserved in snow, as in salt. They were resolved they would try the experiment presently. They alighted out of the coach, and went into a poore woman's howse at the bottome of Highgate hill, and bought a hen, and made the woman exenterate it, and then stuffed the bodie with snow, and my lord did help doe it himselfe. The snow so chilled him, that he immediately fell so extremely ill, that he could not then return to his lodgings (I suppose then at Graye's Inne), but went to the earle of Arundell's house at High-gate, where they putt him into a good bed warmed with a panne, but it was a damp bed that had not been layn-in in about a yeare before, which gave him such a cold that in two or three dayes, as I remember he [Hobbes, Aubrey's informant] told me, he dyed of suffocation.'

Bacon's heart probably lay in his books, not in other people. As he said of himself: 'I have rather studied books than men'. Pope who felt differently, 'The proper study of mankind is man' delivered an awful rebuke:

> If parts allure thee, think how Bacon shined,
> the wisest, brightest, meanest of mankind.

Bacon would not have cared. In a famous letter written in 1592 to William Cecil, the Queen's Chief Secretary of State, he wrote: 'I have taken all knowledge for my province'. Remarkably, he combined a busy public life with major, spare-time philosophy, but although he married, he never had children, so he could find both the time and the energy (Aubrey said he was a pederast). As Bacon wrote in his essay *Of Marriage and Single Life*: 'He that hath wife and children hath given hostages to fortune; for they are impediments to great enterprises, either of virtue or mischief. Certainly, the best works, and of greatest merit for the public, have proceeded from the unmarried or childless men' (F. Bacon, *Essays,* ed. M. J. Hawkins, London: Everyman, 1994).

Macaulay wrote a hilarious essay in denunciation of Bacon's character.

POSTSCRIPT ON ADAM SMITH

Adam Smith was born in Kirkcaldy, in Scotland, in 1723, to a respectable family, his father being the Comptroller of Customs. Adam Smith was an academic. He studied philosophy at Glasgow University (1737–40), and at Balliol College, Oxford University (1740–6) before returning to Scotland to lecture at Edinburgh University on Rhetoric. Between 1751 and 1764 he was a professor at Glasgow University, first of Logic and then of Moral Philosophy. He then resigned his chair to accompany, as tutor, the young Duke of Buccleuch on a Grand Tour of Europe (1764–6) which enabled him to meet most of the great thinkers of the day (Voltaire, Quesnai, Turgot, Helvetius, *et al*). After a long sojourn in London, Smith returned to Edinburgh in 1778 to assume, with pleasing filial symmetry, the office of Commissioner of Customs. He died in 1790. He never married, nor had children, so he had time to write his two great books.

Smith's career illustrates the value of the private funding of scholarship. His studentship at Oxford was funded out of a private bequest, the Snell Foundation (still extant). His professorship at Glasgow was funded out of the students' fees, and the Duke of Buccleuch settled a life-long pension of £300 a year on him (double his academic salary) in return for his company on the Grand Tour. *The Wealth of Nations*, which was written in London, was therefore made possible by the Duke of Buccleuch's funding.

POSTSCRIPT ON RASHID AL-DIN

Francis Bacon was not the first person to suggest the government funding of science. He was anticipated by the extraordinary Rashid al-Din (1247–1318). Rashid, in a remarkable career that paralleled Bacon's in many ways, rose to become Vizier, or chief minister, to the Persian Empire. A scholarly man, he collected all the knowledge open to him in his vast *Jami al-Tawarikeh* (1302). In language very similar to Bacon's, he defended government support for research: 'There is no greater service than to encourage science and scholarship', he wrote: 'It is most important that scholars should be able to work in peace of mind without the harassments of poverty'. Unlike Bacon, however, Rashid left little lasting influence. He was executed for blasphemy, after which his head was carried through the streets of Tabriz with cries of: 'This is the head of a Jew who abused the name of God. May God's curse be upon him!' His murderers then destroyed all the copies of his books they could find, and although some survived in the libraries of neighbouring Muslim states, his work influenced neither them, nor Europe.

Although circumstances ensured that he would never assume Bacon's importance, he retains one advantage; unlike Bacon, he was a decent man. His letters reveal him to have been humane, wise and humorous and, unlike Bacon's convictions for corruption, his own trial was transparently unjust and the charges against him manifestly false. He was a good, as well as a great, man.

2 Research and Development in Antiquity

Technology is so important that much of our history as a species is named for it: the Stone Age, the Bronze Age, the Iron Age. During our first few millions of years as *Homo sapiens*, the period that is called the Palaeolithic Period or Old Stone Age (from *palaios*, the Greek for ancient, and *lithos*, the Greek for stone), our only tools were crude rocks or sticks (chimpanzees use them too, as Jane Goodall described in *In the Shadow of Man*). But throughout the Palaeolithic Period we became more and more skilled at flaking and polishing stone, and some of the flint arrowheads and knives that have since been excavated are exquisite. No one knows the exact sequence, probably no one will ever know, but Palaeolithic Man then initiated a chain of technological advances. First he learned to control fire (the terms 'he' and 'his' are used here as a generic for both genders; this is not a sexist book). Then he developed the bow and arrow, the harpoon and the spear. On water, his early rafts evolved into canoes; on land, he started to domesticate the animals.

About 20 000 years ago, certain human communities abandoned their nomadic, hunting existence for a settled life depending on animal husbandry and agriculture. Archaeologists describe this as the transition from the Palaeolithic to the Mesolithic to the Neolithic Periods (Old Stone Age to Middle Stone Age to New Stone Age). Technology accelerated. Potters developed the wheel, and soon Man was making pots and baking clay. He moulded bricks. Crops were sown, seeds improved and querns constructed to grind corn. The spinning and weaving of textiles emerged, as did dyeing, fermenting and distilling. The sail was invented and ships were constructed, as were water mills, dams and irrigation canals.

About 5000 years ago, men started to smelt metals. Gold and silver were worked, and copper was extracted from its ore to be alloyed with tin to create bronze – the Bronze Age. And what an age! The late Neolithic Period had spawned some wonderful civilisations – the Sumerian, the Assyrian, the Babylonian – but the Bronze Age flowered with the Egyptians, the Hittites, the Medes, the Persians and the Jews. It also produced the Indian, Chinese and Japanese civilisations.

15

With the appearance of the great Bronze Age civilisations, however, a curious paradox emerges. Those civilisations had only developed because their new technology had liberated Stone Age people from bare subsistence and had afforded them the opportunities required to create a new culture. But having created new civilisations on the back of technology, those cultures failed to develop any more. Stagnation set in, innovation dried up. Consider the Egyptians. By 2500 BC they had built the pyramids – wonderful artefacts – but over the next 2000 years they produced nothing new. It was not as if they did not understand mathematics or science; consider this examination question found on a papyrus dated 2000 BC: 'A building ramp is to be constructed 730 cubits long, 55 cubits wide, containing 120 compartments filled with reeds and beams, 60 cubits high at its summit, 30 cubits in the middle, with a batter of cubits high at its summit, 30 cubits in the middle, with a batter of 15 cubits and a pavement of 5 cubits. How many bricks will be needed?'[1] To answer that question, the examinees needed a good grasp of quadratic equations. The Egyptians had got as far as quadratic equations – no mean feat – but could get no further. For hundreds, if not thousands of years they asked the same questions of their students, creating nothing new. Between Rameses I and Cleopatra nothing changed.

The Chinese also produced a wonderful civilisation, and they spawned some marvellous technology. They were the first to create an efficient harness for horses, to roll paper, and to develop mechanical time keeping. They produced gunpowder, the magnetic compass and printing – discoveries which Francis Bacon was to describe as the 'magic trio that have changed the whole face and state of things throughout the world'. But Bacon meant the late medieval western world, because it was renaissance Europe which seized on those discoveries and developed them fabulously. The Chinese had possessed them for 2000 years, but had done nothing new with them. So stagnant was their culture that some of their textbooks on architecture from around AD 1000 were still in use during the eighteenth and nineteenth centuries when western traders forced themselves in. Nothing had changed.

India had achieved a similarly high degree of technology and organisation. As late as 1750, India and China were still so advanced that they accounted for 73 per cent of the world's manufacturing output.[2] In India, as in China, tens of thousands of peasants could, for centuries, be regimented into complex collective efforts: dams could be built, as could canals, roads, forts or temples. But for generations, the same dams, canals, roads, forts or temples were built. Designs did not change.

Why did technology stagnate? What property did the Bronze Age civilisations share that their neolithic predecessors, the cultures from which they had emerged through the development of new technology, did not?

One major property that the sterile cultures all shared was that they were totalitarian states whose citizens were denied freedom. In no Bronze-Age civilisation did individuals enjoy legal rights, and each person was totally subject to the central authority (generally an emperor and his court). Each civilisation, moreover, elaborated an all-powerful religion which froze all intellectual or cultural development. We saw in Chapter One how certain crucial figures in western science such as Roger Bacon, Galileo and Vesalius were persecuted by the Church, but such intolerance has never been restricted to Christianity. When, for example, during the fifth century BC the experimental philosopher Anaxagoras suggested that the sun might not be a god, but a rock 'bigger even than the whole of the Peloponnese', the devout Athenians were so shocked by his impiety that they prosecuted the unfortunate man, and he would have been executed had his old friend Pericles not helped him to escape to Lampascus, where he died in 428 BC. That Anaxagoras, Roger Bacon, Galileo and Vesalius were persecuted is not remarkable, but it is remarkable that these European thinkers were not executed. In ancient China, Japan, India and Egypt, thinkers died for much smaller intellectual deviations.

Economically, as well as intellectually, these civilisations were equally rigid. Within none of them was a free market permitted. Indeed, it was barely possible; the Egyptians, for example, did not even possess coins. Barter, moreover, was largely irrelevant, because these economies were centrally planned. Officials would appropriate the peasantry's entire crop, leaving the peasants only what they needed for food and for planting, and the surplus would be distributed to the court and church according to social precedence. Private property, safe from the court, did not exist. A few merchants were tolerated, under strict control and heavy taxes, but their trade was largely confined to baubles and luxuries 'the objects of oriental traffic were splendid and trifling' (Edward Gibbon, *Decline and Fall of the Roman Empire*). In the absence of free markets of any size, the few large workshops or mines served the State or the church. Towns were very small, because over 90 per cent of the population worked the fields, and those towns that did exist were very different from those of today's in the western world which are centres of trade and of production. Then they were just centres of administration and of consumption. The spirit of such civilisations is well summarised by this extract of a surviving Egyptian papyrus, on which an official is dictating to his subordinates: 'No one has a right to do what he likes, but everything is managed in the best possible way'.[3]

The lack of innovation within such civilisations becomes easy to understand. There was no scope for it, nor any incentive. No peasant who thought of a new way of doing something would be rewarded financially; any extra surplus would merely go to the King or his officials. Nor would anyone be rewarded socially for innovation. Everyone was trapped by rigid hereditary caste laws that condemned him, and his children, like his father before him, to his status and occupation for life. There was no mechanism of reward or promotion – only death for heresy. Even the emperors and priests were frozen by protocol, and they were regulated in their daily lives by precedent and ritual. The Bronze Age civilisations were like beehives; they had evolved complex societies that were busy, but which trapped every individual into pre-determined intellectual death.

Had the whole world been ruled by the Bronze Age empires, *Homo sapiens* would have stagnated into totalitarian sterility, but on the borders of the empires were occasionally to be found small, non-imperial peoples, and if those people were traders, innovation and new technology would flower. One such people were the Canaanites, a small trading people who clung on to a corner of Palestine, constantly threatened by myriads of enemies including the Egyptians to the south. Yet it was the endangered Canaanite merchants, not the subjects of the great empires, who, needing a proper script to record their transactions, developed during the second millenium BC the first modern, non-hieroglyphic script. The Canaanites made literature possible.

Literature notwithstanding, the Canaanites continued to find life difficult. For centuries they were threatened by their immediate neighbours the Philistines, another small, trading nation, and when in about 1200 BC the Philistines introduced iron into the world, the Canaanites' hold on their native land grew increasingly precarious (note that the Iron Age was introduced by marginal traders, not by one of the great empires). At around the turn of the millennium, the Canaanites abandoned most of their land to the neo-Hittites, Arameans, Israelites and Philistines (the word 'Palestine' is a corruption of 'Philistine') and from their ports of Tyre, Sidon and Byblos, they took to the sea. Cyprus contained copper, Spain contained silver, tin and iron. These were needed by the empires, and the Canaanites (increasingly called Phoenicians), made their living by shipping these goods in exchange for timber, olives, cereals, vines and dates.

The Phoenicians prospered, and their trading grew so extensive that they eventually founded up to fifty colonies all over the Mediterranean basin. Carthage, of course, was a Phoenician colony. As both a consequence and a cause, of their success as traders, the Phoenicians developed a free-market culture very different from those of the great empires. Learning from them were the Greeks, another small, Mediterranean people whose poor soil had driven them to take up maritime trade. The Greeks adopted many of the Phoenicians' customs, such as their alphabet, and they considerably developed their commercial practices to spawn a sophisticated free market.

In a free market, traders specialise. Herodotus wrote that the Lydians were pure retailers, and the first to use coins widely. Greek merchants developed subspecialties. The *naukleros* traded in his own ship, the *emperos* subcontracted the passage of his goods to another's ship, the *kapelos* traded from home. To support such trade, Greek law elaborated the essentials of a commercial code, and the idea of private property was conceived, underwritten by the rule of law. Individuals acquired the right to go to court to defend their legal interests. The notion of contracts, freely undertaken between individuals but enforceable in law, was also created. These concepts were largely alien in a world where everyone was the slave of the monarch, court or priest; but the requirements of a free market had created a culture where individuals had to be free and autonomous under the law, or the market and their livelihood would collapse.

We have explored the economics of the Phoenicians and the Greeks because, coinciding with their advances in commerce and civil law, they produced some wonderful technology. The Phoenicians excelled in textiles, glass making, fine metalwork, quarrying, masonry and carpentry. The Greeks did even better, and the temples, pots and canals which survive speak of their vibrancy. The Greeks revelled in their mastery of metal. Homer's *Iliad* dwells lovingly on it: 'Hebe deftly got her chariot ready by fixing the two bronze wheels, each with eight spokes, on the ends of the iron axle-tree. The felloes of these wheels are made of imperishable gold, with bronze tyres fitted on the rims – a wonderful piece of work – while the naves that rotate on each axle are of silver. The car itself has a platform of gold and silver straps tightly interlaced, with a double railing round it, and a silver shaft running out from the front. To the front of this pole, Hebe tied the beautiful golden yoke and attached the fine gold breast-straps.'[4]

It appears, therefore, that we can derive a great lesson from the early history of technology – it depends on a free market. The technological creativity of neolithic, Greek and Phoenician man, compared with the sterility of the great civilisations, speaks of cause and effect.

We cannot, of course, attribute all the attributes of a modern free market to those Stone-Age tribes that created the new technology that flowered in the great Bronze-Age civilisations, and indeed we will never know much about those tribes, but if contemporary anthropological studies on present-day 'stone-age' peoples are representative, we can say that palaeolithic men would have created an enormous range of tribal cultures, some authoritarian and repressive, but others libertarian, individualistic or fun-loving. Within that range of cultures, there would have been some that tolerated, perhaps even fostered, curiosity and experimentation, and which rewarded, by approbation, those clever or brave individuals who found ways of doing things better.

Tentatively, we can propose a social and technological history of the Stone-Age that goes like this: the first technological advances were made by chance or by trial and error. Perhaps a cave man noted that copper flowed out of a piece of ore that happened to abut onto one of his fires, and he discovered that the copper could be shaped usefully. Encouraged, he and his fellow cave men might have systematically explored other rock outcrops for different, better metals or alloys. Progressively, technology would thus have advanced and, with each advance, life would have become progressively easier; that ease, in turn, would have facilitated the exploration of more technology. In time, the beginnings of an economic surplus will have appeared and, human nature being what it is, certain individuals such as chiefs or priests will have begun to distance themselves from the production of food and to start to live off the labours of others. In exchange, they will have provided leadership (this may have been an ugly business, resembling the creation of protection rackets by the Mafia). In time, the leaders will have assembled armies and will have started to conquer their neighbours and to create the great empires of the Bronze Age. Some of the very early Bronze Age societies, one can assume, must have been pleasant places, but competition will have ensured that the most aggressive of those early societies will have conquered all the rest. Thus did the Bronze Age witness the triumph of the most unpleasant people.

Once a Bronze Age king or priest had established his empire, it would not have made sense for him to foster intellectual, economic or technological freedoms for his people; they might have used them against him (the sort of man who creates an empire will think like that). It was much easier, more sensible and more fun for him merely to oppress his subjects. Such complacency would be dangerous in the face of an enemy who could exploit a technological weakness, but the great empires grew until they were protected by geography: the Egyptians and Babylonians were surrounded by desert, the Indians and Chinese sheltered behind mountains or the steppes, and the Japanese were an island people. Thus did the great ancient Bronze Age civilisations settle into totalitarian rigidity.

For the small mediterranean peoples that ringed the great empires, however, life was very different. Beset on all side by enemies, they could not afford technological or economic complacency. The Mediterranean, moreover, offered its small peoples three great opportunities. First, it made trade much cheaper, because transport on water was so much easier than by land. Second, such trade was freer because it was less policed; the sea is less easily controlled by society than the land. Third, it promoted intellectual freedom. A sailor, free on his boat, hopping from port to port, would grow sceptical of the claims of total, but contradictory, certitude he would receive from the priests of different lands. The Greeks grew highly sceptical, and they evolved a light-hearted, entertaining religion that rarely oppressed anyone (Anaxagoras and, of course, Socrates were unfortunate exceptions).

The sailor, hopping from port to port, buying produce where it is in surplus, selling where it is short, creates by his very presence a free market – even if he can only do so by bribing the local officials of the great empires. And free-market cultures then promote technology because they provide motive. In a free market, the man who invents a better mousetrap or a better ship or a better plough, will make his fortune. The free market also provides the technologist with his intellectual opportunity, because a free market is incompatible with rigid intellectual control; it affords him intellectual freedom. It also affords him economic opportunity because a free market enables a successful trader to build up capital, which liberates him from the daily pressure of subsistence and allows him time and equipment to experiment. Thus it was the small, trading peoples like the Canaanites, the Philistines, the early Greeks and the early Romans, who proved technologically vital.

Curiously, however, first the Greeks, and then the Romans, did what the Bronze Age empires had done; having prospered through the technology that had flowed out of trade, they then abjured trade for imperialism. From around 500 BC, the Greeks increasingly turned to plunder. The Acropolis in Athens, for example, was financed by tribute money and, like all the other engineering works, public monuments and civic buildings of the Graeco-Roman period, it was built by slaves, the Graeco-Roman economies being slave economies. So enamoured were the Greeks of slavery that they even tried to enslave each other. The Helots, for example, were the Greek slaves of their neighbours the Spartans.

As the Greek economy moved away from one based upon free enterprise to one based on exploitation, Greek culture metamorphosed into one that despised trade. Fashionable Greeks abhorred the *banausoi*, men who worked for a living. Bankers, businessmen and most doctors were dismissed as *banausoi*. In his *Politics*[5] Aristotle maintained that it was in poor taste even to discuss the banausic professions, and he explained that their trades rendered them unfit to share in the ruling of society.[6] The classic expression of contempt for the *banausoi* was made by the Athenian general Xenophon: 'The banausic trades spoil the bodies of workmen and foremen who are forced to sit still and work indoors. They often spend the whole day at the fire. The debilitation of the body is accompanied by a serious weakening of the mind. Moreover, the banausic occupations leave no spare time for service to one's friends or the city. Thus the *banausoi* are considered unreliable friends and poor defenders of their country.'[7]

Such a culture was not likely to promote free trade. Neither the Greeks nor the Romans, until the second century AD, actually undid their inheritance of free trade, but they never developed it further. It thus remained desperately primitive. The real problem was that their system of commerce never recognised any economic unit larger than the individual. Ultimately, it was the joint stock company that was to provide the western world with much of its economic and technological thrust, but neither the Greeks nor the Romans conceived of it. Nor did they conceive of independent business capital, separate from private wealth. Nor did they conceive of limited liability. Companies, other than *ad hoc* agreements between individuals, did not exist. No patent laws were framed. Banks were small and cumbersome. Pasion's Bank, in Athens, was little more than a pawnshop with only a tiny working capital, and the later Roman banks were little better. Throughout the Graeco-Roman period, individuals experienced enormous problems in transporting bullion about, just trying to pay for things.

With the decline of the culture of trade came a decline in technological innovation. The early Romans had not been uncreative and, to give two examples, they had developed the round arch (although that might have been an Etruscan contribution) and they had discovered how to carbonise iron to coat their swords with steel, but during the

Hellenistic or Imperial periods, despite their vast wealth, it is hard to identify more than two significant inventions that emerged: the development of harvesting machines for corn, and the development of water mills for its grinding (and even that may be too generous – the water mill was a barbarian invention which the Romans only improved, see Chapter 3). As we shall see, also in Chapter 3, this rate of technological innovation was to be dwarfed by that of the subsequent Dark Ages.

Paradoxically, the ancients did appear to appreciate the value of science and technology, and their governments invested huge sums in promoting it. In his *Achaen Woman*, Alexis wrote:

> Discovery attends on every quest,
> Except for renegades who shirk the toil.
> Now certain men have pushed discovery
> Into the sphere of Heaven. Some part they know –
> How planets rise and set and wheel about,
> And of the sun's eclipse. If men have probed
> Worlds far remote, can problems of this earth,
> This common home to which we're born, defy them?'[8]

The Greek city states prided themselves on their savants. Polycrates, the sixth-century BC tyrant of Samos, collected engineers. Herodotus was most impressed: 'Polycrates's engineers built three of the greatest constructions in Hellas. One was a tunnel running straight through the base of a 900 foot high hill. The length of the tunnel is almost a mile, and the height and width are each eight feet. Along the course of the tunnel there is another cutting, thirty feet deep and three feet wide, through which spring water is piped into the city.'[9]

Philosophers were lured from court to court. Thales, the astronomer, adorned in succession the palaces of the tyrant Thrasybulus, the Lydian King Alyattes and his replacement, the fabulously rich Croesus. Plato served as court philosopher to both Dionysius I and II of Syracuse. The most famous of the fourth century Greek scientists, Aristotle, was successively the *protégé* of the tyrant Hermias, Philip of Macedon and Alexander the Great. Aristotle's studies in natural history were funded by the Macedonian Crown.

States competed in their support for philosophy and science. Athens produced the Academy and the Lyceum, and Pergamum its museum and library, but none outshone the wonderful museum and library at Alexandria. From the third century BC until the era of Christ, Alexandria fostered some fabulous science. Euclid, the master of geometry and the author of the *Elements*, taught at Alexandria during a career that started around 300 BC. Amongst his near-contemporaries were (i) Erastosthenes, who calculated the diameter of the earth correctly to within 50 miles, (ii) Archimedes, the great mathematician of Syracuse, who was educated at Alexandria and who corresponded with its scholars all his life, and (iii) Ctesibius, who discovered the elasticity of air, and who created compressed-air weapons, suction pumps, force pumps and wind-organs. Later, Alexandria fostered Hero, who around AD 60, described the *aeolipe*, the world's first steam engine; and, later still, Alexandria nurtured Ptolemy, the great astronomer.

These scientists, and many more, were paid by the Egyptian State. Under a series of rulers, all called Ptolemy, the Hellenistic Egyptian state poured money into its library and museum (the term 'museum' derives from the 'muses'), and Alexandrian science flourished. But what good did it do the people of Egypt? Why did no useful technology emerge? Why, indeed, did so little useful technology emerge from any Greek or Hellenistic science?

To answer these questions, let us first distinguish between technology and science. Technology is the activity of *manipulating* nature, science is the activity of *learning* about nature. Each employs the same method (the so-called scientific method of systematic observation, hypothesis and experimentation) but each is different in purpose. A technologist, to take an example at random, might try to test different alloys for their durability as components of knives, while a scientist might try to understand the molecular basis for different metals' different properties.

Technology came first. That is one of the great lessons of history. The palaeolithic, mesolithic and neolithic developments of fire, harpoons, spears, canoes, pottery, querns, spinning, weaving, dyeing, fermenting, distilling, sailing, sowing, milling, smelting and irrigation were not based on science. They were the products of empirical experiments aimed at improving Man's lot. Only thousands of years after Man made his first technological advances did the first scientists emerge, the astronomers of Sumeria who, around 3000 BC, were plotting the movements of the stars and planets, and were constructing calendars and predicting eclipses.

By Hellenistic times, as we have seen, science was flourishing, yet no new technology flowed from it. Why not?

Science can be performed for a number of reasons; these include intellectual curiosity, a desire for social status, the support of religion (by calculating the correct date of Easter, for example) and the support of technology. Technologically supportive science is, of course, the one that translates into economic growth. For example, a technologist trying to improve the properties of an alloy may well be advised to study metallurgy, the better to understand his materials. But that sort of technologically inspired science barely existed in Hellenistic or later Roman times. The decline of the free market removed any incentive to fund or develop new technology, which removed any incentive to fund or develop technologically inspired science. The only science left was that pursued for reasons of status, religion or scholarship. The Sumerian astronomers, for example, were priests, whose religion enjoined astrology.

The Greek and Hellenistic kings and states certainly supported academic science for glory and culture. This is what Philo of Byzantium wrote: 'Success in this work was recently achieved by the Alexandrian engineers, who were heavily subsidised by kings eager for fame and interested in the arts',[10] but if the kings also hoped for useful discoveries, they were disappointed. Even at the beginning of the Hellenistic period, Aristotle was complaining that pure science had no practical applications, 'We notice that the geometricians are quite unable to apply their scientific proofs in practice. When it comes to dividing a piece of land . . . those who are concerned with mathematics and with the reasons for these things, while they may know how it is to be done, cannot do it.'[11] The situation never improved. None of the great Alexandrian science bore technological fruit. It is sometimes supposed that Archimedes had, at least, invented a useful water pump or 'screw', but even that, it emerges, was well-established in Egypt on his arrival. He merely transmitted its details to Syracuse.

Alexandrian science was useful militarily, of course. Archimedes was famed for his catapults and weapons, as were Ctesibius and Dionysius of Alexandria (Dionysius even invented a machine gun for arrows). Alexandrian science also bore religious fruit. Hero described how priests employed hydraulic pumps to open and close temple doors as if by magic. But Alexandrian science never bore economic fruit. This, of course, was because the Greeks had effectively stopped doing science for economic reasons. Science had become viewed as a luxury for the rich. Herodotus (485–425 BC) had believed that geometry had emerged in Egypt because of the country's need to control the flooding of the fields. Herodotus was almost certainly right, but the explanation did not satisfy the aristocratic requirements of the later Greek intellectuals. For the snobbish Aristotle (384–362 BC) science emerged from the leisure of the wealthy: 'When all inventions had been discovered, the sciences which are not concerned with the pleasures and necessities of life were developed first in lands where men began to have leisure. This is the reason why mathematics originated in Egypt, for there the priestly caste was able to enjoy leisure'[12] (note Aristotle's assumption that there was no technology left to discover).

The Greek scientists' disdain for any possible economic benefit from science is well illustrated by the famous story about Euclid. When one of his pupils, who correctly answered a question on geometry, asked what use it might be, Euclid tossed him a drachma, saying 'he wants science to be profitable'. But Euclid did not disdain his own salary: like Ctesibius of Chalicis, who, on being asked what he had gained from philosophy, had replied 'free dinners',[13] Euclid and his fellow academics were happy to be paid by the state – but they were not concerned with being useful – nor were they. Neither of the useful technological innovations of the Hellenistic or later Roman periods, the harnessing of water power to grind corn and the employment of harvesting machines for corn, emerged from the State's science; each was the fruit of private enterprise.

So unconcerned with research did the Roman State become, that the Emperors actually suppressed technology. Petronius described how: 'a flexible glass was invented, but the workshop of the inventor was completely destroyed by the Emperor Tiberius for fear that copper, silver and gold would lose value'.[14] Suetonius described how: 'An engineer devised a new machine which could haul large pillars at little expense. However the Emperor Vespasian rejected the invention and asked "who will take care of my poor?"'.[15] So uncommercial had the Romans become, their rulers rejected increases in productivity. In such a world, advances in science were never going to be translated into technology. Thus we can see that the government funding of ancient science was, in both economic and technological terms, a complete waste of money because the economy lacked the mechanism to exploit it.

The fall of the Roman Empire was frightful. The growth of the Empire had always been based on conquest, and the Empire's economy had been fuelled by the exploitation of new colonies. When the Empire ran out of putative victims, its economy ceased to make sense, particularly as the mere maintenance of the Empire, with its garrisons and its bureaucrats, was so expensive. From the beginning of the second century AD, the State had to raise higher and higher taxes to maintain itself and its armies. It was under the Emperors Hadrian and Trajan, when the Empire was at its largest, that residual freedoms started to

get knocked away to ensure that revenue was collected. Special commissioners, *curatores*, were appointed to run the cities. An army of
secret police were recruited from the *frumentarii*. To pay for the extra
bureaucrats, yet more taxes were raised, and the state increasingly took
over the running of the economy – almost on ancient Egyptian lines.
In AD 301, the Emperor Diocletian imposed fixed wages and prices,
by decree, with infractions punishable by death. He declared that 'uncontrolled economic activity is a religion of the godless'. Lanctantius
wrote that the edict was a complete failure, that 'there was a great
bloodshed arising from its small and unimportant details' and that more
people were engaged in raising and spending taxes than in paying them.

The origins of medieval feudalism emerged from the Roman Empire
as it decayed. To ensure that the peasants continued to work under an
economy which had lost its free-market incentives, Constantine promulgated a law in AD 332 which bound all *coloni* to the state as serfs.
Their children were *glebe adscripti*, tied to the soil. To reinforce state
control on all aspects of the economy, the city trade guilds or *collegia*
imposed compulsory, hereditary trades on all. An edict of AD 380 for-
bade children of the workers in the mint to marry outside their caste
or trade. The towns shrank, and the population condensed on the patri-
archal, self-sufficient, isolated estates that adumbrate the medieval
European villages. Indeed, the word 'village' derives from the Latin
villa, indicating that the feudal villages originated as the private estates of Roman magnates. And the Roman Catholic Church, once adopted
by Constantine as the official religion, started to burn heretics. Religious and intellectual freedom, the great gifts of the Graeco-Roman
period, were extinguished. No new technology emerged.

Contrary to myth, the empire did not collapse in the face of unstoppable
barbarian hordes. The numbers of barbarians were always small (a mere
80 000 vandals took the whole of Roman Africa in less than a decade). The empire fell because many of its citizens had emigrated to
the freer, more pleasant barbarian lands (under the late empire, the
population fell from 70 to 50 million) and, crucially, the invading barbarians found themselves welcomed as armies of liberation by vast
numbers of oppressed people. The empire had been warned. In *De
Rebus Bellicus*, published anonymously around AD 370, the author called
for tax cuts, new technology, and political freedoms: 'In the technical
arts, progress is due not to those of the highest birth or immense wealth
or public office or eloquence derived from literary studies but solely
to men of intellectual power . . . [the barbarians] are by no means considered strangers to mechanical inventiveness.' The author blamed the

greed of the rulers for the desperation of the poor: 'This store of gold meant that the houses of the powerful were crammed full and their splendour enhanced to the destruction of the poor, the poorer classes of course being held down by force. But the poor were driven by their afflictions into various criminal enterprises, and losing sight of all respect for the law, all feeling of loyalty, they entrusted their revenge to crime. For they often inflicted the most severe injuries on the Empire, laying waste the fields, breaking the peace with outbursts of brigandage, stirring up animosities, and passing from one crime to another, supported usurpers.'[16] Unfortunately, this very sensible tract was never shown to the Emperor, Valentinian I, even though Ammianus Marcellinus tells us that he was one of the emperors who actually was interested in inventions.[17]

The empire collapsed, not for a lack of Hellenistic science – there was plenty of that – but because it abandoned capitalism. It was a plunder empire, not a market empire. For plunder, it forsook free trade, and it therefore forsook the developments in technology that the free market would have fostered, and it also forsook the development of technologically inspired science. Since new technology is effectively synonymous with economic growth (see the discussion in Chapter 7), we can say that, in modern terms, the empire failed to raise its GDP *per capita*.

The fall of the Graeco-Roman hegemony teaches that the government funding of academic science will not generate useful technology in the absence of an appropriate, capitalist economy. This is so different from the conventional history that we must underline it. A standard textbook like Buchanan's *Technology and Social Progress* emphasises, in the author's own italics, on the very second page, that '*A strong state, in short, is a necessary precondition of industrialization*'[18] but we have shown that, historically, the reverse is true. In antiquity, it was the strong states that suppressed technology, and the weak ones that fostered it, because the weak ones were too weak to rob individuals of their freedom. As we shall see, it took the Dark Ages and their attendant chaos to liberate the human spirit and so fructify commerce, technology and a healthy science.

3 The So-called Dark Ages

In *De Viris Illustribus,* Petrarch described the thousand years between the fourth and the fourteenth centuries as the Dark Ages and, indeed, the fall of the Roman Empire brought little joy to aristocrats, poets, writers, historians or academic scientists. For the common man, however, and for technologists in particular, the so-called Dark Ages represented nothing but liberation. The Roman Empire, like all those before it, had been built on the back of human labour, but by the fourteenth century, men were learning to harness the power of animals and of nature.

We have already noted how, during the terminal centuries of the Empire, its population fell from 70 to 50 millions as its people fled its tyranny or succumbed to its horrors; compared with that imperial nightmare, the world of the so-called barbarians was free and enlightened, as some contemporaries well knew. Salvian of Marseilles, for example, enumerated in his *De Gubernatione Dei,* published during the 430s, the superior economic and personal freedoms of the barbarians. Thanks to their freedom, of course, the barbarians produced much new technology, which came as a nasty shock to the Romans. Tacitus (AD 55–120) had dismissed any threat from the peoples who ringed the Empire: 'Barbarians are ignorant of military engines and the management of sieges, while we Romans excel in such matters.'[1] But when the Visigoths crossed the Hellespont on rafts in AD 400, the Byzantine historian Zosimus could only lament 'the inventiveness of Barbarians'.[2] A century later, the Huns had designed a battering ram superior to that of the Byzantines or the Persians.[3]

For soldiers, the most important barbarian development in imperial times was the saddle. We know that horses were being ridden in the Ukraine around 4000 BC, because archaeologists have excavated horse teeth which show the characteristic signs of wear produced by metal bits,[4] but without a saddle, riding remains painful and uncertain. Despite hundreds of years of Greek and Roman riding, it was the barbarians of Northern Europe who, in the first century AD, invented the saddle.[5] Without a stirrup, however, even the saddle provides only partial control, and the first crude stirrups (which only accomodated the big toe) emerged, not from Rome, but from the mountain peoples of

29

Afghanistan and Northern Pakistan around AD 100 (there is a Kushan engraving in the British Museum of a big toe stirrup hooked to a saddle). Further East, during the fourth century AD, the penetration of China by Buddhist missionaries from Afghanistan and Turkestan generated a cultural ferment which proved creative to technology, and the full foot stirrup emerged, being first mentioned in the Chinese literature in 477 AD. From there, it spread over the whole of the Eurasian land mass.

There was still much work to be done if the horse was to be properly exploited. In *L'attelage et le cheval de selle a travers les ages*,[6] his beautiful and comprehensive study of the horse in antiquity, Lefebvre des Noettes (a retired Commandant of the French cavalry) showed that the Romans, like all their predecessors, used horses with ridiculous incompetence. There were three main problems: (i) there were no horse-shoes, so hooves broke all the time, incapacitating the animals; (ii) horses were harnessed with a yoke, so the moment the horses tried to pull they strangled themselves; and (iii) no one knew how to harness one horse in front of another, so there was no way of using more than a small number of horses together. During the ninth and tenth centuries the Northern Europeans overcame all three problems. The nailed horseshoe, the horse collar (which threw the weight of a load onto the horse's shoulders) and the tandem harness all appeared. The economic consequences were enormous; the horse was to the Dark Age what the steam engine was to be to the Industrial Revolution, and men had learnt to extract more than four times as much power from the horse. In consequence, the use of the heavy plough was facilitated, the switch from two- to three-field crop rotation was eased, and the ox was relegated. Agricultural productivity doubled; no wonder Europe suddenly seemed more vital from the turn of the millenium.

One of the most important of the early barbarian advances was the harnessing of water power. Archaeological excavations of two dams in Jutland which date from around the time of Christ have revealed the first known water wheels.[7] Even the Romans, technically sterile though they were, could not ignore such an advance and, by the first century AD, Vitruvius was describing Roman water mills.[8] Remarkably, these were an improvement on the original barbarian models, which were horizontal mills; the Romans built vertical ones. Nonetheless, as Marc Bloch noted,[9] the Romans built very few mills, and it was only after the Empire collapsed that the water mill spread rapidly throughout Europe. After the fall of the Empire, the Northern Europeans continued to develop water technology; the world's first lock gates were installed at Bruges in 1180.

Soap was a barbarian invention. Pliny states that it was: 'an invention of the Gauls for giving a reddish tint to the hair'.[10] It was prepared from tallow and the ashes of the beech and yoke-elm. By the second century AD, Aretaeus reported that the Gauls were using soap as a detergent, washing their clothes with it (Aretaeus himself used it in the bath; see Galen's report in his *Works*[11]). Galen says that the Gauls' soap was made from ox, goat or sheep tallow, and a lye of ashes with quicklime. Despite the personal experience of Aretaeus, F. W. Gibbs, the historian of soap,[12] doubts that any other Romans used it, but after the fall of the Empire its use and manufacture spread widely, and by the seventh century there were enough soap makers in Italy to support a craft guild or '*arti*'.[13] Under Charlemagne, the soap maker was counted as one of the craftsmen of the large domains.

Following the western imperial collapse, other technologies flowered. The power of the wind was increasingly harnessed. Consider the sail: Roman square-rigged ships could not tack, and sailors could only move upwind under oars (although T. R. Holmes has suggested that they could sometimes sail slightly upwind).[14] None the less, the situation improved dramatically with the invention of the lateen sail, the characteristic fore-and-aft sail of the Mediterranean, Nile and Red Sea (the Arab dhow). Representations of lateen sails first appear in Greek manuscripts of the ninth century, but whether they were invented by the Byzantines or Saracens is not known.[15] Perhaps it was the juxtaposition of the different cultures that promoted its development, but one thing we do know is that the lateen sail is not Latin in origin, despite its misleading name. That name was a piece of Italian cultural theft.

The next technology to harness the power of the wind, the windmill, was clearly a Northern European invention. The first record of a windmill is in a charter of St Mary's at Swineshead, Lincolnshire, dated 1170, and it spread so rapidly through Europe that Pope Celestine III, 1191–8, decreed that windmills should pay tithes.[16] The windmill then definitely spread East, and not the other way round; in his account of his experiences during the Third Crusade, Ambroise wrote:

The German soldiers used their skill
to build the very first windmill
That Syria had ever known.[17]

More fundamental than the windmill was the crank. There are two types of mechanical motion, reciprocal and rotary, and it takes the crank to convert one to the other. Without a crank, therefore, only the crudest, simplest machines can be built, and the crank is second only to the wheel in importance to technology. Yet the Romans did not have it, and the first reference to the crank in the West is as late as an illustration in the Utrecht Psalter, dated to between 816 and 834. (Although the Chinese also had the crank, the Europeans almost certainly invented it independently, contact between China and Europe being then very tenuous; there are no contemporary records of the crank in countries between China and Europe.)[18]

In conclusion, the Dark Ages witnessed wonderful improvements in technology. In addition to the ones mentioned here (the saddle, the stirrup, the horseshoe, the horse collar, the tandem harness, the water mill, the fore-and-aft sail, soap and, perhaps most importantly, the crank), Lynn White Jr, the historian of technology, has also chronicled the introduction in the Dark Ages of *cloisonné* jewelry, felt, trousers, skis, butter, rye, oats, hops and the making of barrels.[19] In consequence, men entered the fourteenth century with a greater mastery of animals and machines than ever before. In consequence of that, men could again produce the economic surplus off which academics, writers, scientists and poets – men like Petrarch – could feed. Sadly, the consequence was ingratitude, a denunciation of the era of advancing technology as a Dark Age.

Charitably, one might excuse Petrarch's superciliousness as ignorance, but for one consideration: everything in which Petrarch believed – academies, poets, historians, science, etc. – flourished in Hellenistic and Imperial times. The consequence was degradation. Little of Petrarch's precious humanistic culture flourished between the fourth and fourteenth centuries, and the result was a glorious leap into lovely technology and better living standards. Rather than address the obvious question that this raises – is academic science important to economic growth? – Petrarch opted for abuse. This is very suspicious.

4 The Commercial Revolution

We saw in the previous chapter how, up till about AD 1000, technology accelerated in the hands of the barbarians and the inhabitants of Northern Europe. From the turn of the millennium, however, the acceleration in technology did itself accelerate. The rate of change increased markedly. This can be easily seen just by walking through the old cities of Northern Europe.

The departure of the Romans left architecture in a poor state. The Anglo-Saxon churches that survive in England, for example, show crude workmanship and a copying of Roman forms such as round-headed windows and arches. The Normans introduced higher standards of workmanship, but Norman romanesque was still a primitive architecture, requiring vast pillars and hugely thick walls. Around 1135, however, Abbot Suger,* building S Denis, near Paris, inaugurated a superior architecture known as gothic. Gothic has at least three advantages over romanesque. First, its arches and windows are pointed, not round (the weakest point in a round-headed arch is the centre, but a pointed arch transmits the vertical weight to the supporting wall, so it is stronger). Second, gothic roofs are light because they are vaulted around individual ribs, so allowing the supporting walls to be thin and generously windowed (the romanesque barrel vaulting was immensely heavy). Third, the walls in a gothic building can be made thinner still by using flying buttresses. The consequence of these developments, coupled to improved standards of workmanship, was that gothic walls became so strong that they could consist almost entirely of windows if luminosity was desired (see, for example, the chapel of King's College in Cambridge) or that gothic cathedrals could be built immensely tall if that was the aim (see the cathedrals of Amiens, Beauvais or Cologne). The contrast between a Norman building (dark and heavy) and a late gothic one

* It was Suger, Regent of France for seven years, who claimed that: 'The English are destined by moral and natural law to be subject to the French and not the other way around.' That was a perfectly proper attitude for a Frenchman, but it is odd that the contemporary British political caste, in its adherence to Brussels, should now agree.

33

(luminous and sublime) speaks of the considerable medieval advance in technology.

Technology advanced on all fronts during the middle ages. Consider the fifteenth century: Johannes Gutenberg printed the first book with movable type, a Vulgate Bible, in 1448; a large number of advances in naval technology, such as the stern rudder, allowed Christopher Columbus to cross the Atlantic in 1492; blast furnaces hot and durable enough to actually melt iron, such that it could be poured into moulds (cast iron) emerged during the fifteenth century (previously, it had had to be beaten into shape); and mining for metals and the creation of a coal industry advanced so rapidly that by 1556 Georgius Agricola could publish his encyclopaedic *De Re Metallurgica*.

Despite the technological ferment, however, living standards remained low. It is commonly supposed that, before the Agricultural and Industrial Revolutions, Man inhabited a rural Arcadia. Bucolic, pink-cheeked, besmocked peasants ambled plentifully through green and pleasant fields, counting their sheep and plucking at apples – or so runs the myth. Wickedly, the Agricultural and Industrial Revolutions changed all that; ruthless landowners enclosed the fields and forced the peasants into urban slums, where cruel mill owners drove them fourteen or sixteen hours a day – all the while forcing their children up chimneys and befuddling them with gin. This makes a good story – it has heroes and villains and it updates the Garden of Eden – but it is just that, a story. The truth is more complex, yet more hopeful.

In 1688 the first systematic survey of living standards in England, *Natural and Political Observations and Conclusions upon the State and Condition of England*, was published by Gregory King, the pioneer statistician. As Secretary to the Commissioners for the Public Accounts, King had access to official statistics such as the Hearth Tax and excise returns; he also conducted his own surveys. In 1688 he found that there were 5 450 000 people in England and Wales but, of those, no fewer than 2 825 000 starved. Horrifyingly, over half of all Englishmen or Welshmen would have died from malnutrition and poverty, King found, but for charity or poor relief. Some of those people – common soldiers or common seamen – earned a wage, but one too small to feed and clothe a family; it needed to be supplemented. Huge sections of the population, however, could not even find a wage (Table 4.1).

A pre-Agricultural, pre-Industrial Revolution society was a terrible place for most of its inhabitants. Some people, of course, lived well, but for the majority life was bitter. As Table 4.1 shows, the fundamental problem was not one of distribution – had the aristocrats' in-

TABLE 4.1   *The population and wealth of England in 1688*

Rank	Numbers of people	Average annual income per person	Average annual expenditure per person
Aristocrats and gentry	153 520	£40 18s	£37 16s
Professionals and officers	308 000	£11	£10 3s
Freeholders and farmers	1 730 000	£9 16s	£9 6s
Tradesmen and craftsmen	484 000	£13 12s	£12 6s
Common soldiers and sailors	**220 000**	**£6 16s**	**£7 10s**
Labourers	**1 275 000**	**£4 5s 7d**	**£4 12s**
Cottagers	**1 300 000**	**£2**	**£2 5s**
Vagrants	**30 000**	**£2**	**£3**

SOURCE: These figures have been calculated from Gregory King's table as reproduced by P. Mathias (*The First Industrial Nation*, London and New York, 1983). The numbers of people increasing the wealth of the country were 2 675 520, but the numbers reducing it were 2 825 500 (**in bold**).

come been equally distributed it would not have made much difference – the fundamental problem was one of poverty. There simply was not enough wealth around. Yet, grim though life was in England at the end of the seventeenth century, it was significantly better than it might have been. Maddison has calculated that the average income per head in England in 1700 was around $200 a year (calculated as US dollar purchasing equivalents in 1965), while as late as 1965, the average for Africa was only $130, and some African countries managed less than $100.[1]

England was richer in 1700 than present day Africa for two reasons. First, England enjoyed all the technology we have described above, while much of Africa has yet to leave the Iron Age. Second, England, like much of North Western Europe, had acquired a crucial social discipline – the control of fertility.

The fundamental problem of population was posed by the Reverend

Thomas Malthus in his *Essay on Population* published in 1798: human populations will always tend to grow faster than will food production; it will always be easier for a couple to produce four children than to double its agricultural productivity. Undisciplined, therefore, Man will always starve. We see that in Africa today. Across vast tracts of Africa, the majority of women may never experience a menstrual cycle. Sexual intercourse in many tribes starts very young, and the first evidence of sexual maturity for most young women is pregnancy. Thereafter, women produce child after child until the menopause (there is a gap of around two years between each child because breast feeding is a natural contraceptive). Such fecundity must ineluctably condemn a society to grinding poverty, because however quickly technology might develop, it can never outstrip the rate of reproduction.

Western European society, however, learned to control fertility. By 1700, in England, over 98 per cent of births occurred within marriage.[2] And, by 1700, couples were marrying late. The average age at first marriage was 28 for men and 27 for women.[3] Moreover, parish records show that the age at marriage depended upon wealth; in hard times people married late, in good times early.[4] The birth records also show that, even after they had married, couples must have controlled their fertility, because family sizes rarely approached the biological potential, although we can only speculate how conception actually was regulated (marital sodomy was apparently the favoured technique in renaissance Italy).[5] This contrast between European and many Third-World cultures (where people marry early to produce as many sons as possible) accounts for much of the relative wealth of early Europe. It has also accounted for some amusing literature. In his comic novel *England their England* (1933) A. G. Macdonell, a Scot, teased the English over their materialistic and unromantic approach to marriage:

'You see, if we got married, we couldn't be so badly off', said the man. 'I've got three hundred [pounds a year] of my own, and I get four hundred from the shop; that's seven hundred. Do you think you could sting the old man?'

'Might squeeze another hundred', said the girl, cool, precise, steady. 'Not more. He doesn't part easy.'

'That's rather grim', said the man.

'It won't go far', pointed out the girl. 'You can't get a service flat under five hundred'.

'That's a rather grim prospect', said the man gloomily. 'And I'm damned fond of you, Slick.'

'And I'm damned fond of you, Crabface'.

'But still – '.

'No good', she said decisively.

Despite its poverty, England in 1688 was significantly richer than every other contemporary European nation (but one) and the explanation can be found in Gregory King's table; England in 1688 was remarkably commercial. There were 50 000 families of merchants, tradesmen or shopkeepers; 60 000 artisans and craftsmen; and 55 000 professionals. Only one other country in Europe was more commercial – Holland – and that was the only country that was actually richer than England. Every other European country was predominantly agricultural, with populations of peasants engaged in subsistence (i.e. growing food solely for their own consumption, not to sell), but in England and Holland most people engaged, to some extent, in the market economy, buying and selling produce as well as producing or consuming it. (England's early move away from an agricultural, subsistence culture is reflected in the fact that it, alone in Europe, does not retain a national peasant costume). England and Holland enjoyed vast foreign trade and their imports and exports each accounted for 10 per cent of the national product of both countries, then remarkable figures.[6]

The two richest countries in seventeenth century Europe, therefore, were the two that boasted the largest merchant class, and which embraced the market most enthusiastically. Was this a coincidence? This is an important question because it was precisely these two countries of England and Holland that were later to lead the world through the Agricultural Revolution of the eighteenth and nineteenth centuries – and this revolution inaugurated a new era in human history, the era in which countries first learned to feed all of their people. To unravel the relationship between the market and the series of technological changes that led to the Agricultural Revolution, we will have to go back to the feudal period.

The decline of the Roman Empire bequeathed to Europe a social and economic structure that presaged feudalism. We saw in Chapter 2 that as the Empire forsook free trade, free association and the rule of law, its peoples coalesced on the great estates. Each estate was effectively autonomous. It grew all the food it needed, and its craftsmen supplied its local wants. Such an estate constituted a rational response to the chaos, tyranies and the destruction of free trade of the declining Empire. But rationality is never enough in human affairs and feudalism acquired, from the Church, a profound moral underpinning.

At one level, feudalism was a continuation of late imperial society. Each estate – now called village – continued to belong, as it had under the Empire, to the local chief or count or lord; and his workers, now called serfs, belonged to him too – again, just as they had under the Empire. But serfs, as Christians, were not slaves; they had certain rights, even if they were highly restricted. Very importantly to the medieval mind, serfs and lords were bound together by a network of mutual obligations. Each serf was allocated some land and some of his own time for his own use; the rest of his time was spent on the lord's land, growing the food off which the lord and his retinue would feed. In return, the lord and his retinue would protect the village from marauders.

Feudal society was self-consciously moral. Land was distributed between peasants with scrupulous fairness, and each peasant worked a number of separate strips of land, distributed between different fields, so that good and bad soil was fairly shared. Nowhere did feudalism adhere more closely to Christian morality than in its economic relationships. Feudal thinkers abhorred money. Human relationships under feudalism were based upon loyalty, duty, respect or love. The village was an almost cash-free zone. Peasants donated time, not money, to the lord, and in return he gave them land and protection. This morality even pervaded urban society. Intellectuals like Geoffrey Chaucer criticised lawyers for defending clients for money rather than from a personal belief in the client's case. Doctors, too, were attacked for charging fees. Johannes de Mirfield of St Bartholomew's Hospital wrote: 'But the physician, if he should happen to be a good Christian (which rarely chances, for by their works they show themselves the disciples not of Christ but of Avicenna and of Galen) ought to cure a Christian patient without making even the slightest charge if the man is poor; for the life of such a man ought to be of more value to the physician than his money.'[7]

Free trade was abhorred. In AD 301 the Emperor Diocletian had declared that 'uncontrolled economic activity is a religion of the godless', and feudal society agreed. Medieval Europe believed that the State should impose 'just prices' and 'just wages'. Consider this passage from *English Parliamentary History* for 1315: 'to settle the price of provisions, which were grown excessive dear at that time, the King summoned a Parliament to meet at Westminster on the 20th December. When the prelates, peers, and commons were there assembled, they took into consideration the sad condition of the kingdom, and how to abate the excessive price of victuals, which by reason of the late bad years was grown so scarce that the ordinary people had much ado to live. The archbishops, bishops, earls, barons etc. presented a petition to the king and his council, praying, that a proclamation might be issued out settling the Price of Provisions in the manner following: "Because, say they, that oxen, cows, muttons, hogs, geese, hens, capons, chickens, pidgeons, and eggs, were excessive dear, that the best ox, not fed with corn, should be sold for 16s and no more: and if he was fed with corn, then for 24s at most. The best live fat cow for 12s. A fat hog, of 2 years old, for 3s 4d. A fat wether or mutton, unshorn, for 20d, and shorn for 14d. A fat goose for $2^1/_2$d. A fat capon for 2d. A fat hen for 1d. Two chickens for 1d. Four pidgeons for 1d and 24 eggs for 1d. And those who would not sell the things for these rates were to forfeit them to the king". Proclamation was made in every county of England accordingly.'

Where trade was tolerated it was tightly regulated, because free markets were viewed as, literally, the work of the devil. How could it be right that, of two men working equally hard, one should earn more than the other simply because of vagaries of surplus or shortage? Usury was especially abhorred (Pope Clement V's fierce prescription on usury in his bull of 1312 is still remembered) for how could it be right for a man to enrich himself without labour or social obligations just by lending money at interest? In his *Inferno*, Dante stated that *usura offenda la divina bontado* (usury offends the divine goodness).

Feudal societies were rigid. The conditions of service in the fields – the amount of time the peasants had to donate to the land, or the amount of land they received in return – were determined by custom, ancient usage or law, and they obtained over generations. Feudalism, in short, presaged the huge, stable but sterile empires of the past. Like them it enmeshed all humanity in a complete theocratic, cash-free, social tyranny. And, like them, it promoted no new technology: it provided no opportunity for it, nor reward. But feudal Europe differed in one crucial respect – it was not politically stable. Whereas all of Egypt or Rome had been ruled from one centre, medieval Europe was divided against itself. Village against village, province against province, duchy against kingdom – even pope against pope – all of Europe was at war. Progressive thinkers deplored the endless conflict, and intellectuals dreamt of a new Charlemagne, a Holy Roman Emperor who would unite Europe under the guidance of Mother Church. Yet war created capitalism.

A lord, defending his village, would desperately need chain mail; but few medieval villages, however self-sufficient in feudal theory, boasted of deposits of iron or of blast furnaces. So medieval lords had to trade; they had to sell food or wool in exchange for breast-plates or swords. Medieval lords, therefore, had to foster a merchant class. Medieval kings, too, initially attempted to fight wars in the proper feudal spirit, under which nobles and villeins served in their armies from loyalty or duty, but even medieval kings needed to purchase services that feudalism could not provide. Such services might include the use of a fleet or the buying of sophisticated siege weapons or the long-term, long-distance provisioning of an army. So medieval kings, too, had to foster a money economy and a merchant class.

Medieval kings never knowingly meant to foster capitalism; a merchant class, possessed of independent wealth, might threaten the royal power and imperil feudal ideology. None the less, kings grew dependent on the merchants, not only for their supply of *materiel*, but also because merchants, even those engaged in non-miltary trade, grew rich, and so provided a ready source of tax or loans. Those kings who refused to accommodate the merchants found their countries impoverished and their war chests depleted. They tended, therefore, to lose wars. Thus did capitalism spread by a form of social Darwinism, red in tooth and claw.

Once established, capitalism tended to grow. Traders concentrated in towns which then promoted their trade by purchasing 'liberties' or 'freedoms' from the King. Such liberties or freedoms amounted to little more than the right to hold markets or the right to self-government, but they fostered trade (and their sales put money into the King's coffers).

The first European country to break away from feudalism was Italy.* Italy was well placed. It enjoyed a good climate, it straddled the trading crossroads between the Mediterranean and the North, and its mountains fostered the local autonomy of city states, many of which became so autonomous that they shrugged off their feudal overlords to become republics. Those republics were generally ruled by the leading merchants as mercantile oligarchies, although sometimes individual merchants would emerge as supreme rulers.

Some mercantile republics grew very rich. Trade enriches in two ways, short-term and long-term. The short-term way is the wealth that emerges from individual transactions. When, for example, it was discovered that salt preserved meat, traders transported vast quantities of

---

* The early Italian dominance of trade and commerce has left its traces throughout Europe. The Bank of England, in London, is situated close to Lombard Street; and until it was discontinued, the two shilling coin in Britain's pre-1970, pre-decimal currency was called the florin – a residue of the time when Florence's coin dominated European finance. Indeed, the symbol of the British pound, £, represents the 'l' of lira.

salt from mines and coasts to peasants throughout Europe. The peasants were enriched, because they could eat meat regularly, which made them stronger; this enabled them to grow more food, through which they could pay the traders for the salt, and perhaps purchase a few more goods. Thus was everyone immediately enriched. But trade also enriches in the long-term by providing incentives, in the form of increased profits, for innovation. Thus the medieval trading Italian cities invented new instruments of commerce. Insurance, for example, is a vital spur to trade: a merchant is more likely to risk a ship if he can insure it against loss. A primitive form of marine insurance had been developed by the Greeks; known as a bottomry and respondentia bond, it consisted of a loan, repayable at a high premium if the ship returned with goods, but not repayable if the ship was lost. But by the twelfth century the Italians had invented premium insurance, the modern variety, under which a ship is insured for a fixed price. By 1400, two-thirds of Venetian maritime trade was underwritten in Florence.

The Florentines invented double-entry bookkeeping. In itself, this was only an aid to accounting. Under double-entry bookkeeping, every transaction is entered into a ledger twice, first as a credit or debit of money, and secondly as a credit or debit of goods. The initial purpose of double entry bookkeeping was to act as a check on accounting: at the end of a period of time, the debits and credits should balance; but the long-term effect of the system was to invent the concept of the trading organisation or firm as an entity separate from that of any individual. If an enterprise can accumulate debts or credits in its own right, even if only on paper, it assumes an entity of its own. As Sombart wrote: 'One cannot imagine what capitalism would be without double-entry bookkeeping.'[8] Once the concept of the firm had been created, capitalism ceased to be restricted by the capacity of lone individuals and could assume the shape of a corporation, handling vast sums of money, managing armies of employees, and persuing long-term strategic goals (including research and development).

The first cheque in the world was written in Pisa during the fourteenth century. Before then, the mule trains that crossed Italy or Europe from one end to another had to return with gold bullion or coins, and these would cross other mule trains passing in the opposite direction, also burdened with gold. Each, moreover, would be harassed by bandits. But the invention of cheques, bills of exchange and deposit banking facilitated trade wonderfully. Great banks like those of the Bardi, Peruzzi and Acciaiccioli spread branches throughout Europe. The Florentine Medici bank, in the time of Cosimo (1399–1464) had branches in Rome, Milan, Pisa, Venice, Avignon, Geneva, Bruges and London.

And the Italians invented the patent. Italian merchants increasingly devoted themselves to developing new products, but they soon encountered a disincentive: why should a man devote time or money on developing a better mouse trap, if a competitor can then sell an identical copy? Why, indeed, should a man devote time or money on developing a better mouse trap, if a competitor can then actually undercut him because the competitor would not need to recoup the costs of invention? Medieval Italian inventors, therefore, soon started to demand grants of monopoly on their own inventions – grants that were known as patents. The world's first patent was awarded to an inventor by the Republic of Florence in 1421, and the world's first systematic patent laws were tabled in Venice in 1474.

It can be seen, therefore, that the technological innovation that flows from capitalism is but one aspect of the innovation that capitalism inspires. The insurance policy, double-entry bookkeeping, the cheque or the patent were just as innovative as the lock gate, the flying buttress, movable type or the blast furnace. The patent, however, is of especial importance here, because it provides the formal link between capitalism and technological innovation. It represents, for the first time in history, the recognition of commercially inspired research. We saw in Chapter 3 that, in Hellenistic and Imperial times, Kings and tyrants were often very generous to academic science, but nothing useful emerged. We can see with the invention of the patent, however, that capitalism had generated both a motive and a mechanism for the development of useful new technology.

Early Italian capitalism also provided much pure science. Consider mathematics. We saw in Chapter 2 that the Egyptians had solved linear and quadratic equations, but in the intervening millenia there had been no advance in solving cubic equations. In 1510, however, Scipione del Ferro, a professor of mathematics at the University of Bologna, discovered the solution to the cubic equation of the form $x^3 + mx + n = 0$ where there is no simple x term. In 1534 Niciolo Tartaglia, a teacher of mathematics in Venice, found the solution to the cubic equation $x^3 + px^2 = n$ where there is no simple x term. In 1545 Gerolano Cardona, a lecturer in mathematics in Milan, published his *Ars Magna*, the major mathematics book of the era, which incorporated the cubic solutions of Del Ferro and Tartaglia, the solution to the quartic equation discovered by his own servant, Lodovico Ferrari, and his own fundamental contributions to algebra.

These discoveries were financed by the free market. Cardona, the greatest of the mathematicians, taught in Milan at the Piattine schools, which had been founded by a bequest of Tommaso Piatti. In turn, he taught his own servant, Ferrari. Tartaglia was a freelance teacher, while Del Ferro taught at the University of Bologna which, though it enjoyed papal and royal recognition, survived on students' fees. Nor did the free market ignore these discoveries. Tartaglia, Ferrari and Cardona became famous throughout Europe for their genius (Del Ferro never published) and all were offered appropriate rewards. Ferrari, for example, was made the supervisor of tax assessments in Mantua by Ercole Cardinal Gonzaga, which made his fortune.

Early Italian capitalism transformed the world of art. Surviving medieval and renaissance Italian buildings, paintings and books constitute one of the wonders of the world. When England, for example, was still building crude Anglo-Saxon churches of rough stone, thick walls and tiny windows, the Florentines were building beautiful churches like S. Miniato al Monte (1018–62). Long before King's College chapel, the flower of English gothic, was even being built (1446–1515), Brunelleschi was not only inventing a new style of architecture (the Foundling Hospital, Florence, 1419–26) and not only helping to invent perspective, he was also inventing new ways of building domes and arches: the traditional way of building a dome or an arch was to first build a frame in wood, then to lay the stones or bricks over the frame, and then to remove the frame, leaving the final arch or dome as a self-supporting structure. The dome of Florence Cathedral, however, was so large that no trees could be found long enough to provide the frame. Brunelleschi solved the problem by inventing a way of laying bricks such that they supported themselves as the dome was being built (1420–34).

We can see, therefore, how capitalism not only creates the commercial motivation for technological innovation, but also how it will produce wealthy men like Piatti who will, freely, support artists, architects, scientists and schools because their importance is so obvious.

Certain aspects of feudalism might reasonably be mourned. Many contemporaries, men such as John Gower, Geoffrey Chaucer, John Clanrowe, Eustache Deschamps and, of course, Petrarch – intelligent, thoughtful, kind men – regretted the supersession of a society whose relationships were based on honour, trust, loyalty and duty by a new order, one based upon money. Modern romantics such as William Morris or Eric Gill have also regretted the invasion of a money economy, and they even tried to lead medieval lives in medieval-like communities.

The dream of a planned, fair, moral, ethical, cash-free society remains strong, particularly amongst socialists and liberals. It clearly represents a fundamental human instinct. But feudalism just did not work very well, if only because powerful people will not always obey moral imperatives.

Feudalism was believed to embody the highest morality. Ever since Plato, people had understood that society needed to be divided into rulers, guardians and peasants, and it seemed natural for the peasants to supply food in exchange for defence and spiritual guidance. As Master Thomas Wimbledon preached in a sermon at St Paul's Cross: 'These three offices are necessary: priesthood, knighthood and labourers. To the priesthood it falls to cut away the dead branches of sin with the sword of their tongues. To the knighthood it falls to prevent wrongs and thefts being done, and to maintain God's law and those that are teachers of it, and also to save the land from the enemies from other lands. And to the labourers it falls to work with their bodies and by their sore sweat to get out of the earth the bodily sustenance for themselves and the others.'[9]

But, for all the theory of mutual obligations, in practice the lords exploited their peasants – despising their 'villeins' as 'villains' – who more than once rose in rebellion (as in the Peasants' Revolt in England of 1381). Moreover, the knights rarely came to the peasants' defence if their own interests dictated otherwise. During the 1370s and 1380s, for example, England was repeatedly invaded by French and Scottish armies. In 1377 the French attacked Winchelsea and burned Hastings. In 1380 a French fleet sailed up the Thames and burned Gravesend. These raids were perennial, and were believed to presage invasion. Parliament feared that the French intended 'to destroy the English language and occupy the territory of England, which God forbid, unless a remedy of force is found against their malice.'[10] In 1386 the French assembled a huge armada at Sluys; Froissart wrote: 'I believe since God created the world, there was never so many great ships together as was that year at Sluys.'[11] The Government planned to abandon London where, according to Walsingham, the mass hysteria drove its citizens to ripping down the houses that bordered the city walls.[12] But where were the knights? Time and time again, just when they were needed at home, English knights left for more lucrative battles abroad. In his *Mirour de l'Omme* (1376–8), John Gower complained:

> Oh knight, who goeth far-off
> Into strange lands and seeketh only

Praise in arms, know this:
If your country and your neighbour
Are at war themselves, all the honour
Is in vain, when you flee from
Your country and estrange yourself.
For he who abandons his duties
And does not wish to fulfil his obligations
But rather fulfils his own desires
Has no right to be honoured
No matter how mighty he may be.[13]

In the late fourteenth century, the other orders of society had to learn to defend themselves. The Abbot of Battle Abbey, Hastings, for example, led his own men to the defence of Winchelsea in 1377, while John Philpot, a wealthy London merchant (he was a grocer) equipped his own fleet to capture a notorious Scottish pirate in 1378. He criticised the absent knights, explaining that he had acted out of pity for: 'the afflictions of the common people and of our native land which now, through your indolence, has fallen from being the most noble kingdom and mistress of nations into such wretchedness that it lies open to plundering by whom it pleases of the most ruthless people; so long as none of you applies his hand to its defence'.[14]

Feudalism did not work economically either, if only because moral principles are contradictory: if prices are to be 'just' they will be inefficient as a signal of supply or demand, yet that signalling is itself a highly moral activity. A society needs to know when essentials are scarce so that resources can be appropriately devoted to increasing them. Equally, a society needs to know when a good is in surplus, so that scarce resources can be diverted elsewhere. This transmission of information constitutes the moral value of a free market, but it is one that feudal societies generally ignored. On occasion, even feudalism had to recognise surplus and shortage, yet its shifts could be comic. For example, the price of a loaf of bread was always fixed – the staff of life could not be tainted by economic exploitation. But sometimes corn was plentiful, and sometimes it was scarce, so the size of the loaf was allowed to vary. It was always the same price, mind.

The transition from a feudal to a capitalistic society required a vast moral and ethical shift as people moved from a world rooted in personal bonds to one regulated by contract. Many people, even within capitalist societies, have not fully made that shift, and many cultures have yet to embark on it (visitors to the West from the Third World are often appalled by its apparent immorality). These issues have been discussed in great depth by many writers, including Max Weber in *The Protestant Ethic and the Spirit of Capitalism* (1904). It is enough for us to note here that the replacement of feudalism by capitalism was historically a precondition for the resumption of technological growth. The Commercial Revolution preceded, and caused, the Agricultural and Industrial Revolutions, and it also preceded and funded the early revolutions in science, art and architecture.

5 The Agricultural Revolution

The commercial and technical leadership of Europe passed away from Italy. Poor Italy, everybody conquered her. The Spanish invaded from the west, the French from the northwest and the Austrians from the northeast. On a *per capita* basis, none of these countries could match Italian wealth, but each was stronger militarily because each was united and each was big. Italy was warned: Machiavelli devoted the last chapter of *The Prince* to urging the Italians to unite, but the republics, kingdoms and duchies continued to squabble, and the predators took their opportunity. In 1527, for example, the Spanish sacked Rome (they were to rule over half of Italy for the next two centuries) and in the same year the French conquered Lombardy; for centuries afterwards the French and the Austrians were to compete for the remaining choice bits of the peninsular.

The commercial and technical leaderships of Europe did not, however, settle on any of the conquerors. For all their power, neither Spain nor France nor Austria advanced standards of living very much. It was sixteenth century Holland and seventeenth and eighteenth century England that led the way through the Agricultural Revolution. Why?

The clue comes from restating the historical facts, using fuller titles for the countries. The commercial leadership of Europe passed from the Florentine Republic, Venetian Republic and Genoese Republic to the Dutch Republic. And from the Dutch Republic it passed to England, which was effectively a republic after the execution of Charles I in 1649 and the Glorious Revolution of 1688. (Ultimately, of course, the commercial leadership of the world was to pass to the republican United States of America; although the richest *per capita* country in the world today, and the long time centre of banking, is the Swiss Republic.) In a near-feudal, monarchical Europe, when republics were rare, it is startling that the commercial leadership of the age should have been monopolised by the handful of republics.

How did the republics arise? Monarchs did not relish the loss of prosperous provinces or cities, but two features seem to have facilitated the appearance of the early republics in feudal Europe. First, they were protected from their hereditary rulers by geography. The Venetians were an island people, while the Swiss and North Italian republics were protected by mountains. The Dutch Republic, though geographically vulnerable, was protected by its distance from its nominal rulers in Madrid. Second, the republics were highly commercial entities, whose wealth enabled them to proceed so far down the route of purchasing 'liberties' and 'freedoms' that they achieved outright independence. A battle with the feudal overlord, however, was generally necessary before he finally acknowledged his loss of sovereignty.

The emergence of a republic, therefore, represented the triumph of the merchant classes over the feudal lords. Republics were rarely democratic in the modern sense, although as oligarchies they were often more democratic than the kingdoms, but they were emphatically capitalist. And the importance of capitalism to technological innovation is borne out by the historical observation that the technical leadership of Europe followed the commercial leadership, passing from the Venetian and Florentine Republics (Commercial Revolutions) to the Dutch Republic (Agricultural Revolution) to the English near-republic (Agricultural and Industrial Revolutions). It was, therefore, capitalism that spawned the most important of all technological revolutions, the Agricultural Revolution (important, as noted above, because it enabled society, for the first time, to feed all its citizens). Let us examine how capitalism succeeded where feudalism failed.

Under feudalism, farming practices were determined by intricate mutual obligations that penetrated to the very core of society. They were, therefore, effectively fixed. Re-negotiation or change was appallingly complex; any development would damage a vested interest. Nothing could apparently dent feudal relationships: after the Black Death (1348) had nearly halved the population of England, some of the remaining peasants threatened to leave for the towns unless their conditions of service improved. This would, actually, have made good commercial sense, which is why the exiguous free-market forces were attracting peasants off the land, but the English Parliament responded in 1350 with the Statute of Labourers to force servants and labourers to stay where they were, with no change to their conditions of service. After 1348, some marginal land might have been allowed to lie fallow for longer, but otherwise farming practices remained unchanged. (The difficulties in trying to modernise a feudal village's farming practices are well described by Tolstoy in *War and Peace*, whose hero Levin tries in vain to persuade his own serfs to adopt contemporary methods.)

With the growth of trade, however, and with the introduction of the money economy, feudal landowners increasingly paid the serfs with money for their work. This proved to be highly efficient; under feudalism, peasants owed their lord a certain amount of time on his land, but people work harder if they work to fulfil a contract rather than to watch a clock. Feudal landlords who paid wages found that they could recoup the costs through selling the produce at market, while the extra productivity of the workers would generate an additional profit. Later, landlords would even sell their agricultural land to the peasants, which would not only generate capital (which might provide valuable investment elsewhere), it would also maximise the nation's agricultural productivity, because farmers were now working for themselves. Alternatively, a landlord would 'enclose' common land. Under enclosure, a lord would acquire the peasants' traditional land for himself by buying them out, only retaining a minimum number as paid labourers.

The commercialisation of agriculture, therefore, replaced the inflexibility of feudalism with a system driven by personal motivation; let us examine how such human motivation led directly to the Agricultural Revolution in seventeenth and eighteenth century England.

The eighteenth century in Britain really did witness an acceleration in agricultural productivity. Whereas the sixteenth and seventeenth centuries each enjoyed total increases in agricultural output of about 30 per cent a century, the 18th century enjoyed 61 per cent.[1] Between 1700 and 1800 the amount of land under cultivation rose by no more than 5 per cent (because almost all available land was already being exploited) and between 1700 and 1800 the agricultural work force only rose by about 8 per cent; so it can be seen that the eighteenth century enjoyed an increased output per acre of 44 per cent and an increased productivity per person of 47 per cent. Although some of these increases can be attributed to increases in capital (in the sense of more farm buildings or more equipment) the overriding conclusions that emerge from any survey of the eighteenth century is that it witnessed real improvements in technology – the Agricultural Revolution.

The greatest of the technological changes was the rotation of crops. In medieval times, fields had to lie fallow for one year in three to recover their productivity, but the agricultural innovators discovered that certain crops, such as clover, could not only enrich the fields, they could also sustain cattle. The Norfolk four-course system, which emerged during the seventeenth century in East Anglia, involved the planting of wheat the first year, the planting of turnips the second, the planting of barley (undersown with clover and ryegrass) the third, and the grazing of the clover and the ryegrass by cattle during the fourth year. Thus no years were wasted to fallow, and crops that impoverished the soil were rotated with those that both enriched it and also sustained cattle whose droppings would further fertilise the fields.

Such developments did not emerge by chance – British agriculture was a ferment of 'improvements' as they were called. New types of cattle such as the Herefords were carefully bred, new fodder crops were introduced, new machines such as Jethro Tull's seed drill were invented. Agriculture had its heroes, many of whom farmed in Norfolk. One such was 'Turnip' Townsend (1674–1738) who combined a life at the highest levels in politics (he was Secretary of State between 1721 and 1730) with improvements of crop rotation. Another Norfolk politician, aristocrat, landowner and improver was Thomas Coke (1752–1842) who inherited an estate at Holkham with a rent-roll of £2200, and who by his efforts increased it £20 000. He converted a sandy rabbit-warren into a model farm which agriculturalists came from all over Europe to visit.

The 'New Husbandry' had its propagandists such as Arthur Young (1741–1820), another East Anglian whose *Farmer's Kalendar* went into its 215th edition in 1862. Young was like Abraham Lincoln in that all his early efforts failed. He performed 3000 experiments on the farm he inherited from his mother in 1763, and not one worked. Between 1766 and 1771 he worked a good-sized farm in Essex, but he experimented that one into bankruptcy too. For a time he was forced to farm in Ireland until his writings rescued him. He was better at explaining what to do than at doing it.

These men, great though they were, were but the prominent manifestations of a profound social movement. British agriculture revolutionised itself. From the seventeenth century onwards, British farmers in their thousands sought new technology, and they banded together to provide the appropriate institutions. Innumerable farming societies such as the Highland and Agricultural Society of Scotland (1785) or the Royal Agricultural Society of England (1830) were created to promote the best technology. The Gordon's Mill Farming Club, for example, which was active around the Aberdeen area, helped introduce the Norfolk four-course rotation to North East Scotland during the 1760s.[2] The dissemination of new knowledge and of new strains was taken very seriously; here, for example, is an extract from the 1799 edition of Arthur Young's *Farmer's Kalendar*: 'Swedish Turnips: very fortunately for the agricultural interest of the Kingdom, Lord Romney last winter presented the Board of Agriculture with four sacks of the seed of this invaluable root, which was distributed in parcels of four or five pounds to a great number of gentlemen, and to all the provincial societies in the Kingdom.'

Agricultural shows and fairs were held regularly throughout the country, and their competitions helped promote the development of better breeds. Individual landowners diffused the best practice: The Earl of Haddington (1680–1735) wrote 'I got a farmer and his family from Dorsetshire in hopes that he would instruct my people in the right way of inclosing, and teach them to manage grass seeds.'[3] Farmers knew how to judge advances scientifically: in June 1793 a Basingstoke farmer, Thomas Fleet, advertised a sheep drench to cure liver rot. He offered to test against any infected herd, suggesting that the herd be divided into two equally sick groups, only one of which would receive the treatment: 'all the Sheep shall be together during the Experiment, and live in the same manner, the *Drenching* only excepted'.[4]

By the mid-nineteenth century, it became obvious that empirical technology was not enough to advance agriculture; new science was needed. The major scientific societies in the Kingdom had long devoted much time and effort to the systematic study of agriculture. Thomas Bate, for example, published his paper on cattle diseases in the *Philosophical Transactions* of the Royal Society in 1718. The Royal Society of Arts and the Lunar Society* of Birmingham were as active in agricultural research, but the time had come for organisations to be dedicated solely to agricultural research, education and practice, so academic institutions like the Royal College of Veterinary Surgeons (1844) or the Royal Agricultural College at Cirencester (1845) were created. Individual landowners, too, experimented in science. In 1843 the wealthy farmer J. B. Lawes, in association with his scientific colleague G. H. Gilbert, turned his farm at Rothamsted into a laboratory and discovered, amongst other good things, superphosphate, the marvellous fertiliser that is produced when bones are treated with sulphuric acid.†

---

* The Lunar Society, an association of scientists, engineers and industrialists (Wedgwood, for example, was a member) was so called because it met every full moon, to allow its members to travel home by moonlight.

† Lawes and Gilbert worked together for over half a century, examining a host of topics in soil chemistry, plant and animal nutrition and crop production. Their Rothamsted Experimental Station survived them, celebrating its 150th anniversary in 1993. It has continued to host wonderful research, including Fisher's transformation of statistics, Bawden and Pirie's fundamental work on the tobacco mosaic virus (which was integral to the development of molecular biology) and the development of the pyrethroid insecticides. Rothamsted has specialised in long-term studies, the oldest of which is the 150-year-old Broadbalk study on winter wheat, which has been sown and harvested on the same plots since 1843.

Farmers were quick to exploit the new science: in his *Patriotism with Profit; British Agricultural Societies in the Eighteenth and Nineteenth Centuries*, K. Hudson described how, during the nineteenth century, the agricultural societies started to employ chemists to develop new manures, to test established ones, to perform soil tests and to teach. In 1857 one society reported 'a marked change . . . in the average value of samples of manure offered for sale. . . . The fraudulent dealer is disappearing or hiding in obscure corners, well knowing that the member of our society has an infallible test at hand.'[5]

The net result of this ferment of technology, education and science was marvellous: by 1850, agricultural productivity in Britain was increasing by 0.5 per cent a year, an unbelievable rate when viewed historically.[6] No wonder the quarter-century between 1850 and 1875 was called the 'Golden Age of British Agriculture'.

But where was the Government? Surely all these improvements in machinery, agricultural technology, fertilisers, crop rotation, the development of new breeds, the introduction of new strains and the creation of the agricultural societies and colleges could not have occurred without the control and strategic guidance of politicians and civil servants? Where were the policy documents, the mission statements and the sub-committees? Where were the central government laboratories, the agricultural research stations, and the councils for planning and co-ordination? Could a society like that of eighteenth- and nineteenth century Britain have initiated a profound technological revolution, have led the world in agricultural self-sufficiency, and then grown so well-fed that the French came to call the British '*rosbifs*', in the absence of government-funded research and development?

Remarkably, that is exactly what happened. Eighteenth- and nineteenth-century England subscribed to *laissez faire*. Taxes were minimal (there was no income tax in peacetime, for example) and governments rarely intervened in the economy. There was no government support for agricultural research and development. For a short time, the government did create a Board of Agriculture, but all that did was to sponsor, between 1803 and 1813, an annual course of lectures by Humphry Davy on agricultural chemistry. So dwarfed was the Board by the activities of the privately funded societies for agricultural research and development that, in 1822, it was disbanded – to general indifference. Thus ended the government's sole contribution to the world's most important technological revolution.

People are not stupid. If it is in their own interest to invest in research and development, they will. If their own individual efforts are insufficient, they will club together in cooperative schemes like the Royal Agricultural College at Cirencester or the various Royal Agricultural Societies (the epithet 'Royal' does not imply government funding in Britain; it is purely honorific). Enlightened self-interest, working through the free market, will ensure that people pay for research, training and education if these are needed. Intervention by government, moreover, during the eighteenth and nineteenth centuries would have seemed especially gratuitous. Britain was then predominantly agricultural, with perhaps 70 per cent of the workforce employed in the fields. Farmers ran the country, since most lords and most MPs owned land. Taxes to support agriculture, therefore, would have fallen largely on agriculture, simply because that was the largest and most productive element of society. It would clearly have been absurd for the farmers to have enacted legislation to tax themselves so that bureaucrats could then dictate which research would best suit agriculture; the farmers would have known how best to control their own money. Imagine how J. B. Lawes, the wealthy discoverer of superphosphate, would have felt if someone had suggested that he be taxed so that agricultural experiments could then be funded by the State. Imagine, then, how Lawes would have felt if he had been told that he might possibly do some experiments of his own but only if his applications for research grants were judged acceptable by peer review! Special interests only struggle for government funding if they suspect that the tax load will fall on others.

It might be argued that, had governments intervened in agricultural research, their judgements might have proved superior to those of the gentleman farmers. But who was going to advise the government, if not the gentleman farmers? It was they, after all, who most understood agriculture and its needs. It might also be argued that governments would have focused greater collective resources on research and development than did the free market. Yet some other countries' governments (notably the French) did indeed support agricultural research very considerably (see Chapter 6) but it was *laissez faire* Britain that spawned the Agricultural Revolution. Indeed, it is just as probable that had governments intervened in Britain, *less* money would have been spent on agricultural research and development. British agricultural research was supported to a degree that has amazed economists. Their amazement has been directed at the phenomenon of 'hobby farmers'.

During the eighteenth and nineteenth centuries it became fashionable for gentlemen to sponsor agricultural research and development. King George III, for example, ran a model farm at Windsor, and he rejoiced in his epithet of 'Farmer George'. Of all his titles, it was the one of which he was most proud, because it was the one conferred upon him spontaneously by the collective British aristocracy, much of which was also trying to improve agricultural productivity.* The economists' amazement at this activity has derived from the lack of private return, since it appears that very few gentlemen hobby farmers could have personally benefited from their research. As Professor E. L. Jones of La Trobe University wrote in the standard textbook *The Economic History of Britain since 1700*, 'The central puzzle is the emergence of a British taste for hobby farming and agricultural improvements.'[7] It is a puzzle for which Professor Jones could provide no answer.

The apparent difficulty of recouping one's personal investment in agricultural research is illustrated by the problems encountered by improvers who were not hobbyists and who wanted some return. Consider this advertisement which appeared in *The Country Magazine* for 15 December, 1788, where an inventor offered to reveal 'an entirely new mode of Winter Feeding Sheep' which could increase profits by over 50 per cent, if one thousand persons, each subscribing 20 guineas, would club together to buy the secret from him. We do not know if one thousand subscribers were found, but we do know that British agricultural research and development benefited from vast private resources, despite the apparent lack of private return. Why did people do it?

There are two answers to this question. First, there actually was private return to an agriculturalist who did research or development; the return to others was, admittedly, greater, and this has obscured the benefits to the initial researcher, but that has always been true of all research, and it remains so today. This very important issue will be more comprehensively covered in the sections on 'first movers' and 'second movers' in Chapter 10, but to illustrate it with reference to agricultural R&D in a poor society, let us examine the economics of agricultural research in today's Third World; it is easier to collect detailed data on present-day activities than on those of England's past, yet the agricultural circumstances are similar, and they confirm that the relative poverty of Agricultural Revolution England was no inhibitor of private return on agricultural research.

The two most important crops in India today are wheat and rice,

---

* The long-standing interest of the British aristocracy in agricultural improvement persists, in life but also in art. The most famous pig in history is the Empress of Blandings, Lord Emsworth's prize animal in P. G. Wodehouse's *Summer Lightning*.

and because of their importance they have been intensively improved by the international scientific community – the so-called 'green revolution'. Almost as important to India, however, in terms of acres planted, are sorghum and pearl millet, yet these have not attracted international interest. They are drought-hardy crops, and they are, therefore, well suited to non-irrigated dry land. They are known as 'poor peoples' crops' because they are generally sown by farmers who are barely removed from subsistence agriculture.

In the words of Carl Pray, the agricultural economist, there is a: 'widespread belief that small-farmer subsistence agriculture in developing countries cannot sustain a commercial private breeding industry for food crops'.[8] Belying this common belief, however, Pray and his colleagues have found that the major seventeen private seed companies in India invest seriously in R&D. They spend an average of 4 per cent of seed sales on R&D, and between them, they employ 31 scientists with PhDs, 45 with MScs, and many more with BScs or technical qualifications. The private sector spends as much on seed R&D as does the Government of India through its National and State Seed Corporations, but the new hybrids produced by private research are markedly better, in terms of yield, than those produced by the State's scientists. This is because the private sector directs its research more carefully than does the State. It is only the artifically low prices of the State's seeds that ensure that they still find buyers but, as Pray and colleagues show, they are a false economy: farmers would do much better to pay the higher costs of the private seeds since they are better.

The economics of agricultural R&D are almost absurdly altruistic. Pray *et al.* found that, of the benefits that accrue from private seed research, no less than 85–94 per cent are reaped by farmers or consumers in terms of higher yields, higher profits and lower retail prices; dealers and distributors capture up to 9 per cent, and the seed manufacturers only capture 6 per cent of the benefits of their own research – yet they are still happy because their internal rate of return on their investment in R&D is no less than 17 per cent. No wonder Pray *et al.* entitled their paper *Private Research and Public Benefit*[9] – but, and let us re-emphasise this crucial point, the private benefit is excellent at an annual return of 17 per cent on investment in R&D.

To extrapolate back to eighteenth- and nineteenth-century England, therefore, it is probable that the hobby farmers did personally benefit from their hobbies, even if the bulk of the benefit was enjoyed by others. But an important secondary lesson emerges from the work of Pray and his colleagues. They found that the State's research was not only less good than that of the private sector, it also inhibited that of the private sector by artificially reducing the private sector's market. Even though India's sorghum and pearl millet agriculture is a near-subsistence economy, private money for high-quality research is readily available, and even more would be available if the State did not intrude into seed research and displace the private sector. Similar processes at work during the eighteenth and nineteenth centuries would explain why it was *laissez faire* Britain that led the world through the Agricultural Revolution despite France's huge State-funded programme in agricultural research.

There is a second reason for the rise of the hobby farmers in Britain that so amazes the economists. Economists work on a very narrow model of humanity; they suppose that people are only motivated by immediate economic self-interest. But people are complex, and one of their motivations is curiosity. Another is altruism. These issues will be more comprehensively discussed in later chapters, but let us note the historic fact of post-renaissance Europe that, whenever a class of persons have possessed economic power – either because they have inherited money or because they have made it or, because as bureaucrats and politicians they have taxed the populace – then that class will spend money on research and development. The desire to support research appears to be as basic to humans as the desire for art, say. Rich people will always support art and science unless they perceive that the State has fulfilled the national need. Since governments rarely support art generously, we are still accustomed to private patronage in that area; since governments now generally perceive science to be in the national interest, they usually fund it, and so they have displaced the private patron and thus made the idea of a private patron in science seem bizarre. But in the absence of the State's funding, private individuals will always fund science, as the eighteenth- and nineteenth-century hobby farmers show. In later chapters in this book, we will try to determine who would be more generous today, the State or the private patron.

POSTSCRIPT ON ENCLOSURES

The enclosing of land has always attracted criticism, of which the most telling has been Oliver Goldsmith's poem *The Deserted Village* (1770). Here are lines 51–62:

Ill fares the land, to hast'ning ills a prey,
Where wealth accumulates, and men decay;
Princes and lords may flourish, or may fade;
A breath can make them, as a breath has made;
but a bold peasantry, their country's pride,
When once destroyed, can never be supplied.
A time there was, ere England's griefs began'
when every rood of ground maintain'd its man;
For him light labour spread her wholesome store,
Just gave what life requir'd, but gave no more;
His best companions, innocence and health;
And his best riches, ignorance of wealth.

Who was this Oliver Goldsmith (1728–74)? What did he know of a contemporary English peasant's life? Goldsmith himself was comfortably middle class, his father being a clergyman. Born in Kilkenny in Ireland, and educated in Trinity College, Dublin, he spent his youth in Ireland. He then spent his adult life in European cities, first studying law in London, then medicine in Edinburgh, Leyden, Louvain and Padua, before settling in London for the rest of his life. He was buried in the very heart of London in the Temple Church. One must conclude, therefore, that he had little personal experience of English villages, especially when we discover that everyone in Goldsmith's deserted village is a paragon. Of the vicar Goldsmith wrote:

Even children follow'd with endearing wile,
And pluck'd his gown, to share the good man's smile.

Here is the school master:

Yet he was kind; or if severe in aught,
The love he bore to learning was in fault.

Goldsmith wrote: 'what life requir'd, but gave no more'; but what did he actually know of a subsistence economy? Goldsmith, in fact, knew nothing about any sort of economy. The £50 that was meant to see him through his legal studies in London had all been lost by 1752 on gambling. During his two years as a medical student in Edinburgh, he lost all his money gambling. He did the same at Leyden and everywhere else he visited on the Continent. Back in London, in permanent debt, he was constantly threatened with prison, and he died owing the amazing sum of £2000. This was the man who, having lived off others all his life, praised the peasants for their 'ignorance of wealth'.

Goldsmith never held down a proper job, nor succeeded in achieving a single respectable qualification, having been rejected as unemployable by the Church, and the legal and medical professions in succession. Yet he wrote in *The Deserted Village*:

> How wide the limits stand
> Between a splendid and a happy land.

There are no such limits, as anyone who has produced anything knows. A splendid or prosperous land is a happy one. Before the enclosures, as Gregory King showed, half of all the people in England – many of them peasants – starved. A subsistence economy very often did *not* give what 'life requir'd' (look what happened to Goldsmith's native Ireland during the famine). The enclosures made the rational farming of land in England possible, and so ushered in that most lovely of technological revolutions, the Agricultural Revolution.

To the extent that they employed coercion, the enclosures were clearly wrong, even if they did increase the supply of food. The enclosures were legal – individual acts of enclosure required the consent of Parliament and the financial compensation of the peasants (the Highland clearances were uglier) – but Parliament was controlled by the aristocracy and its laws would rarely have harmed the aristrocracy's interests. Yet, as wrong-doing, the enclosures were hardly unique; they were merely one of a long line of wrongs perpetrated on the peasantry by the aristocracy. In 1350, for example, the aristocracy had used the Statute of Labourers to keep the peasants on the land. The crucial point, which is generally overlooked, is this: the enclosures were the last of the long line of wrongs perpetrated on the peasantry by the aristocracy. The enclosures were only possible because of the aristocracy's feudal powers, but the enclosures reflected the adoption of commercial values by the aristocracy – values that had been embraced by some individual peasants before 1350, and which the feudal aristocracy had then tried to stifle. Consequent on the enclosures, the relationships between landowners and farm workers evolved into those of any other employers and their employees, i.e. regulated by contract, supply and demand, and equality before the law. The wrongs that the enclosures embodied, therefore, were the wrongs of the feudal society they themselves were destroying. By increasing the supply of food, and by substituting commercial relationships for feudal ones, they ushered in a better world.

Goldsmith praised the peasants for their 'innocence', a cant word for ignorance, but it is surely better to have developed a society where no class need be 'innocent', and where everyone has the knowledge and power to defend their own rights rather than be vulnerable to feudal lords. The economic processes of wealth creation and wealth redistribution that the enclosures accelerated would ensure that never again could an aristocracy oppress the people of England.

6 The Industrial Revolution

For one thousand years, between AD 500 and AD 1500, *per capita* wealth in Europe did not increase. Thanks to the new technology, there was a consistent increase in absolute wealth (about 0.1 per cent a year) but this was absorbed by population growth (also about 0.1 per cent a year).[1] From 1500 onwards, however, coinciding with the stirrings of the Commercial Revolution, rates of wealth increase started to outstrip rates of population increase and standards of living started to creep up. At first, rates of increase of *per capita* wealth crept up desperately slowly, but as the Commercial Revolution bred the Agricultural Revolution, which in turn bred the Industrial Revolution, so the rates accelerated. Overall, across Europe, rates of growth still remained very low, but those countries that initiated the great revolutions enjoyed much greater rates. Indeed Britain, between 1780 and 1860, enjoyed the then stupendous rate of over 1 per cent a year.[2]

Poor Holland. In a just world, she would have gone on to lead the world through the Industrial Revolution as she had led it through the Agricultural, but like Italy which had seen her lead through the Commercial Revolution dissipated by invasion and war, so Holland, too, lost her lead through invasion and war. She was enmeshed in religious and colonial wars with Spain, commercial wars with Britain, and divisions at home. It was Britain, therefore, safe behind its English Channel, that outstripped Holland during the eighteenth century, to lead the world through the Industrial Revolution.

Classically, the British Industrial Revolution is dated from 1780 to 1860. Although those dates must be arbitrary, those eighty years in Britain did witness some of the momentous events in human history. First, the population tripled. There were 7 500 000 Britons in 1780, and 23 130 000 in 1860. This, alone, is a tremendous fact. Until the British industrial revolution, any society that experienced great population growth had invited degradation, poverty and famine. When Ireland, for example, saw its population rise from 4 200 000 in 1791 to 8 295 000 in 1845, the consequence was the terrible famine of 1845. Nearly 2 000 000 million people either died or emigrated desperately, and by 1851 the population of Ireland had been reduced to 6 559 000 (5 813 00 in 1861). Yet when mainland Britain's population tripled

60

between 1780 and 1860, not only did standards of living not fall, they rose terrifically. Deane and Cole have shown that, between 1780 and 1860, real wages more than doubled from £11 per head to £28 per head[3] (when measured at 1850 prices; 'real' wages corrected for inflation or deflation). This doubling in income *per capita* was shared by all social classes (see the first postscript to this chapter). Britain's huge population growth between 1780 and 1860, therefore, was a consequence of its new-found ability to create wealth at such a rate as to more than sustain the rate of population increase. That new-found ability we call the Industrial Revolution.

The mechanism that underpinned this increase in wealth can, at a first approximation, be identified. In his *Principles of Political Economy* (1848) John Stuart Mill defined the four fundamental sources of national wealth – labour, capital, land and productivity (which he calls productiveness): 'We may say, then . . . that the requisites of production are Labour, Capital and Land. The increase of production, therefore, depends on the properties of those elements. It is a result of the increase either of the elements themselves, or of their productiveness.' Economists still use Mill's assumptions, and they agree that a country can enrich itself if it: (i) builds up its capital investment in machinery or factories, (ii) increases its labour force, (iii) brings more land under cultivation, or (iv) if its population learns how to exploit those three fundamentals of capital, labour and land more effectively to improve productivity. There is a simple formula that expresses this statement mathematically:

$$Y = K + L + T + p$$

where: Y = total national wealth; K = capital ('Kapital'); L = labour; T = land ('terre'); p = productivity. (For a deeper analysis of the equation, see the second postscript to this chapter.)

The question, therefore, is this: which of these four factors was responsible for the vast increase in wealth that Britain enjoyed between 1780 and 1860? In Table 6.1, we see that the average annual increase in British GNP was 2.42 per cent, but as the population only grew at 1.25 per cent a year, there was a real increase in GNP *per capita* of 1.17 per cent a year. So we can discount increases in the labour force *per se* as the source of Britain's increased wealth: we are looking at an increase in *per capita* wealth. (Admittedly, Adam Smith did show that increases in population do, in themselves, increase wealth, because larger populations allow for greater specialisation of labour, more competition, and a relative lowering of travel costs, but these advantages are not nearly great enough to account for the increase in wealth seen in Britain between 1780 and 1860.) We can also discount increases in land because, although a little extra land was brought under cultivation, it was only marginal and of poor quality. Increases in capital cannot be completely discounted because there was an average annual increase in capital per labourer of 0.35 per cent (see Table 6.1 on p. 87), but this falls far short of the 1.17 per cent average annual increase in wealth *per capita*. The bulk of that, therefore, must be attributable to p, increases in productivity.

What caused this increase in productivity? From Table 6.1 we see that there were general increases in productivity across all areas of the British economy between 1780 and 1860, which implies that some of the advances in wealth were caused by general improvements in health, education and motivation across the nation. But from Table 6.1 we also see that the greatest increases in productivity were concentrated in certain industries, and as it was precisely those industries that also enjoyed the greatest technological advances, we can reasonably conclude that Britain's Industrial Revolution was largely a product of new British technology.

Consider cotton – This industry enjoyed such fabulous increases in productivity (2.6 per cent annually between 1780 and 1860) that a piece of cloth that sold for 80 shillings in 1780 only cost 5 shillings in 1860. One man in 1860, exploiting the same capital investment as he might have done in 1780, could produce more than ten times as much finished cotton per hour of work. We know, moreover, the technology that underpinned these fabulous advances. At the beginning of the eighteenth century, cotton was a manual trade, the two major processes being each performed by hand. The first process, spinning, which involved the twining of myriad short individual cotton strands into one endless thread, was performed on a spinning wheel, generally by unmarried cottage girls (the 'spinsters'). The second process, weaving, was similar to knitting, although it was performed on a simple loom. It was that which witnessed the first great advance, in 1733, when John Kay created his 'flying shuttle', a mechanised loom whose vertical threads were held apart while the shuttle was shot, mechanically, to lay the horizontal weave.

Spinning was then mechanised when James Hargreaves, between 1764 and 1769, invented the spinning jenny and when Richard Arkwright, in 1769, linked the jenny up to a water mill to remove the need for muscle power. In 1779, Samuel Crompton linked the spinning jenny and the flying shuttle in a hybrid machine called a 'mule' which Edmund Cartwright redesigned as a power loom in 1785. All these advances were British, but in 1793 Eli Whitney, an American, created his cotton 'gin' (a corruption of 'engine') which mechanised the separation of cotton threads from the seed of the plant. From these high points flowed a vast number of smaller improvements which, in association with the development of new institutions such as factories, fed the ever-greater advances in productivity that characterised the eighteenth and nineteenth Century cotton industry.

Whence did the major technological advances come? What enabled Kay, Hargreaves, Arkwright, Crompton, Cartwright and Whitney to make their revolutionary contributions? This question can be answered at two major levels: intellectual and financial. Where did these inventors acquire their ideas, and who paid for it?

INTELLECTUAL ADVANCE

The Industrial Revolution is widely believed to have represented the successful application of science to technology. Some scholars have disputed that simple view, but the conventional story, as told, for example, in the standard school textbooks or in the *Encyclopaedia Britannica*, assumes it. Yet it only takes a moment's reflection to doubt that simple view. Consider the flying shuttle or the spinning jenny. These were admirable machines, but did they incorporate anything more intellectual than ingenuity or common sense? Where was the science?

Since the flying shuttle or the spinning jenny clearly did not embody much science, the myths about the Industrial Revolution have skirted around the cotton industry to attach themselves to the steam engine. Ah, now there surely is science in action! Those huge pistons, those jets of smoke and steam, that vast power – oh, the ingenuity of man in harnessing science. During the twentieth century, a number of historians such as J. D. Bernal have used the steam engine to persuade an unsuspecting public that the Industrial Revolution did indeed embody the appliance of science, so let us study the steam engine and explode a few myths.

The first commercial steam engine, Thomas Newcomen's, was at work in 1712 at Dudley Castle, Worcester. It was huge, expensive, and inefficient, but it clearly met a need because, by 1781, about 360 had been built in Britain, most of them devoted to pumping water out of coal mines. There may have been as many on the Continent, too (one was built at Königsberg in Hungary as early as 1721). Newcomen's engine was, in essence, a very simple machine. It consisted of a chamber or cylinder, in which a piston rode up and down. Steam was admitted into one side of the piston but, unlike later steam engines, the motive power did not come from the pressure of the steam, because the technology to contain high pressures was not available in 1712. The steam was admitted at atmospheric pressure. The motive power came from creating a vacuum. Once the chamber was full of steam, Newcomen cooled it with a jet of water. The steam condensed, which sucked down the piston. Since the piston was connected to a rocking beam, the condensation of the steam could be converted into work. The cycle was then repeated.

Whence did Newcomen get his idea? The *Encyclopaedia Britannica* has no doubt. Here is its introduction to the steam engine: 'The researches of a number of scientists, especially those of Robert Boyle of England with atmospheric pressure, of Otto Von Guericke of Germany with a vacuum, and of the French Huguenot Denis Papin with pressure vessels, helped to equip practical technologists with the theoretical basis of steam power. Distressingly little is known about the manner in which this knowledge was assimilated by pioneers such as. . . . Thomas Newcomen, but it is inconceivable that they could have been ignorant of it.'[4] Inconceivable or not, pioneers like Newcomen *were* ignorant of it. The historian D. S. L. Cardwell has established that Newcomen, who was barely literate, was a humble provincial blacksmith and ironmonger who, stuck out in rural Devon, had never had any contact with science or scientists.[5] Newcommen did, however, have a lot of contact with the tin mines in the neighbouring county of Cornwall, and he knew that they were frequently, and disastrously, flooded. There was, unquestionably, a market for an effective pump.

Being an ironmonger, Newcomen knew all about pumps. He spent his working hours repairing and making cylinders, pistons and valves. These were his trade. Newcomen's idea, though unquestionably brilliant, was also very simple. He reversed the process, using the entry and exit of fluids through a valved pump to turn it into an engine. Newcomen's idea was the stuff of intuitive genius (although it took him ten years of exhaustive experimentation to create it) but it was no more than that: the intuition of a natural genius familiar with pumps and the domestic steam kettle. No theoretical science was involved, nor was it necessary.

Newcomen, moreover, had not been the first engineer to discover, empirically, that cooled steam could be harnessed. Thomas Savery, an English military engineer, had patented a crude steam engine in 1698 (it did not work very well), but even earlier De Caus (1575–1626), a gardener who designed fountains, used the condensation of steam to create the very first steam pump. To suck up water from wells, De Caus would light a fire under a boiler that contained a small amount of water. Attached to the boiler was a pipe that fed into a well, and when the water in the boiler had boiled into steam, De Caus would extinguish the fire, and as the boiler cooled, it would suck water up from the well.

De Caus's success disproves the suggestion that the exploitation of steam power by engineers depended upon developments in the science of vacuums because De Caus (1575–1626) died long before von Guericke (1602–86), the Father of Vacuum Science, had started his researches. (Otto von Guericke, Mayor of Magdeburg and Quartermaster to Gustavus Adolphus, was the first scientist to study vacuums. He created such powerful pumps that, in one famous demonstration before the Emperor and his court, he showed that it took sixteen horses on each side to pull apart two hemispherical vessels that contained a vacuum.)

Not only were the engineers of the Industrial Revolution not dependent on the scientists, the opposite was actually true. Denis Papin, the leading gas scientist of the day, used to explain that he was prompted into studying vacuum steam engines because of Newcomen's success. It was the progress in technology that prompted the scientists to catch up. But, if science had not prompted the initial development of the steam engine, could it at least be credited with facilitating its improvement?

The first significant improvement was made in 1764 when James Watt invented the separate condenser. Before that, to cool the steam in his cylinder, Newcomen had had to cool the entire cylinder, including the wall. This was very wasteful, because much of the energy of each cycle of steam was dissipated in re-heating the wall. Watt built a pipe into the cylinder, regulated by a stop cock, and at the appropriate point in each cycle the steam passed down the pipe into a separate cylinder or condenser. This separate condenser was kept continuously cold, so that it was only the steam, as it travelled from one cylinder to another, that was cooled.

Where did Watt get his idea? The conventional view, as propagated by J. D. Bernal, is that Watt was applying Black's discovery of latent heat: '[Joseph Black] reflected on the fact that snow and ice took time to melt – that is absorbed heat without getting hotter – and that the heat must be hidden or *latent* in melted water. He next measured the large latent heat of steam. . . . The first practical application of the discovery of latent heat was to be made by a young Glasgow instrument maker, James Watt . . . [which] was crucial for the development of the steam engine.'[6] Even in Watt's day, those who took a Baconian view of science believed that he must have been inspired by Joseph Black's discovery of latent heat. Did not Black and Watt both work in Glasgow? Did not Black's discovery precede Watt's? Surely, therefore, the one prompted the other? Yet Watt himself explicitly denied it, and he stated in public: '[Professor Robinson] . . . in the dedication to me of Dr Black's "Lectures upon Chemistry" goes to the length of supposing me to have professed to owe my improvements upon the steam engine to the instructions and information I had received from that gentleman, which certainly was a misapprehension; as, although I have always felt and acknowledged my obligations to him for the information I had received from his conversation, and particularly for the knowledge of the doctrine of latent heat, I never did, nor *could*, consider my improvements as originating in those communications".'[7]

Indeed, Watt then went on to explain that his inventions: 'proceeded solely on the old established fact that steam was condensed by the contact of cold bodies, and the later known one that water boiled *in vacuo* at heat below 100°[F], and consequently that a vacuum could not be obtained unless the cylinder and its contents were cooled [at] every stroke to below that heat'.[8] Moreover, as Cardwell has pointed out: 'The greater the latent heat in steam the *less* the incentive to invent a separate condenser, for since the heat required to warm up the cylinder is independent of the properties of steam it follows that the greater the latent heat the higher the proportion of useful steam supplied per cycle.'[9]

Watt's advances, therefore, owed less than nothing to contemporary science; they proceeded on an 'old established fact'. In any case, Watt had not been formally educated in science; he worked at Glasgow University as a technician. Moreover, the next major advance in steam engine technology, the use of high pressure steam to push the piston, was made by a man in Newcomen's mould. Richard Trevithick, whose engine at Coalbrookdale in 1802 achieved the unprecedented pressure of 145 pounds per square inch, was barely literate. Born in Cornwall to a mining family, Trevithick received no education other than that provided at his village primary school, whose master described him as 'disobedient, slow and obstinate'. But Trevithick addressed a problem. The Cornish tin mines were a long way from the nearest coal fields, so their Watt steam engines were expensive to run. Could they be made more efficient? Unlettered and ill-educated though he was, Trevithick thought so, and he introduced steam under high pressure to push, not suck, the piston. His high-pressure steam engines were twice as efficient as Watt's, and they were so light that they could be transported around the Cornish mines. Nicknamed 'puffer whims' from the steam they emitted, over 30 were built, and some were converted to hoist ore as well as water.

In 1801, Trevithick built his first steam carriage, which he drove up a hill in Camborne, Cornwall, on Christmas Eve. In 1803, Trevithick built the world's first steam railway locomotive at the Pendaren Ironworks, South Wales. On 21 February 1804, that engine hauled 10 tons of iron and 70 men along 10 miles of trackway. He built two more steam engines in Gateshead (1805) and London (1808). His engines were used to drive iron-rolling mills (1805), paddle wheels on a barge (1806) and threshing machines on a farm (1812).

Trevithick, marvellous engineer and lovely person though he was, was hopeless with money, and he died a pauper on 22 April 1833. He was buried in an unmarked grave in Dartford, Kent. His introduction of high pressure steam had been bitterly opposed by Watt and others as too dangerous, and this is what he wrote shortly before his death: 'Mr James Watt said ... that I deserved hanging for bringing into use the high pressure engine; this so far has been my reward from the public, but should this be all, I shall be satisfied by the great secret pleasure and laudable pride that I feel in my breast from having been the instrument of bringing forward and maturing new principles and new arrangements, to construct machines of boundless value to my country.'*

The very next major advance, too, was made by an ill-educated, barely literate, barely numerate, self-taught artisan called George Stephenson. Light though it was, Trevithick's locomotive was still too heavy for the cast-iron rails of the day. His engine, therefore, needed further improvement before a practical, commercial railway could op-

---

*Many great scientists and engineers have disregarded money. In a famous phrase Michael Faraday, who was always rejecting lucrative consultancies and directorships, said: 'I am too busy to make money.' Modern-day scientists are as unmaterialistic. Michael Mulkay and Anthony Williams, two sociologists who have studied physicists, have concluded that: 'Without exception, however, our respondents appeared to value money less highly than professional recognition' [M. Mulkay and A. Williams, 'A Sociological Study of a Physics Department', *British Journal of Sociology*, vol. 22, pp. 68–82 (1971)].

erate. But on 27 September 1825, a steam engine designed by George Stephenson drew 450 people from Darlington to Stockton at the terrifying speed of 15 miles per hour. Stephenson went on to build the Liverpool to Manchester line, for which he then designed the 'Rocket', an engine which could attain 36 miles per hour! Yet Stephenson was unschooled. The son of a mechanic, he followed his father in operating a Newcommen Engine to pump out a coal mine in Newcastle. He only learnt to read (just) at the age of 19 when he attended night school, and he never really acquired mathematics. So unsophisticated was Stephenson, and so dense was his Geordie brogue, that he needed an interpreter when talking to educated men from London. Yet it was the educated men – from all over Europe – who consulted him, not the other way round.

It will be seen, therefore, that the development of the steam engine, the one artefact that more than any other embodies the Industrial Revolution, owed nothing to science; it emerged from pre-existing technology, and it was created by uneducated, often isolated, men who applied practical common sense and intuition to address the mechanical problems that beset them, and whose solutions would yield obvious economic reward.

Looking back at the Industrial Revolution generally, it is hard to see how science might have offered very much at all to technology, because science itself was so rudimentary. Chemists who subscribed to the phlogiston theory, or to the view that heat was a substance, or who tried to build perpetual motion machines, were not likely to be of much use to engineers. Indeed, during much of the nineteenth century, the reverse was true; scientists scrambled to catch up with engineers. Carnot's descriptions of the laws of thermodynamics, for example, emerged from his frustration with Watt's improved steam engine, because that steam engine broke all the rules of contemporary physics. Watt's engine was more efficient than theory stated it could be, so Carnot had to change the theory.[10]

Very often, engineers turned scientist. Torricelli was trying to improve the pump when he performed the experiments that demonstrated the weight of air in the atmosphere.[11] Joule was trying to improve power generation in his father's brewery when he discovered the law of conservation of energy.[12] Breweries have long stimulated science. It was because he was trying to understand what brewers had been doing, empirically, for centuries, that led Pasteur to discover bacteria, and so create the science of microbiology. Leopold Auenbrugger (1722–1809), the physician who developed percussion to help diagnose diseases of the chest, was inspired by watching his father, a publican, tap down the sides of his barrels to determine how full they were. Brewing was so important to the development of science that at the end of the nineteenth century there were more biochemists working in Burton on Trent, the capital of the British beer industry, than in the whole of the rest of Great Britain. But successful brewing came first, science followed.

Perhaps the most eminent example of science following age-old practice was that of Mendel. Farmers have been breeding new strains of animals and crops for centuries – no, millenia – with great success. But it was only towards the end of the nineteenth century that Mendel sought to systematise the biological laws that underlay the practice. The practice came first, the science later.

The irrelevance of academic science to technological or economic development during the eighteenth and nineteenth centuries can be best illustrated by comparing Britain and France. The Industrial Revolution happened in Britain at a time when its universities were notoriously stuporous. England only had two universities, Oxford and Cambridge, and their condition can be judged by these extracts from the *Autobiography* of Edward Gibbon (1737–94), who had been an undergraduate of Magdalen College, Oxford. He described his tutors as: 'monks . . . easy men who supinely enjoyed the gifts of the founder'. Of his own tutor, Gibbon wrote that he 'well remembered he had a salary to receive, and only forgot he had a duty to perform'. Gibbon described his fourteen months at Oxford as 'the most idle and unprofitable of my whole life'.

It is often said that the Scottish universities were not so bad, and certainly Glasgow did employ James Watt as a technician for a time; none the less, they were not much healthier. When Dr Johnson visited St Andrew's University in 1773, he found that one of its three colleges was being demolished because no-one attended it. When he tried to visit the library of another, no one could find the key. (Dr Johnson had a shrewd understanding of the relationship between universities and wealth. As this passage from his *Journey to the Western Isles* (1775) shows, he knew that universities prospered as a consequence of national wealth, and not the other way round: 'It is surely not without just reproach, that a nation, of which the commerce is hourly extending, and the wealth increasing, . . . suffers its universities to moulder into dust.')

How badly all this compares with France! There, technical and scientific education were fostered by a generous State, to produce such men as Lavoisier, Berthollet, Leblanc, Carnot, Monge, Cugnot, Coulomb, Lamarck, Cuvier, Saint-Hilaire, Gay-Lussac, Arago, Ampere, Laplace and Chaptal. Their institutions, the *Ecole des Ponts et Chausses*, the *Ecole du Corps Royal du Genie* at Mezieres and, after the Revolution, the *Ecole Polytechnique*, provided the best-equipped scientific research laboratories in the world. By the early nineteenth century, when it was still only a skilled craft in England, France had established engineering as a profession, with schools, formal examinations and, after 1853, its own research laboratories in the *Conservatoire*. Yet it was Britain, not France, that produced the Industrial Revolution. To understand why, let us examine the economics and history of science funding.

THE FUNDING OF RESEARCH

The British and French governments started off with similar policies for science. Each country's king founded a similar body, the Royal Society in London (1622) and the *Academie des Sciences* in Paris (1666). The draft preamble of the statutes of Royal Society stated: 'the business of the Royal Society is: to improve the knowledge of natural things, and all useful arts, Manufactures, Mechanic practices, Engynes and Inventions by experiment'. The *Academie de Sciences* had similar aims. But thereafter, the course of the two countries diverged. The British adopted *laissez faire*. Whole classes of taxes were abolished, central government withdrew from almost any function other than the defence of the realm and the maintenance of law and order, and organisations like the Royal Society were left to sink or swim (it nearly sunk). France embraced *dirigisme*. Following the direction of Colbert (Chief Minister, 1661–83), the State ran every aspect of French society, controlling trade, industry, education and science, amongst other functions. The government did everything it could to promote economic growth; it raised tariffs to deter imports, it raised taxes to pay for bureaucrats to direct the economy, and it raised yet more taxes to pay for the schools of science and technology that were intended to promote economic growth. Colbert created the *Ecole de Rome* (to teach arithmetic, geometry and draughtmanship), the *Academie Royale de Peinture* (from which emerged the great design school the *Ecole Royale Graduite de Dessin*) and the *Academie Royale d'Architecture*. He fostered the great workshops of the Savonnerie and the Gobelins, the mint, the royal press and royal manufactory in the Louvre. The *Academie des Sciences* received generous State aid, and the world's first scientific journal, the *Journal des Savants*, was created by the state. The *Jardin du Roi* was reorganised in 1671 to conduct research and to teach in botany and pharmacy. Three chemistry research laboratories were created by state, one in the King's library, one in the Louvre and one at the *Observatoire*.

Successive administrations continued Colbert's policies. A magnificent school of civil engineering, the *Ecole des Ponts et Chaussees* was created in 1716, the splendid *Ecole du Corps Royal du Genie* in 1749, and the lavish *Ecole des Mines in 1778*. The *Ecole Polytechnique* was founded in 1795. Trade schools or *ecoles des art et metiers* were founded all over the country to train artisans . . . yet it was *laissez faire* Britain that fostered the Industrial Revolution, not France. Britain grew rich; France remained poor. Why?

There are two reasons for this, both of which were embodied by the popular reaction of Colbert. By the time he died in 1683, Colbert was universally loathed. The people did not cry out 'long live Colbert, he gave us the *Academie des Sciences* which made us prosperous' they cried 'damn Colbert, he made us poor'. And one hundred years later, the impoverished and embittered people of France created a Revolution.

The *academies* and *ecoles* of France were fiendishly expensive. Chemical laboratories do not come cheap. If they are run by the government, they depend on taxes, and the French Government was so anxious to run a scientifically based economy that it had to raise enormous taxes. Yet the collection of those taxes was horrible. Consider indirect taxation, which accounted for about a third of the State's revenue. The collection of these taxes was delegated to Farmers-General (so called because the state 'farmed out' the collection of taxes to private individuals who were allowed to keep a proportion of the revenue). In 1760, there were over 30 000 Farmers-General. Out of a total population of 25 million, therefore, a significant proportion was engaged in the unproductive task of screwing money out of unwilling donors (22 000 of the Farmers-General were paramilitary police, uniformed, armed and possessed of vast powers of entry, search, detention and confiscation). Moreover, because the duties on goods were so high, smuggling thrived. Necker (Director-General of Finance, 1776–81) estimated that up to 60 000 people were engaged in smuggling salt alone. These smugglers needed to be caught, tried, imprisoned and executed. Between 1780 and 1783, for example, in just the one region of Angers, 2342 men, 896 women and 201 children were convicted of smuggling salt. Five times as many had been arrested, only to be freed due to insufficient evidence.[13]

The economic and social cost of all this was horrific, but worse was to come. Salt, for example, was ten times more expensive than it needed to be, because of the Farmers' control and taxing of the trade. The primary producers from the marshes of Nantes and elsewhere were forced to sell to the Farmers at a fixed price. From there, the salt was shipped to coastal depots (located at the mouths of rivers) where it was packed, sealed and registered. Barges took the salt upstream to secondary, riverside depots, where the salt was inspected and re-registered. There, via wagon, the salt was taken to a third set of warehouses, to be re-inspected and re-registered. Finally, the salt was transported to a fourth set of warehouses, the retail outlets or *greniers a sel* where the consumer was allowed to buy it, after a further massive mark-up for tax.

The appalling social and economic costs that France bore as it raised its taxes might not have mattered had the money been well spent, but unfortunately it was not. That is the trouble with *dirigisme*; a centrally planned economy can only work as well as the plans, and France subscribed to the wrong plans. The rulers of France believed in Bacon's model: they believed that science bred technology which bred wealth. Indeed, practically everyone in France, a nation of Cartesians, believed in Bacon – even the political dissidents believed in Bacon. Diderot of the *Encyclopedie*, whose ideas led to his imprisonment, was such a convinced Baconian that he wanted yet more government-funded *academies* and *ecoles*. He was even more Cartesian than the King. But, and this is the crucial fact, the model was wrong. As we have already seen, the Industrial Revolution did not represent the application of science to technology, it represented the development of pre-existing technology by hands-on technologists.* The important factor common

---

* In his *Road to Serfdom* (London, 1944) the Nobel laureate in economics, F. A. Hayek explained that all government economic planning must fail. In a free market, each individual player makes his own plans. There are, therefore, hundreds of thousands of competing different plans, from which the market will select the best. But a centrally planned economy will be committed to a single plan. Moreover, a planned economy loses the information provided by the market, and instead it has to rely on officials to collect, summarise and transmit the details of the success or failure of different economic strategies to a central committee. This then has to transmit orders down again, so the potential for misinformation is infinite. (In a free market, since the decision makers are also the key economic players, they are close to the economic activity.) A planned economy, moreover, lacks the objective measure of success provided by the market, and has to depend on the capacity of officials to make strictly objective observations. Invariably,

to Newcomen, Trevithick and Stephenson, apart from that each was essentially illiterate and innumerate, was that each worked on pumps or steam engines as an artisan. They were not academics. Only Watt was employed in a university, but even he was only employed as a technician and, significantly, he was working in a commercial capacity when he invented his separate condenser, because it was the mending of a Newcomen machine that stimulated his idea.

Had Newcomen or the other English artisans been French, they would have been so harried and taxed by policemen and Farmers that they would never have had been the time or money to put aside for technological development. Even worse, their intelligence might have been recognised, and they would have been whisked away to an *academie* or *ecole*, which would have been just as disastrous economically, because they would have spent their time doing academic science while the peasants starved for lack of technological development.

The Baconian model, therefore, that envisaged scientists in a laboratory dispassionately throwing up knowledge, which would then trickle down into new technology, did not, in practice, match reality. The industrial revolution was created by men looking for solutions to very particular problems – men who had the economic freedom and the economic incentive to invest their time and resources in experimentation and development. The particularity of the problems and their solutions is illustrated by the cotton gin. This was one of the few important inventions of the Industrial Revolution not to have been developed in Britain, it was developed by Eli Whitney, an American working in Georgia. Why? Because it was in Georgia that the cotton was picked. The Industrial Revolution did not represent a trickling down of knowledge from academic science to local engineers, it represented the success of local engineers in solving their immediate problems.

We can see, therefore, why Britain – not France – pioneered the Industrial Revolution. Intellectuals on both sides of the channel cham-

---

personal bias, ideological bias, political bias, corruption, laziness or stupidity intervene.

For science, the consequence is this: in a free market, some companies will invest heavily in pure science, others will only invest in applied science, and others will do neither or a bit of both. The market then determines which approach works (and the market in nineteenth century England showed that a judicious investment in applied science was generally best). But a centrally planned economy can only make one choice; the French opted for a huge investment in pure science, because all the intellectuals recommended it, yet it was the wrong decision.

pioned Bacon and state funding, but intellectuals had no power in free-enterprise Britain, where only the market ruled, so Britain invested its money appropriately, in technology. The French listened to their intellectuals, and dissipated their money on science. (One French intellectual who did get it right had to emigrate to England to publish safely. Voltaire published his poem on French history *L'Henriade* (1728) in London. Later, in his *Letters Concerning the English Nation* he wrote that taxes in England were such that: 'no one is tyrannised over, and everyone is easy. The feet of the peasants are not bruised by wooden shoes; they eat white bread, are well clothed and are not afraid of increasing their stock of cattle nor of tiling their houses, from any apprehension that their taxes will be raised the year following.')

One curious consequence of Britain's early but appropriate neglect of science was that its less successful competitors accused it of stealing their own, just as we all now criticise Japan. A Swiss calico printer complained in 1766 of the English: 'they cannot boast of many inventions, but only of having perfected the inventions of others; whence comes the proverb that for a thing to be perfect it must be invented in France and worked out in England'.[14] Yet, when the time was ripe, when purely technological development had been exhausted, the British did turn to science.

THE FUNDING OF BRITISH SCIENCE UNDER *LAISSEZ FAIRE*

Laissez faire Britain created a flourishing science – witness such names as Davy, Faraday, Maxwell, Dalton, Kelvin, Darwin, Huxley, Lyell, Cavendish and Rosse. How was this funded? There were at least five separate sources of funding for science under the free market in Britain: (i) hobby science; (ii) industrial science; (iii) university science that had been endowed by industry; (iv) university science that had been endowed privately; and (v) university science that was funded out of fees or commissions. Let us consider each in turn.

Hobby Science

This term is borrowed from the hobby farmers, rich men who improved their farms for fun. There were a large number of hobby scientists, of whom at least three are still remembered, Cavendish, Darwin and William Parsons, the third Earl of Rosse.

(i) Henry Cavendish (1731–1810), the grandson of the Duke of Devonshire, was for a time both the richest man in England and also England's greatest physicist. He identified hydrogen as an element, he characterised the atmosphere and he determined the density of the earth. For his delicate torsion balance experiments, he would get up at four in the morning when vibration from horse-drawn traffic was minimal (he lived in Clapham). He was the first person to synthesise water from its elements.

(ii) Charles Darwin (1809–82), the grandson of Erasmus Darwin was, of course, the author of the *Origin of Species by Means of Natural Selection.*

(iii) William Parsons, the third Earl of Rosse (1800–67), followed his father, the second Earl, into politics (the second Earl had been joint Postmaster-General for Ireland in 1809) but the third Earl never left the back benches because his true interests lay in astronomy. By 1827 he was improving casting specula for the reflecting telescope and experimenting with fluid lenses. In 1845, at a cost of £30 000, he built his Great Reflecting Telescope, the largest in the world at 58 feet long, through which he discovered, amongst other new things, spiral nebulae. He was President of the Royal Society between 1848 and 1854.

These three men were but the most prominent of thousands of hobby scientists, whose economics was straightforward. The increasing wealth of the technology of the Agricultural and Industrial Revolutions bred increasing numbers of rich men, and the lack of taxation or death duties bred increasing numbers of those who inherited wealth. A remarkable number of these men took up science with a passion.

Industrial Science

As industries developed, so the technology required for further innovation became more and more complex. Increasingly, therefore, industrialists themselves became scientists, or were drawn from the ranks of scientists. William Armstrong (1810–1900), for example, of the famous Armstrong Engineering Works, first trained as a lawyer, but he turned hobby scientist and, in his spare time, experimented with hydroelectric boilers to make them more efficient at generating electricity from steam. (Cragside, his house, was the first in Britain to be lit by electricity.) For his research on electricity he was made a Fellow of the Royal Society (FRS) in 1846, but he went on to exploit his science to become an industrial magnate, creating his Engine Works at Elswick, Newcastle upon Tyne, which developed the hydraulic crane, the Armstrong gun (a rifled breechloader) and accumulators.

Armstrong's most famous employee was Charles Parsons (FRS, 1898), the youngest son of the third Earl of Rosse of the telescope, who invented the steam turbine and who went on the build his own engineering works at Heaton, Newcastle upon Tyne. Both the Armstrong and the Parsons works fostered considerable research and development laboratories, and such laboratories progressively became characteristic of technological industries throughout Britain. By 1900, for example, there were 200 – 250 graduate chemists alone employed in British industry.[15]

(Parsons did a naughty thing during a Review of the Fleet. To celebrate the Queen's Jubilee in 1897, the Royal Navy assembled 173 warships off Spithead to be inspected by the Prince of Wales. His Royal Highness was meant to sail in state down the vast lanes of warships in the Royal Yacht *Victoria and Albert*, followed by a flotilla including the *Enchantress*, which contained the Lords of the Admiralty, the *Wildfire*, which contained the colonial premiers, the *Eldorado*, which contained the Diplomatic Corps, the *Danube,* which contained the House of Lords, and the *Campania* which contained the House of Commons. Parsons, trying to interest the Admiralty in his turbines, stole the show by roaring around the fleet in *Turbinia*, his turbine-powered yacht which, at 2000 horsepower and a top speed of 34 knots, was faster than anything the Navy could send in chase. It was, in fact, the fastest boat in the world. *The Times* wrote: 'Her speed was simply astonishing, but its manifestation was accompanied by a mighty rushing sound and by a stream of flame from her funnel at least as long as the funnel itself.' The Admiralty promptly ordered two turbine-powered ships.)

University Science Endowed by Industry

As industrial research became more scientific, so industrialists needed increasingly to recruit trained scientists. In the early days, men like Armstrong could, effectively, train themselves. Others, such as Charles Parsons, were trained by their father. Indeed, industrial research dynasties were born. David Mushet (1772–1847), for example, was a distinguished metallurgist who discovered how to prepare steel from bar-iron. His son, Robert Mushet (1811–91)* followed his father into

---

* A sad thing happened to Robert Mushet. He failed to patent his method for producing cast steel, and it was his rival Henry Bessemer who made huge sums of

metallurgy, and he discovered how to produce cast steel. But with the increasing complexity and volume of industrial research and development, industrialists needed systematically to recruit trained scientists, and they could no longer depend on self-taught or family-taught researchers; they had to create institutions to educate potential recruits in the most advanced science and engineering. We call such institutions universities. Throughout the industrial heartland of Britain, therefore, local industrialists created universities: Manchester, 1851; Newcastle upon Tyne, 1852; Birmingham, 1900; Liverpool, 1903; Leeds, 1904; Sheffield, 1905 (these are the dates of the Royal Charters, but these institutions generally originated in older colleges). These universities stimulated considerable research, because good teachers could not be attracted without offering them the freedom, time and resources required for research. Armstrong, for example, was one of those who generously endowed Newcastle University to support its research.

The greatest industrially endowed academic science emerged from London. Some of this came from the University of London (1836) whose constituent colleges, such as Imperial College, became world-famous, but overshadowing even the university was the oldest and best of all the industrially endowed research foundations, the Royal Institution (1799). The Royal Institution was created by Count Rumford, an exciting man. Born Benjamin Thompson at Woburn, Massachusetts in 1753, he took a commission in a New Hampshire regiment but, in 1776, proving himself an honest patriot loyal to his oath to his King, he fled the revolting American colonies for England. His experiments on gunpowder for the artillery prompted his election to FRS in 1779. He returned to America in 1782 as a lieutenant-colonel in the King's service, but the peace left him with little to do (although King George did knight him). So in 1784 he entered the service of a monarch who

---

money from it without having to pay royalties. As L. T. C. Rolt, the historian, has recounted, this influenced Robert Mushet's later style of research: 'Robert Mushet became a pioneer worker in the field of alloy steels. In 1862 he formed the Titanic Steel and Iron Company . . . and the Titanic Steel Works was established in Dean Forest. This was a small crucible steel-making plant where, in great secrecy, Robert Mushet carried out his alloy steel experiments. Mushet's greatest success was a special self-hardening tool steel of immeasurably superior cutting power to the carbon steel used previously . . . and R. Mushet's Special Steel, or 'RMS' soon became famous in machine shops on both sides of the Atlantic. Mushet took the most extraordinary cloak-and-dagger precautions to keep his RMS formula secret. The ingredients were always referred to by cyphers and were ordered through intermediaries. The mixing of the ingredients was carried out in the seclusion of the Forest by Mushet himself and a few trusted men' [L. T. C. Rolt, *Victorian Engineering* (Allen Lane, Penguin Press, 1970)].

did need competent assistance, the King of Bavaria. As Bavarian minister for war, Thompson reformed the army, established a cannon foundry and military academy, planned a system of poor law, introduced the potato, founded an institute for plant and animal breeding, and laid out the English Garden in Munich. For these services, he was made a Count of the Holy Roman Empire. By 1796, he was running the country as President of the Council of Regency, the King having fled the approach of the French and Austrians.

In his cannon foundry, Thompson, by then Count Rumford, had studied the remarkable amount of heat generated by the boring of cannon, and he published his studies which disproved the caloric theory of heat (the notion that heat was a substance). On his return to London in 1799, Rumford perceived that British industry needed more scientific research if it was to develop, and he had little difficulty in finding a group of industrialists who agreed with him and who provided the funding. It was an obvious idea whose time had come. Rumford and his fellow industrialists, therefore, created their research institute, the Royal Institution, in 1799. From the first, it pioneered important, fundamental, basic science. Consider Humphry Davy. Humphry Davy was appointed lecturer in 1801 (having been lured from another privately funded research laboratory, the Pneumatic Institute at Clifton, where he had discovered the anaesthetic effects of nitrous oxide or laughing gas) and, before he finally resigned the chair of chemistry at the Royal Institution in 1813, Davy had discovered six new elements: potassium, sodium, barium, strontium, calcium and magnesium, and he had delivered the famous lecture *On Some Chemical Agencies of Electricity* (1806). Davy resigned the chair in 1813 because he had married a rich woman, but he continued his researches as a hobby scientist, inventing the safety lamp in 1815, and assuming the presidency of the Royal Society in 1820.

Davy's great pupil, Michael Faraday, worked all his life at the Royal Institution. He was such a great scientist that a mere list of his achievements is awesome; let us just mention here that in 1831 he discovered electromagnetic induction, the process by which electricity can be generated by rotating a coil in a magnetic field, and the process which has underlain the commercial generation of electricity ever since – indeed, the process that made it possible. The Royal Institution continued to foster great men and great science, witness such names as John Tyndall (FRS, 1852), Edward Frankland (FRS, 1853), James Dewar (FRS, 1875; he invented the vacuum or Dewar flask), William Odling (FRS, 1859), Lord Rayleigh (Nobel Laureate, 1904), Sir William

Bragg (Nobel Laureate, 1915) and Sir Lawrence Bragg (Nobel Laureate, 1915). The Royal Institution still fosters great science; until recently its Director was Lord George Porter, the Nobel Laureate and President of the Royal Society.

Privately Endowed University Science

The importance and romance of science ensured a considerable private endowment of academic science. This tradition was centuries old. Many of the pioneers of the Royal Society occupied academic positions that had been endowed by rich men. Henry Lucas, for example, the MP for Cambridge in 1640, endowed the Lucasian Professorship of Mathematics at Cambridge to which Isaac Newton was elected in 1669 (Stephen Hawking is the current holder). In 1619 Sir Henry Savile, Queen Elizabeth's tutor in Greek and mathematics, endowed the Savilian Professorships of Astronomy and Geometry at Oxford (Christopher Wren was elected to the Savilian Professorship of Astronomy in 1661). Sir Thomas Gresham (1519–79), the founder of the Royal Exchange, created Gresham College in London, to whose professorships were elected such men as Robert Hooke (Professor of Geometry, 1665) and Christopher Wren (Professor of Astronomy, 1657). By the nineteenth century, this rate of private endowment had vastly accelerated.

University Science Funded by Fees or Commissions

The expansion of industry created a need for trained scientists which individuals met by attending university to obtain qualifications. Such individuals paid fees, from which the university could pay the salaries and some of the research of its lecturers. Moreover, industry would commission research from universities, and this also generated fees which would fund independent research.

Under the influence of these five separate sources of finance, namely hobby science, industrial science, industrially endowed science, privately funded science and fee-funded science, the research base of Britain grew dramatically. Roy Macleod, the historian of science, has calculated that Britain's science base doubled every twelve to twenty years during the nineteenth century.[16] The membership of the metropolitan science societies, most of whose members were amateurs or hobby scientists, doubled from about 5000 in 1850, for example, to 10 000 in 1870. The ranks of the professional scientists had a shorter doubling time of around twelve years, the numbers of university lecturers in science rising, for example, from 60 in 1850 to 400 in 1900. By 1914, according to *Who's Who in Science*, there were 1600 leading scientists who had found 'congenial employments' (appointments from which unhindered research was encouraged and funded). In 1894, money accrued from the profits of the 1851 Great Exhibition was devoted to the creation of research fellowships (Ernest Rutherford was an 1851 Exhibitioner) and by 1914 there were 170 privately endowed research fellowships in British universities (excluding those in Oxford and Cambridge).

Britain funded pure science because, as the Industrial Revolution developed, industrialists increasingly perceived that they were exhausting technological development, and they needed to explore science more profoundly. Since the Government refused to do it, and since the industrialists were not anxious to pay any extra taxes to enable the Government to do it, they invested their own money in science. The Government's reluctance to fund research was strengthened by its unfortunate experience over Charles Babbage. A great myth has attached to Babbage, namely that he was a heroic genius and pioneer, the Father of the Computer, whose brilliance was denied us by the meanness and short-sightedness of the Government. The facts are rather different.

Babbage was born in 1792, a year after Faraday, but few contemporaries were more different. Babbage could never finish a job. The moment things got difficult he would spin off into public controversy (his entry in the 1974 edition of the *Chambers Biographical Dictionary* concludes 'in his later years he was chiefly known as a fierce enemy of organ grinders' and that sums him up: he was a man who poured intense energy into bombastic conflict against trifles, while eschewing the quiet dedicated labour of true achievement). His was a conventional academic career, winning the usual honours (FRS, 1816) and culminating in the chair of mathematics at Cambridge (1823–39), but during the 1820s, unfortunately for everybody's peace of mind, he conceived of a calculating machine or 'Difference Engine'. The subsequent importance of the electronic computer has cast a rosy glow on Babbage's machine, but in reality it was little more than a mechanised abacus. Yet few men possessed Babbage's mastery of public controversy, and by dint of pamphleteering, lobbying and the issuing of dire threats, he eventually squeezed no less than £17 000 out of the Government to build his Engine (when the average wage was £2 a week).

After spending his £17 000, Babbage had still not built his Engine, but rather than finish the job, he announced that he now wished to create a completely new machine, the Analytical Engine, which employed punched cards. The Government, having spent £17 000 on nothing, refused to fund this new machine, which then prompted Babbage into doing the one original thing in his whole life. He inaugurated the tradition by which prominent scientists denounced the government for neglecting science. His *Decline of Science in Britain* was published, with true Babbage incompetence, in 1830, the year before Faraday made the most important of Britain's contributions to science, the discovery of electromagnetic induction in 1831.

Those who still believe that Babbage was a misunderstood genius should consider the subsequent history of the Difference Engine. Two Swedish engineers, Pehr Georg and Edward Georg Scheutz, obtained a grant from their own government to build Babbage Difference Engines and, being determined men, they had succeeded by 1853 (for a cost of only £566, a fraction of the money Babbage had wasted). Georg and Scheutz then tried to sell their machines, but despite much hard work, they only managed to sell two, each only to a government. Quite simply, contemporary mathematical tables did a better job. Private investors had long anticipated the Engine's failure, which is why its only sponsors had been governments – organisations that are notoriously careless with taxpayers' money.

Babbage's wasted £17 000 were, curiously, a blessing, because they warned successive British administrations off science. France's tragedy was that its government-funded scientists did nice work, but without a free market the French could never discover what a bad economic investment it all was.

SUMMARY

In conclusion, therefore, the industrial Revolution was born in Britain *because* Britain was *laissez faire*. In the early years, the free market correctly directed investment into technology, but as the revolution accelerated, it correctly directed more and more investment into pure science. Britain's industrial and scientific success, therefore, evolved out of its commercial success, not only historically but also dynamically.*

---

*That Britain's technological and scientific success was predicated on its commercial success is counterpointed by the now-celebrated observation of the historian William Rubinstein that almost all the millionaires in nineteenth-century Britain were financiers, not manufacturers – let alone scientists. Of more than 160 millionaires in nineteenth-century Britain (excluding landowners who inherited wealth),

But have the economic laws of science funding since changed? Would *laissez faire* now fail? Has *dirigisme* mutated out of error? Or would the free market still do better if it were given a chance? We will explore these questions in later chapters.

POSTSCRIPT ON *PER CAPITA* INCOME

One of the great joys of capitalism is that it redistributes wealth in an egalitarian fashion. Under feudalism, barons were hundreds of times richer than serfs, but under capitalism, the discrepancy of income between a director of a company and a member of the shopfloor may be less than ten-fold. For two centuries, romantics and socialists have tried to portray capitalism as a cause of inequality and of unprecedented degradation. Peter Gaskell's seminal *The Manufacturing Population of England* (1833) described pre-Industrial Revolution England as a rural Arcadia, bathing in a golden age of craftsmanship and prosperity which, tragically, the factories destroyed. But Peter Gaskell was a romantic who only saw pretty cottages and who failed to understand that rural poverty can both be utterly squalid and yet look picturesque. In the seventeenth century, agricultural labourers were the poorest section of society as Gregory King had shown (see Table 5.1). In the nineteenth century, agricultural labourers still starved: in his *Plenty and Want: A Social History of Diet in England from 1815 to the Present Day* (London, 1966) J. Burnett concluded that: 'the agricultural worker was, in fact, the worst fed of all workers in the nineteenth century'. He subsisted on a meagre diet of cheese, bacon and some milk.

The agricultural labourer was also the worst housed of all labourers: 'herded together in cottages which, by their imperfect arrangements, violated every sanitary law, generated all kinds of disease, and rendered modesty an unimaginable thing' [R. E. Prothero, *Pioneers and*

---

only one was a Manchester cotton manufacturer, but no fewer than 82 worked in the city of London [W. D. Rubinstein, *Capitalism, Culture and Decline in Britain, 1750–1990* (London, 1993)].

Japan teaches the same lessons. The richest man in the world in 1987 according to *Forbes* magazine was Yoshiaki Tsutsumi, whose personal fortune was £13 billion, the equivalent of Israel's GDP (the richest American could only muster $2.5 billion, poor dear). The Tsutsumi dynasty's founding father was Yasujiro Tsutsumi whose philosophy was 'borrow and buy but always avoid manufacturing' [Lesley Downer, *The Brothers* (London, 1994)].

Progress of English Farming (London, 1888)]. Labourers were drawn from the country into the cities by the higher wages, which is why those northern agricultural workers who stayed on the land enjoyed wages that were double those in the south, because the demand for labour from the mills and factories was so intense [A. L. Bowley, *Wages in the United Kingdom in the 19th Century* (London, 1900)]. We now view the cities of Victorian England with horror, but the rural degradation from which their labourers migrated is illustrated by the findings of the 1842 Royal Commission on Children's Employment which reported that living conditions in the new Lancashire mill towns were much better than they had been in the cramped cottages and rural workshops of pre-capitalist England.

The two men most concerned to prove that capitalism made things worse for the workers were Engels and Marx. This was because they believed that early capitalism would develop into late or monopoly capitalism, where a handful of rich men would control all the means of production, and everybody else would starve. Because of these internal contradictions, revolution would sweep the workers into power. Unfortunately for Engels and Marx, all the evidence showed that the workers were getting richer in capitalist England, so they cheated. W. O. Henderson and W. H. Challoner, two careful scholars, showed in 1958 that almost every 'fact' in Engels's *Condition of the Working Class in England* (1845) was, actually, fictitious or distorted [W. O. Henderson and W. H. Challoner (trans. and eds), *Engels's Condition of the Working Class in England* (Oxford, 1958)]. Indeed, as early as 1848 the German economist Bruno Hildebrand had exposed Engels and his book as fraudulent [*Nationalokonomie der Gegenwart und Zukunft* (Frankfurt, 1848, pp. 155–61, 170–241)]. Engels would, for example, describe the conditions in pre-Industrial Revolution rural workshops, and then omit to mention that these were the very workshops that were being displaced by the new factories. Engels would quote evidence of poor living conditions published in 1810 or 1818, but not give the dates, to mislead his readers into thinking they still existed.

Marx systematically lied as well. As early as 1885, two Cambridge economists, commenting on Marx's use of the British Government's economic statistics which were published in the so-called Blue Books, wrote: 'he uses the Blue Books ... to prove the contrary of what they really establish' [quoted in L. R. Page, *Karl Marx and the Critical Examination of his Work* (London, 1987)]. Marx's most celebrated lie was his deliberate distortion of Gladstone's Budget speech of 17 April 1863. In that speech, Gladstone said: 'The average condition of the

British labourer, we have the happiness to know, has improved during the last twenty years in a degree which we know to be extraordinary, and which we may almost pronounce to be unexampled in the history of any country and of any age.' In his *Inaugural Address and Provisional Rules of the International Working Mens' Association* (1864) Marx claimed that Gladstone said: 'This intoxicating augmentation of wealth and power is entirely confined to classes of property' [E. Kamenka, *The Portable Karl Marx* (Penguin, 1983, p. 358)]. For twenty years Marx continued to deny that he had lied, but as David Felix has shown, Marx knew full well that he had [D. F. Felix, *Marx as Politician* (London, 1983)].

Capitalism will never create a completely equal society because it is dynamic. In his *Capitalism, Socialism and Democracy* (New York, 1942), Schumpeter engaged in an elegant piece of intellectual exploration. What sort of a society, he asked, would be produced if capitalism flourished in the absence of technological innovation? Such an imaginary society would be egalitarian, because the competition between employees and employers would settle into an equilibrium of equality. Schumpeter concluded, therefore, that the ultimate source of profits is technological innovation. Innovation generates wealth but, because it disturbs the steady state, it also generates profits. These must be borne stoically.

What is the 'natural' degree of inequality in a healthy capitalist society? A survey in the *Economist* (5 November 1994, pp. 19–21) showed a considerable variation. If one takes the ratio of the share of GDP earned by the richest 20 per cent against the share earned by the poorest 20 per cent, one finds that Japan, Belgium and Sweden are very egalitarian, the ratio being about 4 : 1 (i.e.: the richest 20 per cent of society earned some four time more than the poorest). Australia, New Zealand, the USA and Switzerland are relatively inegalitarian (ratios of around 9 : 1) with Germany, Holland, Norway, Italy, Finland, France, Canada, Britain and Denmark falling somewhere in the middle.

Interestingly, the more egalitarian countries tend to be the fastest growing, economically, which may either mean that rapid economic growth breeds equality, or that inequality breeds slow growth (see the discussion in the *Economist* of 5 November 1994, pp. 19–21).

POSTSCRIPT ON JOHN STUART MILL'S EQUATION AND THE RESIDUAL

The formula $Y = K + L + T + p$ is too simple if we are to ask about *rates* of change in wealth. We have to use a more complicated equation in which we use the expression '$\Delta x/x$' to mean the proportional growth rate in 'x'.

$$\Delta Y/Y = \Delta K/K + \Delta L/L + \Delta T/T + \Delta p/p.$$

The equation has to be made even more complicated, because K, L, T and p have to be weighted against each other. This is because they may not be of equal importance to national wealth. In a highly industrialised society, say, a twofold increase in land may not increase overall wealth very much, because farming may only contribute a tiny percentage of national wealth. To weight these factors against each other, economists use the symbols α, β and γ to represent, respectively, the relative shares of profits (the return on capital), wages (the income of labour) and agricultural rents (the return on land) in national income. The equation, therefore, becomes

$$\Delta Y/Y = \alpha\Delta K/K + \beta\Delta L/L + \gamma\Delta T/T + \Delta p/p.$$

Which of these factors accounted for the Industrial Revolution? Economists do not measure changes in national productivity directly, they do it indirectly. They measure changes in capital, profits, labour, wages, land and rents, and then they measure the overall change in national wealth, and if the two sides of the equation do not add up, they attribute that shortfall to changes in national productivity. Indeed, economists use the term 'residual' to represent proportional growth rate in productivity because that is how they determine it, by calculating the residual shortfall in their measurements. They symbolise this by r^*. The final form of the equation is, therefore

$$\Delta Y/Y = \alpha\Delta K/K + \beta\Delta L/L + \gamma\Delta T/T + r^*.$$

What can we say about the residual? First, it is inaccurate, because small errors in the measurement of Y, K, L, T α, β or γ will lead to big changes in r^*. None the less, it is real. The changes in British national wealth between 1780 and 1860 were so big, and the changes in capital, profits, population, wages, land and rents so small, relatively, that only changes in productivity can be invoked to account for the difference. But what makes up the residual? Or, to put the question differently, what factors make up the changes in productivity? These questions have never been fully answered. Some improvements in productivity, of course, must be very general. A healthy, well-educated, well-trained and motivated workforce will produce more wealth, with the same materials, than a collection of sick, ignorant slackers. But from Table 6.1, as discussed above, we also see that the greatest increases in productivity were concentrated in certain industries, and as it was precisely those industries that also enjoyed the greatest technological advances, we can reasonably conclude that Britain's industrial revolution was largely a product of new British technology.

POSTSCRIPT ON HOBBY SCIENCE

The hobby scientists flourished under *laissez faire*, but *laissez-faire* Britain came to an end in 1914. Before 1914 the Government sequestered less than 10 per cent of the nation's wealth in taxes, but between 1918 and 1939 the Government increased this to about 25 per cent of GNP, and since 1945 the Government has spent between 40–50 per cent GNP. Because of the attrition of inherited wealth and of private means, the hobby scientist is now practically extinct. By the 1930s, for example, half of the lecturers in the Department of Biochemistry at Cambridge University still had private incomes, but today's tax structure has dramatically cut the numbers of people who inherit sufficient private means to do science for fun. One rare survivor is Peter Mitchell who won the Nobel Prize in 1978 for discovering the chemiosmotic hypothesis in his private laboratories in Bodmin, Cornwall. The occasional hobby scientist can still be found in a theoretical subject; Albert Einstein, for example, was working as a clerk in the patents office in Zurich when, during his spare time, he conceived of his theories of relativity.

Even *dirigiste* France could produce hobby scientists, but its harsh taxes so restricted opportunities that, inevitably, its most distinguished hobbyist was a taxman. Lavoisier was indeed a Farmer-General, which ultimately led him to the guillotine (the judge who condemned him to death remarked that 'the revolution has no need for scientists'; Karl Marx would have disagreed).

The loss of the hobby scientists has been unfortunate because the

TABLE 6.1 *British economic statistics, 1780–1860*

(a) Basic statistics

	Population	Income per capita	National capital
1780	7 500 000	£11	£670 m
1860	23 130 000	£28	£2770 m

(b) Dynamic statistics, 1780–1860

Average annual increase in GNP	2.42%
Average annual increase population	1.25%
Average annual increase in GNP/capita	1.17%
Average annual increase in capital/labourer	0.35%

(c) Average annual increases in productivity, 1780–1860

Cotton	2.6%
Worsteds	0.9%
Woollens	0.9%
Canals and railways	1.3%
Shipping	2.3%
Iron	0.9%
Agriculture	0.45%
All other sectors of the economy	0.65%

SOURCE: These figures come from D. N. McCloskey, 'The Industrial Revolution 1780–1860: a survey', in *The Economic History of Britain since 1700*, ed. P. Floud and D. McCloskey (Cambridge University Press, 1981).

hobby scientists tended to be spectacularly good. They were good because they tended to do original science. Professional scientists tend to play it safe; they need to succeed, which tempts them into doing experiments that are certain to produce results. Similarly, grant-giving bodies which are accountable to government try only to give money for experiments that are likely to work. But experiments that are likely to work are probably boring – indeed, if they are that predictable, they are barely experiments at all; rather, they represent the development of established science rather than the creation of the new (though science is so unpredictable that even so-called predictable experiments will yield unpredictable results on occasion). But the hobby scientist is unaccountable. He can follow the will-o'-the-wisp and he is more likely to do original than unoriginal research, because it is original research that is fun.

Most professional scientists spend much of their time doing repetitive work. Science has become a treadmill, and scientists must be seen to be publishing papers, speaking at conferences, getting grants, teaching undergraduates and training PhD students. These activities will not succeed unless they are predictable, and therefore even boring. The hobby scientist need never be bored. He need only do an experiment if it looks fun. The hobby scientist, therefore, will be attracted to challenging science to the same degree that the professional scientist is attracted to safe science.

The hobby scientist, moreover, will be a different sort of human being from the professional scientist. A professional scientist needs to be tough. It is a harsh, competitive world in a modern university, and if a scientist does not drive himself and his students to write the requisite number of papers and to win enough grants, then that scientist does not survive. But a hobby scientist does not have to be any particular sort of human being. Indeed, many of the great hobby scientists would transparently never have survived a modern university. Peter Mitchell, whose chemiosmotic hypothesis changed the very nature of modern biochemistry, took seven years to complete his PhD. In Britain, PhD grants are only for three years, so Mitchell would never have completed his PhD had he depended on public funds (particularly as he was not a good PhD student, and no one would have fought for him). After his PhD, he obtained a lectureship in the Department of Zoology at Edinburgh University, but he found the job intolerable and left after a few years. He bought a dilapidated country house in Bodmin, in Cornwall, spent two years rebuilding it as a form of psychotherapy, and then started on his researches again, in his own way, on family money – to win a Nobel Prize.

Many of the hobby scientists were decidedly peculiar. Cavendish, a bachelor, only spoke to other human beings on Thursday nights when he dined with a coterie of FRSs. Otherwise he lived in solitude, communicating with his servants by notes and letters. Dinner was served to him through a contraption that shielded the butler from gaze, and if Cavendish ever saw a servant, he dismissed that person instantly. Darwin was also odd. He spent his whole life as a semi-invalid, and although it is claimed he suffered from Chagas' disease, Pickering showed in his *Creative Malady* [17] that Darwin probably pretended to be ill to shield himself from the strains of everyday life. Neither Cavendish nor Darwin would have survived in a modern university any better than did Mitchell, yet they were scientific giants (Darwin could not even survive undergraduate life, and he left before obtaining a degree). Another academic failure was Albert Einstein, one of the greatest of hobby scientists. Einstein did not do well as an undergraduate at university, and he failed to obtain a PhD position, so he had to get a job; he chose to clerk in a patents office because it left him with spare energy in the evenings.

When science was a vocation, personal poverty did not frustrate potential researchers. Michael Faraday, for example, was the son of a blacksmith, and he was apprenticed to a bookbinder, but science was his hobby and despite his lack of conventional qualifications Sir Humphry Davy was happy to employ him as a technician at the Royal Institution. It did not take long for his genius and passion to be recognised. (Even a chronic grumbler like Thomas Huxley prospered as a gifted career scientist despite his lack of private means.) Occasionally, a contemporary private scientific body will be as enlightened as those earlier institutions. Barbara McClintock, who won the Nobel Prize in 1983 for her discovery of transposable genetic elements, was employed from 1942 by the Carnegie Institution in Washington, DC. All they asked of her was that she wrote an annual report, which is all that she wrote. She could not be bothered with all the fuss and nonsense that it takes to publish papers in peer-reviewed journals, and anyone who wanted to know what she had done only had to read the Carnegie Institution's annual report. Only a private body could behave so unconventionally. A modern university would have found McClintock wanting, because she would not have been conventional enough to spend her days writing grants, sitting on committees, and driving PhD students, technicians and post-doctoral fellows to write their quota of papers.

The hobby scientists were the most romantic of scientists, approaching the poets in their intellectual purity and richly individualistic personalities. Rich or poor, the hobby scientists were driven by a vocation and a love of research. We are lessened by their extinction. Those who argue for more government funding of science, or of anything else, should never forget the cost of government money, namely the taxes that impoverish society to enable government to impose its particular, narrow, harsh vision of a modern university.

7 Economic History since 1870

God, during the nineteenth century, was unquestionably an English-man. One little island nation, Great Britain, dominated the world. The citizens of that small foggy island controlled the greatest empire the world had ever seen: a quarter of the human race, a fifth of the surface of the globe including Canada, half of Africa, India, Australasia and various other chunks of the remaining continents were all ruled from London. The Royal Navy policed the oceans – Britannia ruled the waves – and the British army preserved world peace. *Pax Britannica.* Underpinning that vast military and political enterprise was the British economy, also the greatest the world had ever seen. British products, techniques and engineers inspired awe from San Francisco to Irkutzk, from Helsinki to Cape Town. Birmingham was indeed the workshop of the world.

Countries are competitive, and when they become Top Nation they like to boast. Half way through the nineteenth century, in 1848, a civil servant called Henry Cole (later Sir Henry Cole) conceived of an international industrial exhibition. He proposed that products from factories all over the world be collected, to celebrate Britain's supreme contribution to industry. The idea was embraced by His Royal Highness Prince Albert, the consort of Her Majesty the Queen. Albert's motives, however, were different from Cole's. Cole was a triumphalist, but Albert wanted to warn Britain of the growing threat of foreign competition. He feared for Britain's supremacy.

Not every one liked the idea of an international exhibition. Many patriots opposed it because it would attract foreigners to the country, and foreigners were not only notoriously immoral but also, after 1848, revolutionary. Moreover, the sceptics argued, the exhibition was a new idea, and therefore certain to fail. Colonel John Sibthorp, MP for Lincoln, prayed publicly in the House of Commons that the exhibition might collapse into bankruptcy.

Fearless, Albert and Cole forged ahead. Their first task was to secure a site, but this presented them with less of a task than it would other people. The obvious location was Hyde Park, which is to London what Central Park is to New York, and which belongs to the monarch – who in Albert's day was his adoring wife, Queen Victoria. The Park secured, Albert and Cole then had to find an architect. They envisaged an enormous building – it eventually covered 20 acres – but they did not want to wait years while one was built by conventional means. They needed a revolutionary technique of construction. Fortunately Joseph Paxton, the Duke of Devonshire's gardener, had perfected the art of prefabrication. Paxton had built a number of large greenhouses on the ducal estates at Chatsworth, and he had trained the local foundries to cast metal casements to be assembled on site. Using this technique, Paxton built so fast that in a matter of months he had erected a huge greenhouse – nicknamed the Crystal Palace – to house the exhibition. In 1851, only three years after Cole's memorandum, the Queen opened the Great Exhibition of the Industry of all Nations.*

The Great Exhibition was a fantastic success. It attracted 13 937 exhibitors, and during its 5½ months, 6 039 195 people paid to visit. No one who was anyone in Europe failed to attend, and a remarkable number of Americans came too. Gratifyingly, England's morals survived the invasion of foreigners. Moreover, the international jury awarded British artefacts most of the prizes – Rule Britannia.

Not everyone rejoiced, however. There were those who, in the midst of triumph, claimed to have spotted the seeds of British industrial decay. Professor Lyon Playfair FRS, Professor of Chemistry at the Royal School of Mines, Jermyn Street, London, published in late 1852 his *Industrial Instruction on the Continent* in which he claimed that European industry would overtake Britain's unless British industrialists changed their approach. Such warnings were not original; as early as 1835 Cobden had written 'our only chance of national prosperity lies in the timely re-modelling of our system, so as to put it as nearly as possible on an equality with the improved management of the Americans';[1] but these and other warnings were generally ignored – until the frightening Paris Exhibition of 1867.

The 1867 Paris Exhibition was organised similarly to the 1851 London Great Exhibition, except that it was larger, covering 41 as op-

---

*The Grand Procession that inaugurated the opening was led by the Duke of Wellington. The Duke had made a crucial contribution to the Great Exhibition by suggesting an ecological solution to the problem of the pigeons. The Crystal Palace had been rapidly colonised by large numbers of pigeons, and their droppings caused much distress. The only obvious solution, sniper fire, was impossible, but during a meeting of the Privy Council, presided over by the Monarch, the Duke suggested 'Sparrowhawks, Ma'am'.

posed to 20 acres. The outcome, however, was very different; in 1851 the international jury had awarded most of the prizes to Britain, in 1867 Britain only won 10 out of the 90, being overtaken by France, Belgium, Germany and the USA. Britain was shocked and, remarkably, that shock reverberates to this day. Almost unanimously, commentators have described Britain as having entered long-term economic and industrial decline at some time around 1867.

The first person to raise the alarm in 1867 was, again, Lyon Playfair, who wrote a famous open letter[2] to Lord Taunton, then chairing the Royal Schools Inquiry Commission. This letter so impressed Parliament, and so inflamed public opinion, that the Government set up a Select Committee in 1868 to examine the problem. It is often forgotten that this Select Committee dismissed Playfair's claim that British industry was in decline because of a shortage of British science: 'Although the pressure of foreign competition, where it exists, is considered by some witnesses to be partly of a superior scientific attainment of foreign manufacturers, yet the general result of the evidence proves that it is to be attributed mainly to their artistic taste, to fashion, to lower wages, to the absence of trade disputes abroad, and to the greater readiness with which handicraftsmen abroad, in some trades, adapt themselves to new requirements. It is owing to one or more of these favourable conditions rather than to superior education or technical skill, that the lace-makers of Calais and the locomotive manufacturers of Creuzot and of Esalingen, are competing with this country in neutral markets, and even at home; Mr Cochrane states that the lower rate of wages alone enabled the factory at Anzin to furnish him with pumping and winding engines constructed according to his own design but of excellent workmanship, for his own collieries in Northumberland and Durham.'[3]

The Select Committee's dismissal, however, has not dented the almost unanimous verdict of economic historians that Britain entered long-term economic, industrial and technological decline at some time during the mid to late nineteenth Century. Consider these comments.

> The worst symptoms of Britain's industrial ills . . . was the extent to which she was no longer in the van of technical change; instead, even the best of her enterprises were usually being dragged in the wake of foreign precursors, like children being jerked along by important adults. (David Landes, *The Unbound Prometheus*)[4]

> Britain's technological defeat was first alarmingly revealed . . . in 1867. (Correlli Barnett, *Audit of War*)[5]

Britain's relatively poor economic performance (1870–1914) can be attributed largely to the failure of the British entrepreneur to respond to the challenge of changed conditions.[6]

At the turn of the century, a new international pattern of technology was emerging. Britain was in decline, Germany was in the ascendant.[7]

One could go on and on quoting such comments from distinguished economists and historians; but how accurate are they? Let us, initially, dismiss the suggestion that the different number of prizes won by Britain in 1851 and 1867 means anything. First, the numbers of British entries were very different. Of the 13 937 exhibitors in 1851, half were either British (6861 exhibitors) or from the British empire (520 exhibitors): The competition was unbalanced towards the host country. Exactly the same imbalance occurred in 1867, but tilted towards France. Half the exhibitors were French, and the Belgians and Germans also found it very easy to exhibit. The Americans, moreover, made a very big effort to be represented. There were relatively few British entries, and so relatively few prizes.

Second, it is important to distinguish between relative and absolute decline. Let us simplify and represent economic strength as oranges. Let us pretend that Britain in 1851 was so rich that she possessed five oranges, while the USA, Germany, France and Belgium each only owned one orange. Britain, therefore, was not only richer than each individual competitor, it was richer than all four combined. Now let us pretend that, by 1867, all five countries had done equally well, and that each had acquired a further orange. Britain would now possess six oranges, but the USA, Germany, France and Belgium would own two apiece. Two consequences immediately flow. First, Britain no longer emerges as richer than everyone else combined, because Britain's six oranges are outnumbered by its competitors' eight; Britain, therefore, is in relative decline. Second, its competitors have grown at higher percentage rates: to double your oranges from one to two is to achieve 100 per cent growth, to increase your number of oranges from five to six is to only achieve 20 per cent growth. Yet, in absolute numbers of oranges, everyone has grown equally, by one. On the orange analogy, therefore, it is absurd to have expected Britain to have won the majority of the prizes in 1867. 1851 hit just that moment when Britain had five oranges while its competitors each only had one, but the situation could not have lasted – nor should it have. In a just world, everybody should have six oranges.

The third reason for dismissing the prizes won at international competitions as indicators of national wealth is that they are not, in fact, very good indicators of national wealth; they are indicators of national technological achievements in certain fields, but these may never translate into wealth. If an international exhibition had been held in 1957, for example, just after the USSR had launched *Sputnik*, the world's first space rocket, the USSR would have been showered in prizes. The world was, indeed, awed at the time (see Chapter 8). Japan, on the other hand, which by 1957 had aspired to nothing more than copying other people's technology, would have won no prizes at all. Whose economy has since flourished?

Who, moreover, were the judges in 1851 and 1867? They were men like Lyon Playfair, a professor of chemistry (he sat on both the 1851 and 1867 juries, and he has been suspected of marking down the British entries on both occasions to reinforce his claims of British technological decline). These men were not judging economic achievements, but technical ones. To judge economic achievements, we must turn to the economists and to the economic historians. In particular, we must turn to Angus Maddison who has collated a vast amount of scholarship in his *Phases of Capitalist Development* from whose tables many of the graphs in this chapter are drawn.

The pattern of long-term economic growth of capitalist countries is characteristic. Figure 7.1 illustrates Britain's GDP *per capita* since 1700. It shows that growth between 1700 and 1820 was slow, but that it accelerated thereafter. Yet it is hard to determine *rates* of growth from Figure 7.1; this is because the left-hand axis is linear. If, however, the left-hand axis is rendered logarithmic, as in Figure 7.2, it can be seen that rates of *per capita* growth after 1820 remained roughly constant*. We can, therefore, conclude that between 1700 and 1820 Britain's rate of economic growth increased gradually, but between 1820 and 1979 Britain enjoyed an essentially steady rate of growth, although there are some wobbles to the line.

---

*Most GDP *per capita* growth graphs in this book are presented with their vertical axes represented as logarithmic scales. This is a mathematical procedure that converts regular rates of growth into straight lines. Consider a pair of rabbits. If they reproduce such that the initial pair will produce two offspring, each pair of which then go on to produce two more, the numbers of offspring will go 2, 4, 8, 16, 32 etc. This will generate a rapidly rising curve if the vertical axis is represented arithmetically (as in Figure 7.1) but if the vertical axis is represented on a logarithmic scale, the curve emerges as a straight line (which informs numerate people that the rate of reproduction is constant, irrespective of the numbers of rabbits breeding at any one time).

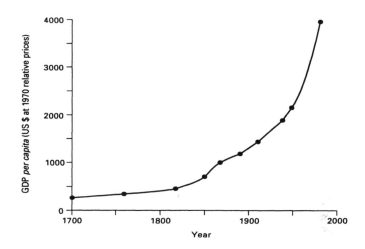

FIGURE 7.1   *GDP* per capita *UK*

GDP *per capita* data for the various years came from Maddison's *Phases of Capitalist Development.*[21]

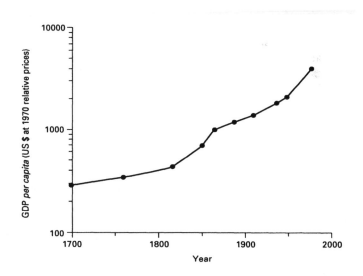

FIGURE 7.2   *GDP* per capita *UK (log scale)*

GDP *per capita* data for the various years came from Maddison's *Phases of Capitalist Development.*[21]

This pattern of national economic growth – a preliminary period as a country emerges from late-feudalism into capitalism, during which growth rates slowly accelerate, followed by a sustained period of constant growth rates – appears to be common to all capitalist countries. In Figure 7.3, France shows the same pattern as Britain, although France started to emerge from feudalism about 50 years later. Sweden's emergence was delayed a further 50 years, and Japan's yet another 50 years, but their patterns of growth are all similar. We can draw up, therefore, a general law of economic growth amongst capitalist countries. As they emerge from a feudal culture, their growth rates start to accelerate until the acceleration into a capitalist culture is complete, whereupon countries enter into the log phase of growth, during which wealth is doubled every 25 years or so, apparently inexorably. The essentially linear nature of sustained capitalist growth is illustrated by the USA, Canada and Switzerland in Figure 7.4.

From Figure 7.3, however, a curious twist emerges: the countries that emerge into capitalism late enjoy higher rates of growth than do the pioneers (thus in Figure 7.3 Sweden overhauls Britain and France by 1939, and Japan is clearly set to catch up). This phenomenon appears to be general, as is shown in Table 7.1. Table 7.1 presents the economic performances between 1870 and 1979 of the 16 countries that are, today, the richest in the world – and it shows that the comparative economic standing of countries in 1870 was very different from today. In 1870, the richest country in the world was Australia; Japan was desperately poor.

Australia in 1870 was widely known as the 'lucky country'; it had vast mineral reserves, lots of sheep, and a British expatriate population which knew how to exploit its resources. The second richest country in 1870 was Britain, for obvious reasons. Almost as rich were Belgium and the Netherlands; we saw in Chapter 5 how those countries led the world during the late Commercial and Agricultural Revolutions, and were only dragged down by religious conflict, and by their wars with Spain and their neighbours. They were not dragged down very far.

There was, in 1870, a band of eight middling economies of similar strength, those of Switzerland, the USA, France, Canada, Italy, Austria, Denmark and Germany (note how poor France and Germany were compared to the UK) and finally, in 1870, the Scandinavian countries were almost as poor as Japan (Japan was a feudal country until 1868, when the Emperor Meiji triumphed over the shoguns and embarked on converting his country into a capitalist one, the so-called Meiji Restoration of 1868).

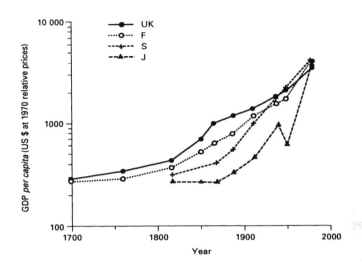

FIGURE 7.3   *GDPs* per capita *UK, France, Sweden and Japan*

GDP *per capita* data for the various years came from Maddison's *Phases of Capitalist Development.*[21]

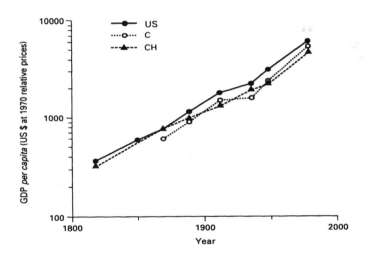

FIGURE 7.4   *GDPs* per capita *USA, Canada and Switzerland*

GDP *per capita* data for the various years came from Maddison's *Phases of Capitalist Development.*[21]

TABLE 7.1   *The economic performances, 1870–1979, of the 16 currently richest countries*

	GDP per capita in 1870	Average annual % growth of GDP per capita	Productivity (GDP man-hour) in 1870	Average annual compound growth in productivity
Australia	1393	1.07	1.30	1.5
Austria	573	1.88	0.43	2.4
Belgium	925	1.51	0.74	2.1
Canada	619	1.97	0.61	2.3
Denmark	572	1.93	0.44	2.3
Finland	384	2.24	0.29	2.7
France	627	1.94	0.42	2.6
Germany	535	2.06	0.43	2.6
Italy	593	1.67	0.44	2.4
Japan	251	2.69	0.17	3.0
Holland	831	1.56	0.74	2.1
Norway	489	2.09	0.40	2.6
Sweden	415	2.31	0.31	2.9
Switzerland	786	1.62	0.55	2.1
UK	972	1.29	0.80	1.8
USA	764	1.89	0.70	2.3

SOURCE: The data comes from A. Maddison, *Phases of Capitalist Development* (Oxford University Press, 1982). The figures represent US dollars at 1970 prices with exchange rates adjusted for differences in purchasing power of national currencies. Productivity is determined as GDP earned per hour worked.

The second column in Table 7.1 provides the average annual growth of GDP *per capita* between 1870 and 1979 for the different countries. Even the least mathematical of readers will immediately note the remarkable phenomenon: the richest countries in 1870 (the ones that embraced capitalism earliest) have grown slowest subsequently, the poorest (who embraced capitalism later) have grown fastest. Australia only grew at 1.07 per cent a year on average between 1870 and 1979, Japan achieved 2.69 per cent, and everyone else grew somewhere between those two rates, depending on how rich or poor they were in 1870. The inverse correlation between wealth in 1870 and rates of growth between 1870 and 1979 is so tight ($r = -0.96$, $P < 0.001$ in mathematical terms) that it can be illustrated on a graph (Figure 7.5).

This principle of economic growth in capitalist countries, that poor ones can grow at faster rates than rich ones, holds over quite short periods of time. Figure 7.6 shows an equally tight relationship between GDPs per capita in 1950 and growth rates between 1950 and 1979. Countries that were poor in 1950 grew faster, subsequently, than

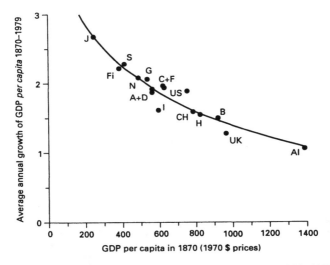

FIGURE 7.5 *GDP* per capita *in 1870 and growth 1870–1979*

GDP *per capita* data for the various years came from Maddison's *Phases of Capitalist Development*.[21] By definition, of course, this graph contains an element of statistical fallacy as it only covers the countries that are now the richest. A graph that included the countries that were the richest in 1870 would be more like Figure 7.10.

KEY: A = Austria; Al = Australia; B = Belgium; C = Canada; CH = Switzerland; D = Denmark; F = Finland; F = France; G = Germany; H = Holland; I = Italy; J = Japan; N = Norway; S = Sweden

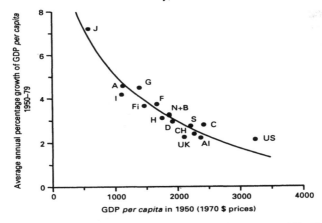

FIGURE 7.6 *GDP* per capita *in 1950 and growth 1950–79*

GDP *per capita* data for the various years came from Maddison's *Phases of Capitalist Development*.[21] See above for key to symbols.

did rich ones. The relationship, however, does not hold over very short periods of time, because national growth rates are very variable over the short term: Figure 7.7 compares national growth rates for the major industrialised countries between 1976 and 1981 with those countries' growth rates for 1981–6. There is no correlation over such short periods of time; one has to consider 25-year periods to determine long-term growth rates.

There are, therefore, two sorts of economic growth, short term and long term. The factors that determine short-term economic growth are the ones that dominate the television screens, the radio and the newspapers; these are the exchange rate, the interest rate, the unemployment rate, the inflation rate, the investment rate, the capital output rates, the expansion of the money supply, business confidence, consumer spending, exports, imports, stock market activity, and all the other factors that exercise politicians, journalists and pundits. These are the factors that determine short-term economic growth rates and so determine the outcome of elections; but in the long term, these factors cancel out. In the long term, there is only one factor that determines the rate of economic growth: productivity.

The third column in Table 7.1 shows the national productivity figures for the 16 countries in 1870, and a dramatic correlation emerges. The poor countries have low levels of productivity, the rich ones have high ones. But the fourth column in Table 7.1 shows the average annual increases in productivity of the 16 countries between 1870 and 1979 and, again, a most dramatic correlation emerges; it is precisely those countries who enjoy the highest rates of increase in productivity who also enjoy the highest rates of economic growth. Figure 7.8 illustrates the very tight correlation between GDP *per capita* growth rates and productivity growth rates that occurred between 1870 and 1979, and Figure 7.9 shows that increases in national productivity since 1870 correlate inversely with national poverty in 1870. These correlations indicate that productivity *is* wealth.

What increases productivity? Let us return to the equation for long-term growth that was presented in Chapter 6

$$\Delta Y/Y = \alpha \Delta K/K + \beta \Delta L/L + \gamma \Delta T/T + r^*$$

where: Δ = the change in; Y = total wealth; K = capital; α = share of profits in national income; L = labour; β = share of wages in national income; T = land; γ = share of rents in national income; r^* = the residual.

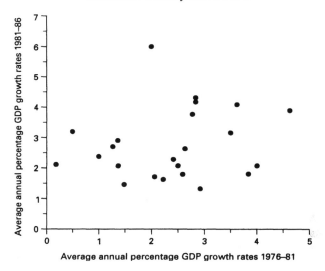

FIGURE 7.7 *Growth rates for 1976–81 and 1981–6*

The 5-year growth rates for the 22 largest OECD capitalist countries came from
OECD Economic Surveys, UK (Paris: OECD, 1983 and 1988).

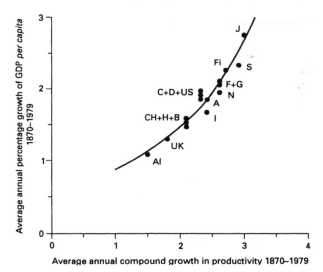

FIGURE 7.8 *Productivity and GDP* per capita *growth rates, 1870–1979*

Productivity and GDP *per capita* data for the various years came from Maddison's
Phases of Capitalist Development.[21] The equation for the graph is
$$y = 0.476 \times 10^{0.247x}, \ r = 0.96 \ (P < 0.001).$$
See Figure 7.5 for key to symbols.

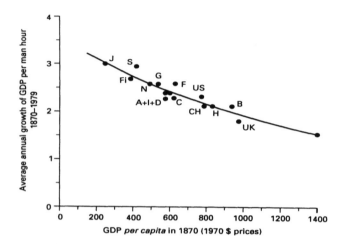

FIGURE 7.9   *Wealth in 1870 and productivity growth rates, 1870–1979*

Productivity and GDP *per capita* data for the various years came from Maddison's
Phases of Capitalist Development.[21] The equation for the graph is
$$y = 3.519 \times 10^{-2.688e - 4x}, \ r = -0.96 \ (P > 0.001).$$
See Figure 7.5 for key to symbols.

This equation was first drawn up in this form during the 1950s when
three different American economists were trying to determine the factors that had been responsible for the increased wealth that America
had enjoyed during the twentieth century.[8, 9, 10] Each of these economists had taken John Stuart Mill's assumption that the four fundamental sources of wealth were labour, land, capital and productivity,
and they had each found that the two major sources for the increase in
American *per capita* wealth during the twentieth century were increased
capital and increases in r^*, the residual.

Increased capital accounted for around a third of the increased wealth,
although different economists come up with slightly different figures.[11, 12]
But increased capital injection will, after a certain point, only yield
diminishing returns.† The rest of the economic growth could not be
accounted for by the measurable indices such as the capital-output rates,
the rates of saving or the rate of growth of the work force. Since it
could not be measured, it was called the residual, r^*, which some econ-

---

† The Law of Diminishing Returns can be illustrated in two ways, anecdotally or
  mathematically. Anecdotally, imagine a carpenter working with a capital of hammer and a saw. If he were to double his capital, so that he then possessed two
  hammers and two saws, he would not double his productivity. At any one time he

omists such as Solow have called 'technical change in the widest sense', but which John Stuart Mill would have called increased productiveness and which we call increased productivity.

What is the residual? By definition, since it is calculated as the difference between all the indices that are actually quantifiable, we do not know for certain. If we could quantify or measure its components, we would. Some factors such as a stronger work ethic or the improved health of the population undoubtedly play a part in it, but the best assessment we have from the economists is that about half of it might represent, in broad terms, the consequence of the improved education of the work force, and about half represents the consequence of technological development. In summary, therefore, long-term economic growth is almost entirely the consequence of increases in productivity, and those increases are caused, in more or less equal amounts, by increases in capital, improvements in education, and advances in technology (a good general review of the residual, and what it means, can be found in Chapter 1 of M. Abramovitz's *Thinking About Growth*).[13]

It may seem obvious to conclude that long-term increases in wealth are the consequence of improvements in productivity, but such a conclusion helps put other factors into perspective. For example, how important are natural resources to national wealth? One only has to compare Japan, say, with the Congo. Japan is an island possessing no minerals and no oil, and it has to import all its raw materials, yet it has transformed itself into a major industrial power. The Congo, on the other hand, is blessed with vast natural resources, yet it remains deperately poor. One can see, therefore, that wealth is fundamentally a consequence of a particular culture, one that promotes improvements in productivity through encouraging capital investment, education and new technology.

---

can only use one hammer and one saw, and even if his new tools were bigger than the old, he would still not produce twice as many products per day. Only if his new tools were different in kind, say someone had invented a screwdriver and a chisel, would his productivity shoot up, but that would represent innovation, not investment *per se*. Increased investment or capital in existing technology will soon produce diminishing returns.

The Law of Diminishing returns can be represented mathematically. The formula for increase in wealth is given by: $\Delta Y/Y = \alpha\Delta K/K + \beta\Delta L/L + \gamma\Delta T/T + r^*$

Now let us assume the term $\alpha\Delta K/K$ doubles. It is obvious that if $\beta\Delta L/L$, $\gamma\Delta T/T$ and r^* remain unchanged, then $\Delta Y/Y$ will not double. Let us create some figures. If $\Delta Y/Y = 4$ because $\alpha\Delta K/K$, $\beta\Delta L/L$, $\gamma\Delta T/T$ and r^* all equal 1, then $4 = 1 + 1 + 1 + 1$. If $\alpha\Delta K/K$ doubles, then $\Delta Y/Y$ only increases to 5 because $2 + 1 + 1 + 1 = 5$.

It should be emphasised that it is not a general economic principle that the poorer a country, the faster it grows. It is only countries that have made the cultural leap into capitalism that obey this principle. Figure 7.10 is the same as Figure 7.6, except that the economic performance since 1950 of the then next-richest 34 countries has been added to the original 16 countries, and it shows that few of them fulfilled their potential. But few wholeheartedly embraced capitalism either, and only those that did (Singapore, Hong Kong and Taiwan) approach the line. Many of the underperforming countries' governments invested in capital, education and technology, but those are not enough – the USSR and the countries of eastern Europe invested in them to excess, yet they did not grow (Soviet science policy is discussed in Chapter 8). The free market is crucial. (The interesting case of Malta is discussed below.)

How, now, do we account for the faster rate of growth of the poorer capitalist countries? We have stated that long-term increases in national wealth are caused by improvements in productivity, and we have also stated that innovation is a crucial contributor to improvements in productivity, so presumably we could show that the poorer countries are precisely those which are the most innovative. Can we test the hypothesis? Can we correlate national growth rates with national innovation? Fortunately we can, and the results are intriguing. Let us, therefore, introduce the economics of innovation.

A SHORT INTRODUCTION TO THE ECONOMICS OF INNOVATION

Innovation is complex. It encompasses the trio of science, research and development, even if it also encompasses more than that. But we can measure aspects of that trio relatively easily, so let us study them first.

A Short Introduction to the Economics of Science

The easiest way of studying different countries' commitment to science is to look at their respective outputs of scientific papers. Over 300 000 papers are now published each year in the world, and Figure 7.11 correlates the numbers of papers published *per capita* with each country's GDP *per capita*. A strong correlation emerges, but it is (perhaps) the

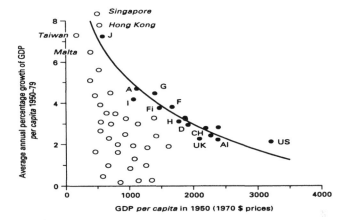

FIGURE 7.10   *GDP* per capita *in 1950 and growth 1950–79*

Data is as for Figure 7.6 with the additional economic data from *The Economist*, 4 January 1992, p. 17. The additional points (O) are only approximations, since the data from *The Economist* is not strictly comparable to that of Figure 7.6, differing in the years studied and in the collection of data. However, the two sets of data for the original 16 countries are so similar as to render meaningful the superimposition of the additional 34.

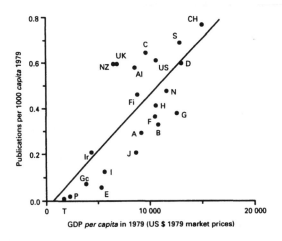

FIGURE 7.11   *National wealth and scientific output*

GDP *per capita* and national populations came from *OECD Economic Surveys, UK* (Paris: OECD, 1981). National publications for 1979 came from Braun *et al.* (1987).[50] The equation for the graph is: $y = -4.29 \times 10^{-2} + 4.99 \times 10^{-5x}$, $r = 0.766$ ($P < 0.001$). See Figure 7.5 for key to symbols. Also: E = Spain; Gc = Greece; Ir = Ireland; P = Portugal; T = Turkey opposite of what one might expect; the richer the country, the more science it does.

Figure 7.12 correlates the quality of different countries' science with their GDPs *per capita*, and a further strong correlation emerges; the richer the country, the better its science. (In this chapter, we only provide a short introduction to the economics of research; Chapter 10 provides the detailed explanation of how the data for figures such as Figure 7.12 was collated.)

A Short Introduction to the Economics of Patents

Although there are well over 100 countries in the world, almost all the patents that are filed originate in the richest 69. There are some 40 or so countries too poor to foster the research that leads to patents. For the 69 patent generators, J. D. Frame has shown a strong correlation between their GNP *per capita* and their published patents *per capita* (Frame's figure comparing patents *per capita* against GNP *per capita* looks very similar to Figure 7.11).[14] We can conclude, therefore, that the richer the country, the more patents it files.

A Short Introduction to the Economics of Research and Development

Not all innovation is published as scientific papers or patents. Much industrial innovation is discreet. Much, moreover, can be described as 'development' rather than as 'research' (development is that process, generally carried out by companies, by which products, technologies or scientific methods are improved for the market; it represents applied rather than pure science). The Organization of Economic Cooperation and Development (OECD) publishes the shares of GDP that its member states spend on civil research and development (R&D) and, again, a remarkable correlation emerges (Figure 7.13). Rich countries spend a higher percentage of their GDP on civil R&D than do poor ones.

This appears to be a general law, not just restricted to 1979 and 1985, the years illustrated in Figures 7.11 and 7.13. The earliest year for which we have OECD statistics for civil R&D, 1965, also shows a positive correlation between national wealth and the percentage of national wealth (GNP in this case) spent on civil R&D (Figure 7.14). So, for example, the European country that spent most on civil R&D in 1963/4 was Britain, then the richest country in Europe. In 1963/4, industrially funded civil R&D ran at $809 million in Britain (at official

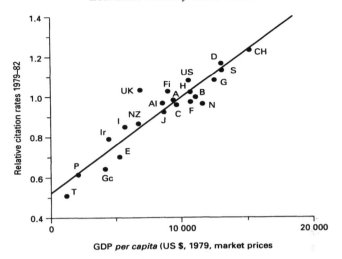

FIGURE 7.12   *GDP and relative citation rates*

The citation data comes from Braun *et al.* (1987)[50] and the economic data from
OECD Economic Surveys, UK (Paris: OECD, 1981).
$$y = 0.525 + 4.77e^{-5x}, \ r = 0.94 \ (P < 0.001).$$
See Figures 7.5 and 7.11 for key to symbols.

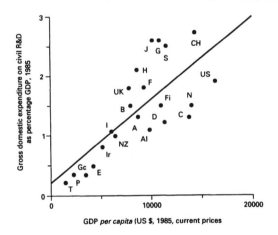

FIGURE 7.13   *GDP and expenditure on civil R&D*

Gross domestic expenditure on civil research and development as percentage GDP,
1981, comes from *OECD Science and Technology Indicators No. 3* (Paris: OECD,
1989). GDP *per capita* data came from *OECD Economic Surveys UK, 1987* (Paris:
OECD, 1987). $y = 0.21 + 1.38e^{-4x}, \ r = 0.739 \ (P < 0.001).$

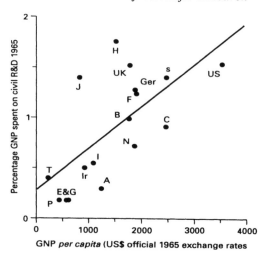

FIGURE 7.14   *Percentage GNP spent on civil R&D, 1965*

The data comes from *OECD International Statistical Year for Research and Development: a Study of Resources Devoted to R&D in OECD Member Countries in 1963/64* (Paris: OECD, 1968). $y = 0.263 + 4.192e - 4x$. $r = 0.677$ ($P < 0.005$). See Figures 7.5 and 7.11 for key to symbols.

exchange rates), while West Germany could only manage $796 million, and France $438 million. Britain's total expenditure on all R&D in 1963/4 was $3170 million (at OECD R&D exchange rates) while West Germany could only manage $1886 million and France $1661 million, and although Britain's figure was bloated by its defence expenditure (as was France's) yet the British government was still outspending the German government on civil R&D.[15] Indeed, as the historian David Edgerton has noted in his comprehensive review of twentieth-century funding for civil R&D, the European nation that throughout the twentieth-century has consistently spent the most on civil R&D is Britain, the country which until recently was the richest country in Europe: 'until the mid-1960s British industry spent more in absolute and relative terms on industrial R&D than the industry of any capitalist country other than the United States'.[16]*

---

*In his writings, Edgerton has explained how the myth arose that during the twentieth century Britain failed to perform enough R&D. There are two major explanations. First most British R&D was privately funded (mainly by industry) and the figures for this sort of funding are hard to collate, but when they are collated, it emerges that Britain was the major European funder of R&D. Second, Britain spent the twentieth century in relative economic decline, and during the 1960s

We can construct, therefore, from these studies, a general economic law: the richer the country, the greater the percentage of its wealth it spends on R&D, the more science it does, and the better that science is. But a paradox immediately emerges. In Figure 7.5 we saw that the richer the country, the slower its rate of economic growth, yet the more science and R&D it does. But is it not innovation that underpins economic growth? This is an old problem, which has long fretted the worthies in this field. In his *Unbound Prometheus*, David Landes worried why, during the early 1960s, Britain was spending four times as much on civil R&D as France, yet it was the French who enjoyed the higher economic growth rates.[17] How do we explain this paradox?

Poor countries enrich themselves by copying. Poor countries are poor because their technology is inferior and their work force inefficient, but they do not need to engage in R&D to correct the deficiencies; all they have to do is to copy the rich countries. Japan illustrates that very well. Until recently, Japan did practically no R&D, but it copied assiduously, and that copying was, for many years, a major arm of state policy. After the Meiji Restoration of 1868, when the Emperor Meiji overthrew the shoguns and embarked on industrialising his feudal country, the Emperor and his government realised that they were so ignorant of Western practices that they even needed to be taught how to copy. So, at vast expense, they imported Western teachers (between 1868 and 1872, no less than 40 per cent of the Japanese government's budget was spent on salaries to foreign experts).[18] One such Western teacher was Henry Dyer, a graduate in engineering from Glasgow University who, in 1873, was invited to create the Tokyo Polytechnic (later to become the Faculty of Engineering of Tokyo University). To lure him, the Japanese paid him four times what he would have earned in Britain. Indeed, the men who hired him earned less: in 1873, a Japanese cabinet minister earned 500 yen a month, but Dyer as the Principal of the Tokyo Polytechnic was paid 660 yen a month.

The Germans, too, copied assiduously, and they too imported foreign experts. The Hamburg and Bergedorf railway, for example which was started in 1838, was designed by an Englishman, William Lindley, because there was no one in Germany who possessed the requisite

---

was actually overtaken by many of its trading partners. So powerful is the widespread assumption that R&D is synonymous with economic growth, that the simple fact that Britain was overtaken has been adduced as evidence for the fact that Britain must have underspent on R&D. This is a circular argument. It is also untrue.

knowledge. The British actually built the railway too, because Lindley found the Germans in 1838 to be lazy and incompetent. He imported British navvies, and although he had to pay them twice as much as Germans, they were at least three times as productive. Lindley, like many other British engineers, stayed on in Germany, designing the Hamburg sewerage and water works, draining and reclaiming the Hammerbrook district, and working as consulting engineer to Frankfurt-am-Main until 1879. (The British also designed and built the French railways, which is why trains in France run, quite properly, on the left. But the British did not build absolutely every continental railway. The Moscow to Leningrad line was engineered by the American Lieutenant George Washington Whistler, the father of James McNeil Whistler, the painter.)

Poor countries, therefore, enrich themselves by copying or by importing expertise, they do not need to do their own R&D. Rich countries, however, have no one to copy. If they are to get richer, then they can only do so by innovation so they invest in R&D. But this creates a twist: copying is much easier than innovation (for obvious reasons) so poor countries can grow faster than rich ones. Much economic history since 1870, therefore, becomes easy to understand. Certain countries, particularly Britain, Australia, Belgium and Holland had, before 1870, through their embrace of capitalism, grown rich. That enrichment had happened because capitalism itself had generated the money for R&D. Other countries, on observing the success of the four lead countries, started actively to copy them. Because copying is easier than innovation, these other countries enjoyed astonishing rates of growth; because innovation is difficult, the lead countries have always grown slowly. It can be seen, therefore, that much of the criticism or praise that has attached to various countries since 1870 has been misplaced.

Consider Japan. It has been widely, extravagantly, praised for its technological genius; MITI, the Ministry for International Trade and Industry, has been ascribed almost magical powers of precognition; and Japanese management techniques, Ah my! On close examination, however, these wonderful supermen turn out to be much like everyone else. Japanese technology, until recently, has been blatantly derivative. Now that Japan is in the technological forefront, it is making the same embarrassing mistakes of all lead countries. Japan's space programme, for example, is years behind schedule as rocket after rocket fails or blows up in trials (the explosion in August 1991 killed a technician). Of Japan's 39 nuclear plants, 12 are currently closed because of accidents or leaks. Indeed, the Japanese do not seem to be very good with nuclear power. Consider the *Mutsu*, Japan's only nuclear-powered ship. Built in 1975, it has leaked radiation since its maiden voyage. For over 16 years it was in dock while engineers tried to repair it. In May 1990, it was finally re-launched for sea trials, but these had to be hurriedly aborted when it became clear that the crew was being exposed to frightening amounts of radiation. The project has now been cancelled, at a final cost of £400 million, every yen of which was wasted.

Japan's civil engineers have their problems too. When the 32-mile tunnel between Honshu and Hokkaido was opened in 1988, it was not only 10 years behind schedule but also, at £1.5 billion, three times over budget. Let us hope that it does not collapse like the chunk of Tokyo subway which, in 1990, swallowed up a whole street, 13 people and four cars. Would those unfortunate victims have been safer on the famous bullet train? A recent rail crash, which killed 42, was caused by faulty signalling equipment.

MITI, far from being a uniquely brilliant leader of government/industrial partnership, has been wrong so often that the Japanese themselves will concede that much of their growth derives from industry's rejection of MITI's guidance. MITI, incredibly, opposed the development of the very areas where Japan has been successful: cars, electronics and cameras. MITI has, moreover, poured vast funds into desperately wasteful projects. Thanks to MITI, Japan has a huge overcapacity in steel – no less than three times the national requirement. This, probably the most expensive mistake Japan ever made in peacetime, was a mistake of genius, because Japan has no natural resources: it has to import everything; the iron ore, the coal, the gas, the limestone and the oil to make its unwanted steel. Undaunted, MITI then invested in giant, loss-making (£400 million losses by 1992) 5th generation supercomputers at the precise moment that the market opened for the small personal computer; and MITI's attempts at dominating the world's pharmaceutical and telecommunications industries have each failed. Nor is this just anecdote. In a meticulous study of MITI's interventions into the Japanese economy between 1955 and 1990, Richard Beason of Alberta University and David Weinterin of Harvard showed that, across the 13 major sectors of the economy, surveying hundreds of different companies, Japan's bureaucrats almost invariably picked and supported the losers.[19] In the words of the *Economist*: 'Much as in Europe, Japan's clever technocrats picked and supported losers. That the economy succeeded for decades is plain enough. But, on this evidence, industrial policy may well have hindered rather than advanced the cause.'[20]

Japanese management techniques, moreover, emerge on examination to represent nothing more than the intelligent application of common sense and experience. In some respects they lag behind Western practice: it should not be forgotten that the Japanese are actually less productive than the Americans or the Western Europeans, as measured by wealth produced per worker per hour,[21] and this is not surprising when Japanese work customs are examined. Decisions in Japan are made agonisingly slowly, junior staff are obliged to spend all day and most of the evening in the office (long past their peak of efficiency) and women's skills are wasted because women are only employed in humble grades to provide male executives (the ones who never leave the office) with access to potential wives. The Japanese are workaholics (80 per cent of them do not take their full holiday allowance, and many take no holidays at all) and this workaholism, as much as Japanese efficiency, management or innovation, helps explain its success.

These words are not written to knock Japan – far from it, it is an admirable and impressive country; these words are written to emphasise that Japan could not fail. It possessed an industrious, disciplined, civilised work force that, by 1868, had lagged terribly behind the UK, the USA and other lead countries. All it had to do, initially at least, was to copy, and its path was so straightforward that it could afford the occasional mistake, however bad. (This is not just the wisdom of hindsight; Japan's economic and technological growth were predicted in detail by Graham Wallas in *The Rise of Japan* published in 1906.) But the secret of Japan's success is simple – *laissez faire*. The Emperor may have initiated his capitalist Restoration in 1868, but after 1884 it became State policy to leave industry to its own devices. Having germinated it, the Japanese government understood that capitalism runs best unfettered. This is ironic, since the popular perception abroad has been that Japan has prospered because of MITI's direction of the economy, whereas the reality is that since 1884 Japan's government has seen itself as the servant of industry, anxious to tax it and its customers as lightly as possible. A 1988 survey by the British Central Statistical Office showed that Japan was one of the three most lightly-taxed industrialised countries, the other two being the USA and Switzerland. Those three countries' governments only sequester 32–35 per cent of GNP compared to the British State's 44 per cent, the West German's 45 per cent, and the French 52 per cent.[22]

Japan's *laissez faire* extends dramatically into the funding of civil R&D, which is almost entirely funded by the private sector (as is Switzerland's, see Chapter 10) and into academic science, which is the most privatised in the industrialised world (the private sector funds more than half of all Japanese university research – see Table 12.1). Thus we see that the historic economic lead of *laissez faire* Britain and *laissez faire* USA has been superseded by that of *laissez faire* Switzerland (now the richest industrialised country in the world) and *laissez faire* Japan (the second richest). That is a vital historical lesson, which has never been understood by those who have failed to note that neither *dirigiste* France or Germany has ever been the lead country economically (and, in as much as Germany has done better than France, it is because it has often been the more *laissez faire*).

The Scandinavian countries, too, had lagged far behind by 1870, and they too only had to copy. Scandinavian cooperative management techniques have attracted much praise during the twentieth-Century, and free-market ideologues have wondered how such cooperative societies have flourished so well under socialist governments; but like Japan, Scandinavia could not fail. It possessed an educated, civilised work force that had fallen terribly behind by 1870 (as early as 1750 the Lutheran Church, concerned that its members should be able to read the Bible, had used its schools to push literacy in Sweden to 100 per cent) and all it had to do subsequently was to copy. Socialism and cooperation do not preclude copying. Now that the Scandinavian countries have caught up, however, they are being forced to shed their socialism, for reasons that are discussed further below.

The case of Malta is instructive. As Figure 7.10 shows, it has grown fabulously since 1950, yet it is neither wholeheartedly capitalist, nor very committed to science – indeed Dom Mintoff, its long-serving prime minister, was a doctrinaire socialist who effectively dismantled the island's one university. But its economy has grown most wonderfully. How? Malta, like the socialist Scandinavian countries, could not fail. It possessed an old, cultured, educated population absolutely determined to copy its way into wealth. Interestingly, the clever people of Malta are now turning to free enterprise, and restoring their university's science base, to meet their new needs as a relatively rich country. (Malta also enjoys the advantage of not being shackled by membership of the European Union.)

The case of Britain is intriguing. It has long been fashionable to criticise Britain's economic performance since 1870, yet it has never been clear by which criteria; no one has stated what Britain's performance should have been. Instead, the typical comment has run along the lines of 'since 1870, Britain's growth rate has only been half of Germany's'; but as Figure 7.5 shows, this is an irrelevant criticism.

Each country's performance has adhered closely to that which was predicted by its wealth in 1870. Indeed, this is not a Marxist book, but economic determinism must be admitted: the rates of economic growth of the capitalist countries since 1870 appear to have been determined by their wealth in 1870, and Britain merely fits the predictable curve. Specific criticisms of Britain's industrial or technological performance since 1870 also melt on scrutiny.

Consider the iron and steel industry. Until 1880, this was Britain's industry. In 1880, Britain produced more iron (so-called 'pig' iron) than the USA, and double the amount produced by Germany. Indeed, in 1880, Britain produced 46 per cent of the whole world's pig iron, and 36 per cent of the world's steel. But 1880 represented Britain's apotheosis, thereafter British production levelled out and, as other countries increased theirs, so Britain's world share fell. By 1913, Britain only produced 14 per cent of the world's pig iron and 10.3 per cent of the steel. Indeed, between 1880 and 1913, Britain moved from being the world's largest exporter of iron and steel to the world's largest importer. Productivity fell relatively, so that by 1929 *per capita* production of iron and steel in Britain was only two-thirds that of Germany, and less than half that of the USA. An industry which in 1871 accounted for 11.6 per cent of British GNP, had fallen to only 5.8 per cent by 1901 and was going down fast.

Criticism has poured on Britain for this apparent failure. Burham and Hoskins, in their *Iron and Steel in Britain, 1870–1913* were almost apoplectic in their denunciations: 'If a business deteriorates, it is of no use blaming anyone except those at the top.'[23] Blame, too, has descended on British education and research. In his *Audit of War*, Correlli Barnett wrote: 'Before 1884 there was no university department of metallurgy anywhere in Britain, no university research, [but] the Germans looked from the very start to organised science and technology, to thorough training at every level, as the necessary instrument of future industrial success. The forging of this sophisticated and elaborate tool began half a century and more before the resulting tempered blade began cutting into world markets: the Berlin Technical Institute for engineers dated from 1821; technical high schools at Karlsruhe from 1825, at Dresden from 1828, at Stuttgart from 1829. By around 1870 there were already some 3500 students in German technical high schools. Since the early 1820s German universities had likewise been developing formidable teaching departments in chemistry, metallurgy, physics and engineering. They established research laboratories as well, to act as path-finders to future technological leadership.'[24]

It all sounds jolly impressive. Let us look at the actual results. The three great technical advances in steel making during the second half of the nineteenth-scentury were Bessemer's invention of the converter in 1856, which made mass production possible, Siemens's invention of the open-hearth furnace in 1866, and Thomas's invention in 1879 of the 'basic' process which enabled steel to be extracted from phosphoric ores. Now, the curious thing about all three inventions was that they were made in Britain. Sir Henry Bessemer was an amateur, untrained in science, who wrote in his *Autobiography* that his knowledge of metallurgy was 'very limited, and consisted only of such facts as an engineer must necessarily observe in the foundry or smith's shop'. Wilhelm Siemens was, admittedly, German in origin; he trained as a metallurgist in Germany, and he was assisted by Pierre Martin, a French metallurgist; but Siemens moved to Britain, where he created his branch of the company and where he did his research, finding the entrepreneurial spirit of the UK highly conducive to creative research and industrialism. But it was Thomas's discovery in 1879 of the 'basic process' which rendered most ores suitable for the Siemens–Martin open-hearth furnace, and Thomas was another amateur, a hobby scientist who worked as a clerk in a police court in Wales, and who experimented in chemistry at night.

It can be seen, therefore, that the many suggestions that Britain's steel industry was overtaken by Germany's because German science was superior are false: it was Britain that made the scientific advances, not Germany. And Britain made the advances for exactly the same reasons it always had; not because it possessed vast, gleaming, government-funded laboratories of white-coated scientists, but because it possessed men like Bessemer and Thomas who were close to the market and who trained themselves in the relevant technology and science to meet the market's need. Moreover, of the continentally trained metallurgists, the most productive researchers were Siemens and Martin, the very ones who emigrated to Britain, where the market was free, state control least and cartels rare.

If British metallurgical science was so good, why did the industry decline? Relative decline domestically, of course, was inevitable. Britain could not continue to devote 11.6 per cent of GNP to iron and steel after 1871, or the island would now be knee-deep in rolling mills. The rest of the economy had to grow. Relative decline *vis-à-vis* foreign steel making was also inevitable; other countries were bound to develop their own steel industries, and so Britain's relative predominance was bound to fall. The absolute decline of Britain's steel industry was not inevitable, however, but that was the consequence of factors out of British control. Let us consider them. First, the USA, Germany and France raised tariffs against British steel, to protect their own industries. Second, Thomas's discovery benefited the Germans and other competitors much more than it did the British, because continental ore, unlike Britain's, is largely phosphoric. Third, Britain's competitors subsidised their industries in ways that the British did not. Consider training: in Britain, workers were trained as apprentices, on the job, on the shop-floor. In Germany, workers were trained in a large number of technical schools, founded across the country. The German system of training has been praised, because the German steel industry did so well, but has the case actually been proved?

The British system of training was actually more economical, more efficient and more rational, because the greatest experts were those already working in the industry, so the apprentices' teachers were the most experienced steel workers in the country. In Germany, in contrast, the teachers were just that, teachers, desk-bound theoreticians teaching at second hand about matters of which they were frequently inexperienced and often outdated.* Learning, moreover, was also more efficient in Britain, because the apprentices produced as they learnt, and their products could be sold; in Germany, apprentices were taught separately in colleges, their products could not be sold, they lost years of potential productivity while they sat in classrooms, and their very classrooms had to be built, equipped and maintained at vast expense. How then did German steel overtake Britain's?

The costs of training were not equally shared. In Britain, the cost of training was fully borne by the industry, but in Germany a considerable proportion was met by the State. Since the costs of training

---

*Curiously, these inadequacies in the much-vaunted German vocational training colleges or *Berufsschulen* persist to this day. In 1991, a team of British school inspectors reported that the *Berufsschulen* were manifestly inferior to their British counterparts, the colleges of further education. Amongst their inadequacies, most *Berufsschulen* (i) lacked central libraries, (ii) they were overcrowded and lacked study space, (iii) a much lower proportion of German staff had recent industrial experience compared with British staff, (iv) there was little project work, and (v) general courses were not sufficiently challenging. The British school inspectors found that the reason the *Berufsschulen* have, for over a century, been supposed to be so excellent, is that they award the qualification of Master Craftsman, for which there is no equivalent in Britain. This qualification carries high status in a nation obsessed with qualifications, but the actual products of the *Berufsschulen* are in practice no better than their British equivalents with their modest diplomas [*Aspects of Vocational Education and Training in the Federal Republic of Germany* (London: Her Majesty's Stationery Office, 1991)].

in technological industries are so huge (the annual costs of training in technological industries often exceed annual profits as a percentage of annual turnover, and may well amount to 15 per cent or more of annual turnover), the consequence was that the German steel industry was heavily subsidised, the British one was not. The cumulative effect, year in year out, of this vast one-sided subsidy was inexorably to undercut British steel (but the rest of the German economy paid a high price for its subsidy of steel – see below).

Because the British Government did not create technical schools, it is often assumed that Britain lacked them, but the free market is perfectly capable of supplying education if it is needed. Between the 1820s and the 1840s no fewer than 700 Mechanics' Institutes were set up, privately, in Britain to offer technical instruction[25] but, of course, their costs were fully met by industry, because the artisans' fees (or loans to meet the fees) were ultimately translated into higher wages.

There were other, less honourable, reasons for the German rise in steel making. The German government actually paid bounties to companies that exported steel and it fostered a domestic cartel so that excessive home profits could further subsidise exports. The *Iron and Coal Traders Review* for the 24 December 1909 wrote: 'Without its vast system of syndication – its almost military-like production and distribution methods – and the organized fostering of export trade by countries, the German iron and steel industries could hardly have obtained their present status.'

The British response to this foreign pressure was strictly rational, because it was market-driven; it gradually closed down its industry. Obviously this was a slow process; phosphoric ores in Lincolnshire and Northamptonshire were exploited, an Industrial Research Council was created, and tariffs were erected, but the comparative advantage in steel making had moved abroad; foreigners enjoyed commercial ores, State-trained employees, bounties, cartels and comprehensive tariff protection. It made better sense for Britain to import cheap foreign iron and steel and to move into other industries.

The point of economic activity is to consume – production is only a means to an end – and if the German taxpayer wanted to subsidise steel exports to Britain, that was Britain's good fortune. For British steelmakers, of course, there was a heavy social cost involved in changing industries, but for the economy at large it was beneficial. The final collapse of the British steel industry's productivity merely reflected an appropriate, terminal lack of investment.

The German taxpayer, however, in subsidising his steel, was not so rational, because he had an ulterior motive. He wanted to build up his domestic steel industry for military reasons. He had a heavy schedule of invasions ahead of him (Denmark, 1864; Austria, 1866; France, 1870; Belgium *et al.*, 1914; Czechoslovakia, 1938; Poland *et al.*, 1939) and for these he needed steel. The German taxpayer perverted normal economic activity to subsidise production rather than to maximise consumption; not that he had much choice – his government was not democratically accountable. Bismarck only answered to the Kaiser.

But the apparent success of Bismarck's policies – his successful invasions of Denmark, Austria and France – corrupted all of Europe, not just the once-civilised Germans, and by the second half of the nineteenth century all of Europe was drunk on national pride and obsessed with expansion. In April 1904, for example, the British imperialist Leo Amery wrote in the *Geographical Journal*: 'The successful powers will be those who have the greatest industrial base. Those people who have the industrial power and the power of invention and science will be able to defeat all others.'[26] Amery wanted the Government to intervene to direct British industry and British science to military ends, but Britain was protected from this perversion by its long-standing attachment to *laissez faire*, very low taxes and free trade. Germany was not so protected and, even earlier than Amery, Bismarck had delivered his famous *Blut und Eisen* (Blood and Iron) speech to the Prussian House of Deputies (28 January 1886): 'Place in the hands of the King of Prussia the strongest possible military power, then he will be able to carry out the policy you wish; this policy cannot succeed through speeches and shooting matches and songs; it can only be carried out through blood and iron.' Well, Germany got its iron (but at the cost of higher taxes and at a price that was for decades kept unnecessarily high by tariffs) and Germany got its blood too (and a theme of the rest of this book will be that governments fund education, science, universities and technology for military, not humanitarian, reasons).

We have devoted a lot of space to Britain's decline in iron and steel because of the myths that have attached to it. Let us consider one more of Britain's 'failures', the chemical industry. The conventional view of the failure of the British chemical industry was restated recently by Professor Denis Noble FRS of Oxford University: 'The first major government intervention in science in the UK, in 1913, was to save a nation approaching a state of war with a country on which we had become almost totally dependent for chemical products.'[27] Indeed, by 1913, Germany accounted for 24 per cent of the world's chemical output, and the USA 34 per cent, but Britain only accounted for 11 per cent, and the outbreak of war in 1914 left Britain alarmingly exposed. Britain had grown dependent on Germany for many chemicals, not least the khaki dye for its soldiers' uniforms, and 1914–18 witnessed desperate shifts as British chemists and industrialists struggled to provide crucial products. Yet criticism of the British chemical industry are, curiously, hard to sustain. First, as H. W. Richardson has shown, the chemical industry grew faster between 1881 and 1911 than any other industry in Britain, except for public utilities, and by 1911 it was employing 2.7 per cent of the manufacturing labour force.[28] Britain did well in detergents (thanks to Lever Brothers, later Unilever), paints, coal tar intermediates and explosives (thanks to Nobel Industries – explosives are useful in war).

British chemical companies, of course, sometimes made mistakes. It is generally agreed by specialist historians that United Alkali Producers retained the Leblanc process in alkali production for too long after 1897, when it should have shifted more of its production into its successful Solvay process, and it is generally agreed that the lead-chamber process in sulphuric acid production should have been replaced by the contact process more swiftly after 1914; but these are the sort of mistakes that occur in all growing industries, and all countries make these sort of mistakes. It is easy for the economic historian, secure with the benefit of hindsight, to criticise the busy, harassed businessmen of an earlier generation, especially as some decisions are not easy to make. It can be difficult to scrap expensive, profitable, functioning machinery for new, expensive, untried machinery. Consider Britain's electronic industry: this too is often criticised, and the economic historians who criticise the chemicals industry for neglecting new technology are often the same ones who criticise Britain's electronics industry for an over-enthusiasm for new technology. The major British electronic company, G. Z. de Ferranti, was obsessed by new technology, and always installed it, even if the older technology was more profitable or more reliable.[29] Clearly, some economic historians' blanket criticism of certain British industries is not very helpful. In any case, the mistakes are not so serious when put into context. Lindert and Trace have calculated that United Alkali Producers' mistake over the Leblanc process only cost it £1.9 million in lost profits;[30] hardly a major blow to the nation from what is generally described as one of the worst examples of British 'entrepreneurial failure' before 1914.

The British chemicals industry, therefore, cannot be criticised for not growing – it grew fast – but only for not growing as fast as

Germany's, particularly in the organic chemicals that underlay the dyes and drugs in which Germany excelled. This must be a general criticism, because nobody – not the USA, or France, or Belgium, or Holland – expanded into organic chemicals in the way that Germany did. The Germans excelled in organic chemistry for the same reasons they excelled in iron and steel – the government poured vast funds in technical schools, universities and scientific research. It is often forgotten that, yet again, the fundamental discovery that underpinned the growth of Germany's organic chemicals was made by an Englishman, in London, when in 1856 William Perkin synthesised mauve, the first of the aniline dyes. There was nothing wrong with British science. But the director of the Royal College of Chemistry, where Perkin worked, was the German chemist Wilhelm Hoffman, and when Hoffman returned home in 1865, he was one of those who used government money to train the generation of young scientists that subsequently made possible Germany's dominance in dyes, fertilisers and drugs.

Since the German taxpayer had reduced the real cost of German chemicals, it made good economic sense for Britain and other countries to import German chemicals. But did this carry the seeds of decay for Britain? Was Germany engaged in a form of intellectual 'dumping' in reverse, undercutting all other countries' science-based industries so that it would then acquire a monopoly on future economic growth because it would be the only country left with scientists? The empirical evidence shows clearly that this did not happen; let us note here, for example, that whereas late nineteenth-century and early twentieth-century Germany dominated the drugs industry, the industry is now dominated again by the British who discovered penicillin, for example, and it is the British who currently possess the world's greatest drug company, Glaxo.

During the nineteenth and twentieth-centuries the Germans, of course, boosted their organic chemicals industry for military as much as for economic reasons and, while it might have been a mistake, militarily and strategically, for the rest of Europe to have become so dependent on German chemicals by 1914, it made excellent economic sense (Germany's neighbours, then being culturally superior, believed that the purpose of economic activity was consumption – in peace). But Britain learnt its strategic lesson very rapidly. During the 1914–18 war, Britain created the British Dyestuffs Corporation, specifically to compensate for previous German imports, and in 1926 this combined with Nobel Industries, Brunner Mond and the United Alkali Company to create ICI, the chemicals giant, which by the late 1930s was employing nearly 500 research scientists, and of whose managers many were also scientifically trained. Chemically, Britain was fully prepared for war in 1939. (After 50 years of peace, ICI is increasingly looking uneconomical, and it has been broken into two. No longer, thankfully, are industries having to be distorted for strategic reasons. Thanks largely to the Germans, the twentieth century has proved to have been quite the most horrid century in Europe since the fourteenth, but the Germans are looking nicer now.)

The German government's subsidy of research and development has often been adduced as a contributor to Germany's economic miracle, but where is this miracle? Germany's economic growth since 1870 has merely fitted the predictable curve, given its initial poverty (Figures 7.5 and 7.6). Because Germany overtook Britain, economically, during the 1960s, people often suppose that Germany's superiority is of long standing, and they make the usual misleading comments. Consider this statement by Margaret Gowing, the Professor of the History of Science at Oxford University: 'Britain, hitherto industrially supreme, had been very obviously outclassed in the Paris international exhibition of 1872, most notably by Germany'.[31] And, being typical, it is a typically incorrect statement. As late as 1914, Germany only enjoyed 75 per cent of the level of industrialisation of Britain.[32] In consequence, Germany only enjoyed 75 per cent of Britain's wealth as determined by GDP *per capita*.[33] This was one reason it lost the Great War.

The Germans were so poor during the nineteenth-century that no less than 4 853 253 of them emigrated between 1841 and 1900,[34] often to the USA, but sometimes to the UK (which received, in addition to William Siemens, Ludwig Mond the chemist, Charles Halle the musician, Ernest Casel the banker, George Goschen the banker and Chancellor of the Exchequer 1886–1892, the Schroeders, Edgar Speyer and Jacob Behrens). Germans found Britain more entrepreneurial, and more pleasant. This is what Jacob Behrens, the banker, wrote in 1832 to explain his emigration: 'I took a liking for Britain, especially because it presented a picture totally different from retrogressive Germany.... Not only did I feel myself a man amongst men, but the times were great.'[35] The Germans were so poor partly because they were taxed so crushingly to pay for their famous universities, technology, metallurgy and chemicals. Germans actually and literally starved; the *dirigiste* economy frustrated agricultural development to foster industrialisation, and a series of terrible harvests from 1845 precipitated mass malnutrition and death.

Germany's unnecessary sacrifice of its agriculture for its manufacturing base (unnecessary because, under *laissez faire*, it could have had both) contributed hugely to its losing its wars. Here, for example, is how Britain's official economic historians compared British and German agriculture for 1939: 'Great Britain employed on the land less than a million people or not quite five per cent of her labour force; whereas Germany, to provide her people with food, was employing on the land 11 millions, or twenty seven per cent of her labour force.'[36] By 1945, Britain was producing over 80 per cent of its food; its greatly more efficient agriculture had liberated literally millions of men for manufacture or military service. (Curiously, successive supine post-war British governments have, under the Common Agricultural Policy, abandoned Britain's successful agricultural policies for the failed ones of Germany.)

None the less, Britain did decline, relatively. The first country to overtake the UK, in about 1890, was the USA. America, of course, grew rich through *laissez faire*. Its governments did not start to fund academic science until 1940 (see Chapter 8) and its engineers followed the British, not the French tradition. We have already noted how the French state poured so much money into its engineering colleges, and paid its engineers so well, that not only was every engineer formally educated, but the profession became very upper class, ennobling men like the Duc de Sully. And we have noted how, in contrast, the British state completely ignored engineers, in consequence of which most were self-taught and of humble origin. Well, the American state was as unconcerned. The early Americans merely imported their technology, the most famous example of that being Samuel Slater who in 1789 built America's first factory, a cotton mill at Pawtucket, Rhode Island, to a design he illegally stole from a British mill (in those days of Navigation Acts and mercantilism, it was illegal to export industrial designs out of Britain, but Slater memorised the details of the latest British spinning machines, and sailed to Rhode Island with the plans safely hidden from view in his brain).

America's first major canal, the Middlesex Canal (1793–1803), was largely the creation of the British engineer William Weston, and other Britons such as Benjamin Latrobe and Hamilton Fulton played crucial roles in early America, passing on their skills directly to their assistants, to initiate the American tradition of on-the-job training. This tradition was consolidated as the famous Erie system by Benjamin Wright, Canvass White and James Geddes. Since we have seen that national wealth is, essentially, national productivity, and since the engineering profession is a major determinant of that, let us examine those three men as a microcosm of American economic growth.

Wright, White and Geddes, who built New York State's Erie Canal between 1817 and 1820, were originally surveyors, and they possessed only limited technical training, but they studied existing canals, they read widely and they experimented. As they built, they learnt. As they learnt, they taught their assistants. These, in turn, rose to become chief engineers elsewhere, and they spread the Erie's system of on-the-job training all over the USA. By 1890, when the USA overhauled Britain economically, the majority of American engineers were still trained on the job. Any 'explanation' of Germany's or France's overhauling of Britain economically which invokes those governments' funding of research or of engineering or of education will, therefore, have to explain how the first country to overhaul *laissez faire* Britain was *laissez faire* America.

But the later evolution of America's engineering profession teaches further lessons; it confirms that technology is wealth, but it also confirms the value of *laissez faire*. Although the majority of nineteenth-century US engineers had been trained on the job, some engineering colleges did exist. The first civilian engineering school had actually opened as early as 1820 at Norwich University, and Rensselaer followed in 1824. But, by 1860, there were still only half a dozen civilian engineering schools in the USA – there was little call for them. But, during the 1890s, the numbers of engineering colleges suddenly expanded, the profession suddenly embraced formal qualifications, and by 1900 the overwhelming majority of new engineers had been scientifically educated.[37]

Why the sudden *volte face*? And why did it coincide with America's assumption of economic leadership? This has never been explained before, but the readers of this book will not be surprised to learn that it was, of course, no coincidence. Until the 1890s America, being poorer than Britain, was also technologically inferior. It was enough for American engineers to copy British practice. But after 1890 Americans, being in the economic lead, encountered novel technological problems. These could not be resolved just by copying – engineers needed to be good scientists. The need for academic training suddenly became pre-eminent, and thereafter the colleges took over the training of the entire profession. Thus did *laissez faire* direct education appropriately, because after 1890 employers would only hire college-trained engineers, who therefore commanded higher salaries to commute their college fees (just as do American doctors today).

Let us dismiss the myth that America overhauled the British during the 1890s through superior commercial skills. Rudyard Kipling lived in Vermont for four years during the early 1890s, and he was appalled by the incompetence he encountered. In his autobiography he wrote: 'Administratively, there was unlimited and meticulous legality, with a multiplication of semi-judicial offices and titles; but of law-abidingness, or of any conception of what that implied, not a trace. Very little in business, transportation or distribution, that I had to deal with, was sure, punctual, accurate or organized.'[38] Kipling's disaffection cannot just be dismissed as resentment over his problems with his brother-in-law; Vermont's technological backwardness in his day is illustrated by the fact that the very first pair of skis to appear in that snow-bound State was introduced by one of Kipling's British guests, Arthur Conan Doyle.

America initially grew rich for the same reasons that Australia had; it possessed vast, virgin lands, which only had to be cleared by shooting the natives, and it received wave after wave of vigorous, determined immigrants from Europe. The USA enjoyed two advantages over Australia, namely that it was bigger and its interior fertile, so it possessed a larger internal market, but each country enjoyed the same protection from foreign aggression that the oceans provided, which kept their defence costs low. But even America is now no longer the richest the country in the world. It, too, has been overtaken. It, too, is in relative decline. What causes countries to decline?

Let us clarify the term 'decline'. We are talking, here, about relative decline. Consider Britain. Ever since 1870, Britain has continued to grow at about 1.3 per cent GDP *per capita per annum*, and if some industries have collapsed as their comparative advantage fell and it made better sense to import, conversely other industries grew to generate that 1.3 per cent annual growth. Compared to other industrialising countries, that growth rate looks small, and it has generated awful warnings like the one repeated by Paul Kennedy in his *The Rise and Fall of the Great Powers*: 'It is hard to imagine, but a country whose productivity growth lags 1 per cent behind other countries over one century can turn, as England did, from the world's undisputed leader into the mediocre economy it is today'; but that statement is incomplete economic history: rich countries grow slowly, poor countries grow quickly. In itself, a productivity growth which lags 1 per cent behind other countries would not matter if those other countries, on catching up, then settled down to the same laggardly growth rates. Why have they not?

Actually, in the main, they have. Comparisons between the major industrialised powers are ceasing to be very interesting. In 1870, there was a fivefold gap – 500 per cent – between the GDPs *per capita* of the richest and poorest capitalist countries. Today, the capitalist countries jostle for 10 per cent–20 per cent advantages over each other. The moment one country develops an advantage, the others hurry to catch up. The previously poor, but rapidly-growing countries have slowed down, and the old rich ones are speeding up again (Britain grew faster during the first half of the 1980s than any other country in Europe; Japan and Germany are currently experiencing economic wobbles). Of course there are differences in wealth between the different countries, but they are small, transient and reversible, and when a country falls behind, it will copy the productive features of its successful competitors, and it will shed its mistakes – as long as trade between countries is sufficiently free to signal when domestic industries are falling behind foreign competition. (In the long term, tariffs harm those who erect them, because they coddle inefficiency – a lesson the French showed during the 1994 GATT negotiations that they have yet to learn; see the postscript on trade at the end of this chapter.)

Because countries are competitive, they agonise over economic leadership. Those agonies are, themselves, very important, because they reflect the determination of countries to better themselves, but the intensity of the agonies can belie the conclusions they breed. In his *English Culture and the Decline of the Industrial Spirit*,[39] for example, Martin Wiener claimed that Britain had experienced a moral decline, which rendered its managers incapable of managing and its workers incapable of working. Curiously, that book appeared at the beginning of the decade during which Britain enjoyed higher economic growth rates than any other country in Europe. Those growth rates did not fall from heaven; Britain recognised that many of its industrial practices such as the closed shop, its vast and inefficient nationalised industries, its poor industrial relations and its high taxes had fallen behind best current practice, and it reformed.

As we have seen, it is generally claimed by commentators like Correlli Barnett or Margaret Gowing that Britain declined because it neglected R&D, and as we have already seen, that simply is not true. Its expenditure on R&D put Britain firmly in the post-war technological lead. The first nuclear power station was British (Calder Hall, commissioned in 1956). The world's first commercial mainframe computer was British (sold by Ferranti in 1951). The world's first commercial jet aircraft, the Comet, was British (in service in 1952). Whatever else caused Britain's post-war decline, it was not a neglect of research, nor of technology (it was, in fact, a compound of Britain's vast wartime debt, its huge post-war tax burden, its enormous and inefficient post-war programme of nationalisation, its horrendous imperial responsi-bilities and its shouldering of post-war peacekeeping: for example, the gargantuan defence expenditure of £4.7 billion in 1951, to pay for the Korean War, promptly turned Britain's balance of payments surplus of £307 millions in 1950 to a deficit of £369 million in 1951).[40] Today, some Americans are also claiming to be in economic decline because their governments have neglected research, particularly basic science.[41] But the facts belie the claim: Figure 7.11 shows that America per-forms a vast amount of pure science; the USA dominates world sci-ence, with Americans now winning most of the Nobel Prizes.

In conclusion, therefore, we have seen how economic history since 1870 has failed to be very interesting. The pioneering industrialised countries have continued to grow economically, but relatively slowly, because they have largely grown through R&D. Those poorer coun-tries that have caught up have done so by copying the technology of the lead countries, which has blessed them with high rates of initial growth. But on enriching themselves, they too have had to settle for lower rates of economic growth and for increasing their investment in R&D.

Countries, like children fighting over toys, are over-sensitive to each others' perquisites. They are too prone to suspect each other of advan-tage, and they are too prone to bemoan their lot. And, like children, they rush to a parent figure, the State, to correct apparent problems – forgetting how parents use their power on other occasions, such as bed time. Worse, again like children, countries do actually fight, which has caused them to confuse productivity and military self-sufficiency with economic growth and wealth. In the next chapter, we shall see how that confusion has corrupted the funding of research.

POSTSCRIPT ON TRADE

We have stated that the prime factor that underpins long-term increases in wealth is productivity. By so doing, we have discounted many of the other concerns of economists and politicians. We have shown by the example of Japan that a nation may possess practically no national resources (apart from raw fish) and yet it can grow rich if its citizens embrace the work and ethical ethos of capitalism and if they import the raw materials they need. Japan is much less exceptional than is

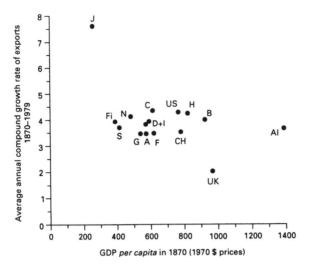

FIGURE 7.15   *GDP* per capita *in 1870 and growth in exports, 1870–1979*

The data came from Maddison's *Phases of Capitalist Development.*[21]
See Figure 7.5 for a key to the symbols.

generally believed: France, for example, has few natural resources in
fuel (no virgin forests, no oil and little coal) yet the French warm
themselves in winter, power their factories and drive their cars on
imported fuels. Many industrialised countries are similarly deficient.
Natural resources are obviously very nice, but if the local population
is not disciplined in the ways of capitalism, those resources will either
be untapped (as in much of Africa) or the easy wealth they bring will
be dissipated in nonsense (as will happen in Arabia). It is productivity
that determines nations' long-term futures.

One other factor that is commonly believed to be very important to
national wealth is trade, in particular exports. In Britain, for example,
the Queen annually distributes special awards to companies that successfully export, and many other countries run similar schemes. There
is, however, no empirical evidence to support the suggestion that exports as such do promote national wealth. In Figure 7.15, we see that
almost all of the major capitalist countries, whether they grew fast or
slowly, increased their exports equally between 1870 and 1979. The
only two exceptions were Japan and Britain, but Japan had practically
no exports at all until 1868, so its apparently huge proportional increase since 1870 is distorted by its peculiarly small export base in

1870 (in pre-1868 feudal Japan, trade could be a capital offence, and those who had any contact with foreigners were routinely executed). Britain's apparent export failure since 1870 is also distorted, but the other way. In 1870, Britain had an almost captive export market in its Empire, so its exports in 1870 were inflated by a lack of competition. For the other 14 major capitalist powers, however, it can be seen that there is no correlation between their growth of exports and their relative growth in GDP *per capita*. None the less, trade is very important to long-term growth, but in more subtle ways than the crude statistics of Figure 7.15 can convey. Indeed, the long-term benefits of trade may not show up in the trade figures at all.

Trade can have long- and short-term benefits. The short-term benefits are obvious. If one country, for example, has no mineral resources, and another no good land, it would be to their mutual advantage to trade food for metals. Thus the Japanese, for example, are not only huge exporters, they are also huge importers; they are third largest importers in the world at, in 1990, $1900 worth of goods per person to the USA's $2050. But the long-term benefit of free trade transcends immediate consumption; the long-term benefit of trade is that it improves domestic productivity (by definition, because all long-term economic benefits are only mediated through improvements in productivity). To illustrate how trade achieves this, imagine two countries, each of which has a steel industry, and between which trade is free. From fear of the competition, the managers of each national steel industry will be constantly seeking ways of improving their own companies' productivity. They will foster management training, optimal conditions for their workforce – and they will undertake research and development. The commercial forces that encourage a company to innovate are simple: the fear that the competition will steal a market with an improved product. This, the major long-term benefit of trade, need never show up in the trade figures. It will be enough for producers to fear their competitors' research and development for them to foster their own research and development. This has been demonstrated empirically in a fascinating study by K. L. Sokoloff.[42] Sokoloff studied the rates of patenting between different counties in early nineteenth-century America, and he found that those counties which had access to navigable waterways enjoyed higher rates of patenting than those that did not. Remarkably, when water transport was introduced into a county, either by building a canal or by dredging a river, the rate of patenting shot up – imports promote R&D.

Paradoxically, this explains how some industries can do so well in some countries behind trade barriers, at least for a time. Because trade is not, in itself, crucial, but only an instrument for fostering incentive, research and development, some national governments have thought of other ways of fostering incentive, research and development. Governments have created state laboratories and they have set productivity targets; in short, they have substituted political will for the commercial will that free trade generates. Such a substitute for the free market may succeed for a time, but home industries generally deteriorate behind tariff barriers because it is hard to sustain political will. Political will represents a kind of civil war, with politicians and managers spurring their workforce against targets. No one likes being driven to work harder, and in time the workers will resent their management. Since few people like being unpopular, the politicians and managers will relax their targets, and productivity will slump. An occasional Lenin or Stalin will keep up the pressure by regularly shooting senior managers (in its early days, the USSR grew remarkably fast) but these are unpleasant substitutes for a free market. In time, even the Soviet Communist Party lost the will to continue. Nineteenth- and early twentieth-century Germany, behind its tariffs, tried to substitute the incentives produced by normal commercial rivalry with an increasingly fevered nationalism, but the logical progression of that collective psychopathology led Germany into a series of wars.

Under free trade, however, companies feel the foreign competition very keenly, so they will unite behind their management, and management will strive to foster the good will and loyalty of their staff. Free trade, therefore, brings people together, tariff protection divides and politicises them (look at the French farmers today). Free trade, moreover, provides powerful incentives for research and development; tariffs do not. A theme of the rest of this book will be that the governments that pioneered the creation of state laboratories were totalitarian; those totalitarian regimes believed that they should control all aspects of society, including trade, and so they were forced to create state laboratories.

An analogy can be drawn with socialism and competition. The early socialists complained that capitalism was inefficient because it duplicated. Thus Haslemere High Street today, for example, has four sweet [candy] shops, each selling much the same sweets. Why not amalgamate them into one, to liberate some of the staff for other jobs? But the Soviet experience has taught that without the goad of competition, the resulting monopoly sweet shop will eventually degenerate. The early socialists wanted to replace competition with planning, which would not only be more 'rational' but would also create a more 'gentle' society. But as Hayek showed, planning is inefficient and, even worse, it divides the planners from those for whom they plan, thus alienating ordinary people to a much greater degree than does capitalism.)

POSTSCRIPT ON AMERICAN ENGINEERING

The American engineering tradition, even in its on-the-job nineteenth-century phase, diverged in two ways from the British. First, American engineers tended to be from the prosperous middle classes.[43] This was because engineers tended to be entrepreneurs. William Sellers, for example, who introduced the standard American screw thread, came from an upper-class (in American terms) Philadelphian family, he skipped college as was then the custom among Philadelphia's elite, and he served an apprenticeship in his uncle's machine shop, before establishing his own company.[44] Because so many American engineers were primarily entrepreneurs, American engineering developed its own style. Where the British emphasised strength, permanency, aesthetics and safety, the Americans emphasised economy of construction and the saving of labour. This was illustrated by Seller's screw thread. The standard thread for screws during the nineteenth-century had been established by Whitworth, the British engineer. His screw had flat sides at an angle of 55 degrees with rounded tops and bottoms. It worked well, but it required a skilled craftsmen to machine it on a lathe. Sellers's screw thread was an equilateral triangle, sides inclined at an angle of 60°, and the top and bottom flattened one-eighth of the thread depth. It worked no better than Whitworth's, but it was much easier to make. The Americans rapidly adopted it, but the British, who had no shortage of skilled craftsmen, stuck to Whitworth.

Economic historians point acccusingly to Britain's attachment to sound craftsmanship. How outdated, no wonder America overtook! But British artefacts during the nineteenth-century were better made, more beautifully crafted and more elegant than America's; Britain produced men like William Morris and, later, Eric Gill, who introduced charm and historical resonance into daily objects. If there was a price to be paid in terms of a few percentage points of GDP *per capita*, it was one that educated Britons chose to pay.

(When Sellers produced his new thread to challenge Whitworth's English standard, the world of engineering had to choose between the two. Sellers, a good *laissez faire* American, was proud that his countrymen's engineers resolved the matter through their own autonomous, professional conferences at the Franklin Institute. He was contemptuous of Europe's *diktats* and *fiats*, whereby government ministers decided engineering matters. As he wrote in his *Report to the Franklin Institute* (1876) on the two threads: 'The government of France has always been in the habit of interfering with the private habits of people, but the American concept of government was that it should do and enforce justice, and that liberty in all things innocent is the birthright of the citizen.' And, as America showed, *laissez faire* was perfectly capable of deciding on national engineering standards.)[45]

POSTSCRIPT ON NATIONAL ATTITUDES TO SCIENCE

In as much as the USA and Britain have declined relatively, could that be blamed on an excessive quantity of pure science? That concept always causes such a shock to American and British scientists that it must be clarified. How could one possibly have too many scientists? Well, consider a society in which absolutely everybody was a research scientist. Who, then, would deliver the food to the shops? – indeed, who then would work in the shops? Who would print and distribute the scientific journals or direct the traffic on the roads? (who would build the cars?). Clearly, it is possible for a society to have too many scientists.

There is at least one set of data which suggests that Britain and the USA may do too much research. The OECD collates, every year, the technological balance of payments of its member states. This balance can be determined because, if companies in one country wish to exploit technology created in another, they have to pay royalties or licence fees to the patent holders. These payments are registered for tax purposes, and so can be collated by the OECD. The results are remarkable.[46] Over the last decade for which figures are available, 1975–85, the two countries in apparent long-term relative decline, the USA and the UK, both enjoyed a surplus in their technological balance of payments ($8486 million for the USA in 1985, $194 million for the UK in 1985) while the countries that are overtaking them, Japan, Germany and France, are all in technological deficit ($247 million, $591 million and $169 million, respectively, for 1985). It can be more economical to import technology than to generate it at home because the purchaser of technology is spared the R&D costs of all the failed technology which is never put onto the market (a lot of R&D projects fail).

Of course, one cannot rely solely on imported technology; one needs in-house technologists to set up and to maintain the technology, even if it is imported, and good technologists will not be retained if they are not allowed to experiment, so domestic investment in R&D is necessary, if only to keep the technologists happy. Moreover, the continued profitability of any technology depends on a myriad of small, incremental improvements which, over a number of years, may yield much more, economically, than any particular imported 'breakthrough'. Arrow, in his famous study, called this 'learning by doing',[47] and this, too, requires in-house technologists. The OECD data on technological balances of payments, therefore, does not prove that a technological deficit is a good thing, or a surplus a bad one, but it does disprove the suggestion that the British or American economies are being overtaken by the Japanese, German or French ones because those latter economies do more useful research.

Academic science is expensive. It does not contribute directly to economic growth, but it is fun. Have some countries suffered relative economic decline because they have opted for fun and for Nobel Prizes, while other countries have concentrated on technology (which does contribute directly to economic growth)? There is a hint of quantitative evidence for that; consider Figure 7.11. All five countries which believe they suffer from relative economic decline, New Zealand, the UK, Australia, Canada and the USA, seem to publish more than their fair shares of scientific papers.

(Figure 7.11 should not be interpreted too simplistically; the five over-producing science nations are anglophonic, and one of the other countries to peep above the line, Ireland, is also anglophonic. The collection of data by the Institute of Scientific Information, from which this figure is ultimately derived, is biased towards journals which are published in English, since almost all important papers now are published in English. The Institute only collects the papers published in the 4000 or so leading international journals, most of which are anglophonic, and many respectable non-anglophonic journals are neglected (*Chemical Abstracts* alone covers 15 000 journals). Moreover, Braun and his colleagues, from whose papers this figure is drawn, included in their analysis of the Institute's data some scientific papers that actually appeared as letters; this is because some anglophonic journals traditionally publish small scientific advances as letters. Since French

and German journals only publish full papers, this will artificially in-
flate the numbers of anglophonic papers (see the discussion by B. R.
Martin in 'The Bibliometric Assessment of UK Scientific Performance', *Scientometrics*, vol. 20, pp. 377–401, 1991). Having said all
that, however, Figure 7.11 eliminates a so-called lack of pure science
as an 'explanation' for the recent relative economic decline of the US,
Canada, UK, Australia, or New Zealand – if there has been one).

There is, moreover, anecdotal evidence to suggest that successful
countries have a more hard-nosed approach to academic science than
those that have, apparently, recently declined relatively, although something as intangible as an approach can be hard to chronicle. But consider the research prospectus published in 1991 by Mitsui Pharmaceuticals
(the pharmaceutical arm of the Japanese company Mitsui, one of the
largest chemical companies in the world). This is what it says about
its R&D activities: 'Mitsui Pharmaceuticals has put its main effort into
the R&D of new potential ethical drugs since its foundation, recognising that research is fundamental to the company. The R&D activity of
the company consists of the following three functions: 1) Exhaustive
survey of the literature, collection and analysis of various information,
establishment of R&D targets, and coordination between various research groups. 2) Chemical synthesis of new compounds as emphasised
above and exploration of new biologically active substances from natural sources. 3) Complete confirmation of effectiveness and safety by
various pharmacological screening tests, clinical trials by medical doctors,
studies of dosage forms and drug stabilities for commercialisation.'

What a remarkable order of priorities! No Anglo-Saxon researcher
or research company would happily proclaim a greater priority for reading
other peoples' literature than for original work. There can be little
doubt that the Japanese order of priorities is commercially sound, but
it runs against the contemporary Anglo-Saxon grain.

Another post-war success story, Germany, also seems to have a different
perspective on research. Germany, like so many other countries recently, has experienced economic difficulties, and one of the government budgets to have suffered is science. When the British and American
governments recently also cut their science budgets, their universities
and researchers exploded in a storm of protests which many opposition politicians actively championed. But the Germans took it more
calmly. This, for example, is an extract from an article written by
Professor Wolfgang Fruhwald, the President of the Deutsche Forschungs-
gemeinschaft (the main government science funding agency): 'Today,
at the most, it is only the competitiveness of certain branches of German

science which is at risk – although that is bad enough! There can be no comparison between the richly organised, still sound, and highly differentiated German system of promotion [of science] today with the desperate situation faced by the founders of the Notgemeinschaft der Deutschen Wissenschaft during the years of inflation and global economic crisis.'[48]

How extraordinarily this reads to the Anglo-Saxon! It is almost complacent about the loss of competitiveness of certain branches of German science, and it actually admits that things have been worse. No Anglo-Saxon scientist or science policy maker would ever view the loss of national competitiveness in any branch of science as anything other than a major crisis and a desperate disaster. Nor would any Anglo-Saxon scientist care if things had ever been worse before – that simply would not matter to him. Indeed, he would probably not even believe it.

Germany and Japan clearly view academic science differently from the UK or the USA. In their *Made in America: Regaining the Productive Edge*, Dertouzos, Lester and Solow, three Nobel laureates in economics, explained how the British and Americans have tended since the war to revere academic science, as both a cultural and economic good.[49] The UK and the USA have assumed that as long as they did good pure research, its economic benefits would trickle down into the economy of their own accord. But the post-war Japanese and Germans have understood that pure science is but one link in the chain of events that leads to economic growth – and a tenuous link at that. So, where Anglo-Saxon universities have lauded science and scientists, the Japanese and Germans have continued to honour good production engineering, say, and production engineers. A *volte face*! The British and Americans, in short, have recently made the same mistake as the Alexandrian Greeks or the Enlightenment French; they have poured government money into academic science, they have neglected production, and now they complain that other people create wealth from the science that they, the British or the Americans, first developed. Meanwhile, the Japanese and Germans have adopted the harder nosed approach of pre-1914 USA or Britain (see Chapter 8).

Britain and America, of course, share a common history of victory in war, but Germany and Japan share a different common history. War has shaped science since 1945 to a much greater degree than is generally recognised. Consider the OECD countries. From Figure 7.11 we saw that there is a group of seven countries (New Zealand, the UK, Australia, Canada, the USA, Sweden and Switzerland) which seems to publish more than its fair share of papers. Those are also the only seven countries of the OECD not to have experienced invasion or civil war during the twentieth-century. We can conclude, therefore, that peace is good for science, but war is bad, and its deleterious effects can last for decades.

War damages science in at least two ways. First, war destroys research laboratories and equipment, but war also disperses researchers, who are killed or emigrate. The effects of migration can be dramatic; many of Britain's Nobel Prizes since 1945 were actually won by *émigrés* like Hans Krebs, a German Jew. The country which now publishes the most scientific papers *per capita* is, curiously, Israel. It is actually quite a poor country, but it receives a constant influx of Russian Jewish scientists whose political strength extracts generous funding from the government in Tel Aviv (a government that is culturally predisposed to fund science, art and other expressions of culture). This phenomenon may explain how the nations that were invaded between 1939 and 1945 were able, subsequently, to keep their science funding down; the dispersal and death of their scientists will have shrunk the science lobby.

But, if war is bad for science, it is good for economic growth. To show that, let us use the data collected by Maddison on the economic history of the 16 countries that are currently the richest in the world. Let us divide those countries into two groups: (i) the war-torn countries that were invaded during the Second World War (Austria, Belgium, Denmark, Finland, France, Holland, Italy, Japan and Norway) and (ii) the non-war-torn countries which were either neutral or which were combatants but escaped invasion (Australia, Canada, Sweden, Switzerland, the USA and the UK). Now let us average the economic performances of the two groups. As Figure 7.16 shows, an intriguing picture emerges. Between 1870 and 1979, the non-war-torn countries enjoyed steady, consistent economic growth, with very little variation in growth rates. The countries that were to be war-torn between 1939 and 1945 started out poorer in 1870, and between 1870 and 1939 they too enjoyed steady growth rates, gently catching up; but after 1950 they raced away. How do we explain this?

War alters a nation's culture. The trauma of invasion will make countries determined to recover economically and socially. One of the expressions of that determination will be a re-evaluation of the role of science in economic growth. Invaded countries will conclude that pure science is a luxury. They will, therefore, postpone investing in it until more immediate goals have been met, and even when they do invest in it, they will not be overly romantic about it – they will squeeze as

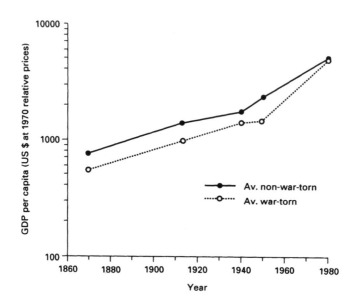

FIGURE 7.16   *GDPs* per capita *for war-torn and non-war-torn countries*

GDP *per capita* data for the various years came from Maddison's *Phases of Capitalist Development.*[21] The war-torn countries that were invaded during the Second World War were Austria, Belgium, Denmark, Finland, France, Holland, Italy, Japan and Norway; the non-war-torn countries which were either neutral or were combatants but escaped invasion were Australia, Canada, Sweden, Switzerland, the USA and the UK. The economic performances of these two groups of countries have been averaged.

much economic benefit from it that they can. It is this hard-headed culture of defeat that has enabled the invaded countries actually to overtake the established leaders like the UK or the USA.

But the invaded countries are beginning to forget their past traumas, new generations are adopting the more relaxed lifestyles of the established rich countries, and their governments are increasingly investing in academic science. This is helping the erstwhile rich countries, now temporarily overtaken, to catch up again. Moreover, neither the UK or the USA is stupid. They do not like being overtaken, and they are learning from the overtakers. Thus it is noticeable how universities in the UK and the USA will now seek to exploit their science commercially, where once they would have disdained such a practice.

Consider, for example, Ernst Chain. Chain, a German Jewish *émigré*

who, working in Oxford, was one of the discoverers of penicillin, tried
during the early 1940s to persuade the Medical Research Council (MRC)
to patent it. The Secretary of the MRC, Edward Mellanby, was shocked,
and told Chain that only tradesmen tried to make money from science
[this is described in G. Macfarlane, *Howard Florey* (Oxford, 1979)].
In consequence, the American drug companies patented penicillin, and
British taxpayers have subsequently had to pay millions and millions
of pounds in royalties to use a drug whose discovery they had themselves helped to fund. The MRC did not learn from this profligate
arrogance, and when one of its other *émigrés*, Cesar Milstein from
Argentina, suggested during the 1970s that his discovery of monoclonal
antibodies be patented, his suggestion was also rejected. British taxpayers are, yet again, now paying royalties for a discovery that they
funded.* Meanwhile, the MRC wonders why governments hesitate over
increasing its funding. It also wonders why, after 80 years of funding
science in the vastly lucrative area of biomedical research, it has not
acquired any substantial independent funds of its own, but still has to
depend on government handouts.

Neither of the scientists, Chain or Milstein, was unworldly – perhaps because each was an *émigré*. Each understood the commercial
implications of his work. It was the bureaucrats – perhaps because
they were British – who failed to consider the taxpayer's rights, and
who viewed the taxpayer as a cow to be milked.

POSTSCRIPT ON THE EUROPEAN COMMUNITY (OR UNION)

The sort of analyses presented in Figures 7.5 and 7.6 allow one to
audit economic policies. For example, propagandists for the EC claim
that it has promoted the economic growth of its members. Is that true?
It is easy to test. If one creates a figure comparing the GDPs *per
capita* of the EEC member states on entry, against their GDPs *per
capita* twenty years later – and then one does exactly the same analyses for the EFTA countries – one discovers that the rates of economic
growth for the members of the two organisations are indistinguishable.
The sacrifice of national sovereignty and national democratic accountability that the EEC demands, therefore, does not accelerate economic
growth. Of course, the change in terminology from EEC to EC to EU

* The lost patent income on monoclonal antibodies has been calculated at £200 million
a year, which is also the annual cost of the MRC to the taxpayer.

reflects the reality that the EU was never primarily an economic organisation; it used economic arguments to disguise the reality that the
EU was always primarily a political construct to enable the Germans
to lose their national identity in a wider, untainted body, and also to
provide France and the Benelux nations with the hope that political
union would stop the Germans from invading them yet again. The EU's
political aspirations are, of course, utterly redundant, because Germany
has experienced a total cultural change, and would no more invade its
neighbours today than paint itself orange. And if Germany did want to
re-invade, just for old time's sake, it would not allow a little thing
like political union to stop it – witness the Serbians, whose union within
Yugoslavia did not prevent them from expansion and invasion.

POSTSCRIPT ON THOMAS EDISON (1847–1931)

Edison is a pivotal figure. Like most of the great pioneers of the first
industrial revolution, he was a barely literate, uneducated artisan, expelled from school after only three months for being 'retarded'. Like
most of the great American engineering pioneers, he learnt on the job,
educating himself through his experience with Western Union and through
Faraday's *Experimental Researches in Electricity*.

But he also realised that as America breasted the technological forefront – one that was growing increasingly complex – it could no longer
depend on lone, self-taught pioneers. It needed to copy Faraday's Royal
Institution to support collaborative, systematic industrial research. In
1876 he created Menlo Park, employing 20 researchers, funded by the
income they generated through the patents they produced.

In 1880 Graham Alexander Bell founded the Volta Laboratory, and
soon AT&T, General Electric, Dupont, Westinghouse and others were
creating their own laboratories. The founding of these laboratories reflects
the capacity of private enterprise to fund appropriate research at the
appropriate time, even if the style of their founding caricatured that of
the bloated plutocrats of socialist fantasy – Edison was a horrible man.

8 Science Policies of the Twentieth Century

There are few activities more important than research or development. Above all, it is innovation that has dragged us out of the palaeolithic mire into light, warmth, comfort and security. Yet even acute observers remain confused as to how best to promote innovation, research or development. Consider Paul Kennedy; his *Rise and Fall of the Great Powers* has been rightly praised for its masterful review of the factors that, indeed, cause the rise and fall of great powers. In it, Kennedy naturally devotes much space to the importance of science and technology. But he cannot decide on a policy. These are the factors to which he attributes the rise of nineteenth-century Germany: 'technical education, an unrivalled university and scientific establishment, and chemical laboratories and research institutes without an equal.'[1] These, of course, were largely funded by government. Now note how Kennedy accounts for Japan's contemporary success: 'Japan's second strength . . . is the fact that a far higher proportion of Japanese R and D is paid for and done by industry itself than in Europe and the United States, where so much is done by governments or universities. In other words, it is aimed directly at the market place and is expected to pay its way quickly. Pure science is left to others and tapped only when its commercial relevance becomes clear.'[2] Within the same book, therefore, a historian as brilliant as Kennedy will attribute Germany's success to the state funding of university science, while praising Japan for avoiding the mistake that Europeans or American governments make in funding university science. Well, which is it to be?

To help answer this question, we will survey over the next few chapters the science policies and economic outcomes of the major science nations. In this chapter, however, we shall largely concentrate on two countries, the USA and the UK, because they share a common and pivotal history. Each was, in turn, the lead country economically, becoming so while pursuing *laissez faire* policies for science. Each, however, is now scientifically *dirigiste*. Let us survey the evolution of their science policies, to explain why they shifted from *laizzez faire* to *dirigiste*, and let us examine the consequences of the shift. Let us also determine

why almost every other major industrialised country is now scientifically *dirigiste*. First, let us start with the USA.

UNITED STATES SCIENCE POLICIES

Science is mentioned in the American Constitution: 'The Congress shall
have power . . . to promote the progress of science and the useful arts'
(Article 1, Section 8), and the Constitutional Convention of 1787 had
discussed the creation of national societies for the promotion of science,
but in practice the American government during the eighteenth-century did nothing for science (although a Patent Act was passed in 1790).
During the early part of the nineteenth-century, a few federal research
institutions were created, but these were of very defined and highly
specialised functions. Thus, in 1800, the Library of Congress was created, in 1807 the Coast Survey was established, in 1818 the Surgeon
General's office and the Army Medical Department were established,
in 1830 a Depot of Charts and Instruments was established (these eventually evolved into the Naval Observatory) and in 1836 the Secretary
of the Treasury was directed to supply the Governor of each state with
a standard set of weights and measures (from this directive evolved
the National Bureau of Standards). None of these institutions, however, existed primarily for the pursuit of science; they were created to
execute functions that were believed to be the proper responsibility of
the Federal Government. Any research they performed was secondary
to their mission. Indeed, research for its own sake was absolutely not
believed to be a responsibility of government. This was illustrated by
the foundation of the Smithsonian Institute.

James Smithson (1765–1829) a distinguished English mineralogist
and chemist (FRS, 1786) left, in his will, over £100 000 'to found at
Washington under the name of the Smithsonian Institution, an Establishment for the increase and diffusion of knowledge among men'.
Smithson was an illegitimate son of the Duke of Northumberland and
his social difficulties had alienated him from his own country. Having
never visited America, he imagined it a haven of tolerance and progress,
so he bequested it his huge fortune. Little did he know the problems
this would cause.

Nineteenth-century America simply did not believe that research, or
knowledge for its own sake, was a proper responsibility of the Federal
Government. There were two major reasons for this: first, many Ameri-

cans subscribed to Adam Smith's view that all government intervention into society should be minimised. Many early Americans had left Europe specifically to escape authoritarian regimes, and they were not keen to pay taxes to empower politicians. These ideological objections to federal power might have been overcome but for a second obstacle: states' rights. Many Americans felt a stronger loyalty to their state than to the Union, and they opposed any strengthening of central power. This was felt particularly strongly in the South, which feared for the survival of its peculiar institution, slavery.

Although Smithson bequested his money to Washington in 1829, it was not actually until 1846, after many bitter debates, that Congress finally brought itself to accept the money to found the Smithsonian Institution. Thereafter, Congress maintained its disdain of science, and no further federal subventions were made. This changed very abruptly with the onset of the American Civil War (1861–5). It took total war to force the federal government into funding science.

For decades, certain prominent American scientists had sighed longingly over Europe. They observed the vast sums spent on research by the French government and they noted the high social prestige of London's Royal Society – and they ached for an American Academy of Science that would channel federal funds and prestige into their own laboratories. No American sighed more than Alexander Dallas Bache, the Superintendent of the Coast Survey. In 1851 he had even created his own scientific group, the Lazzaroni, to foreshadow an Academy, but the peace-time administrations resisted his blandishments. It was the Civil War that enabled him to create a real Academy.

The Civil War presented the federal government with uprecedented technical problems: Ironclads had to be built, gunnery improved and steam engines converted to military use. Bache grabbed his opportunity. Leading a small but prominent group of activists, including Charles Henry Davis (the Chief of the Bureau of Navigation), Joseph Henry (the Secretary of the Smithsonian Institution), Benjamin Apthorp Gould (the astronomer) and Louis Agassiz (a prominent researcher), he lobbied the Federal Government, and in 1863 Congress established the National Academy of Sciences, with Bache as its first President. The Academy promptly organised a series of committees to advise various government departments on military scientific problems. On 8 May 1863, for example, a committee was formed to consider how to operate a magnetic compass in a ship built of iron. On the same day, another committee was formed to report on electroplating ships with copper to prevent corrosion and fouling in sea water. Later that year, committees examined the feasibility of submarines and of ironclad warships, and in 1864 a committee supervised experiments on the expansion of steam (these committees are described by Nathan Reingold in his 'Science in the Civil War').[3]

Much of the work of the Academy was shared with a very similar body, the Permanent Commission of the Navy Department. This organisation was created a month earlier than the Academy, it possessed a similar membership and it did similar things. Why two bodies? This is how the Commission attempted to clear the confusion in the *Scientific American* for 12 March 1864: 'The present members of the Commission are also members of the National Academy of Sciences; and the Commission itself would probably never have been created if the Academy had been in existence at that time, since they both have the same objects, and are designed to perform similar duties.'[4] The explanation actually explained nothing, but it is worth exploring how the two bodies arose, because it illustrates the prevailing attitude towards science policy in the USA.

One of Bache's strongest supporters for the idea of an Academy was Joseph Henry, the Secretary of the Smithsonian Institution. But from his own experiences, Henry feared that Congress would never create an Academy. So, discreetly and privately, he persuaded Gideon Welles, the Secretary of the Navy, to create a Permanent Commission by executive action. In the event, the legislature also obliged, leading to the creation of both the Academy and the Commission, with overlapping membership and similar aims. Fortunately, common sense prevailed, and scientific harmony reigned (although there were conflicts between the scientists of the Academy and of the Commission on the one hand, and some of the scientists employed directly by government departments on the other). Bache, who chaired committees of both bodies, enjoyed himself hugely in writing to himself, as the chairman of one committee to another, asking himself for advice. (Bache enjoyed life: the minutes for the Commission of 10 February 1864 noted, 'Mrs Bache being unwell, the Professor regretfully announced that there were no oysters').[5]

Bache needed a jovial nature because his idea of an Academy offended many scientists. *Scientific American* led the opposition (see its issue for 23 May 1863). Amongst the objections were the following: (i) the Government funding of science would lead to Government control, which would impede the disinterested pursuit of truth; (ii) Academy members were not to be paid, so the Government would obtain its scientific advice for nothing, whereas before it had paid consultancy fees, this would lower the status of a learned profession; (iii) because members would not be paid, the Academy would be largely staffed by scientists who were already employed by the Government rather than by those who earned their living in the market place. It would promote, therefore, the interests of government-employed scientists rather than those of science itself. This featherbedding of vested interests was obvious to contemporary observers; one of the few lobbyists for an Academy who was not a civil servant, Louis Agassiz, wrote to Bache on 6 March 1863: 'How shall the first meeting of the Academy be called? I wish it were not done by you, that no one can say this is going to be a branch of the Coast Survey and the like.'[6]

The Government had its own doubts of the value of an Academy, and although subsequent peace-time administrations recognised it as their scientific advisor – indeed, the Academy was obliged by its constitution to help any government department that asked for it – the government paid the Academy no standing grant; it only paid for the expenses incurred on investigating any particular question. Bache, however, was more than happy. He had wanted an Academy for years, and he had exploited the Civil War to get one.

As we shall see, the manner of the foundation of the National Academy of Sciences was characteristic. There will always be scientists who believe that science should be advanced with government funds, governments will resist the pressure but, time again, war persuades them to create scientific institutions. Once created, institutions are hard to destroy; they invent justifications for their continued existence, and they mobilise political support. One feature of scientific institutions is their intellectual flexibility. In war they emphasise their potential for creating weapons of death, in peace they explain how they can create wealth. (This has been recently illustrated by the ten national laboratories that were created in support of the Manhattan Project, which employ no fewer than 48 500 staff and which cost a staggering \$5.9 billion a year to run. In the wake of the Cold War these laboratories no longer serve a useful function, but rather than close them and alleviate the poor American taxpayer's burden, the Garvin Report for the Department of Energy thrashed around for any and every possible invented justification for their continued support. See the reports in *Nature*, 9 February 1995, vol. **373**, pp. 475–88 and the *Economist*, 11 February 1995, pp. 110–111.)

The American Civil War precipitated a further federal intervention into society: the Morill Act of 1862. This Act, which created federally funded agricultural colleges across the country, was bitterly opposed, both by proponents of the free market and, more powerfully, by advocates of states' rights. An earlier, fiercely contested, Act had actually passed both Houses of Congress in 1859, only to be vetoed by President Buchanan. The Morill Act eventually became law in 1862, under President Lincoln, the Civil War having removed from Washington the strongest advocates of states' rights, the southern representatives. (The Morill Act is important because it represented the first direct federal intervention into higher education, albeit agricultural.) In the same year, 1862, the Department of Agriculture was established with, as one of its aims, the application of science to agriculture.

Post-war, federal intervention in science decelerated. The Office of Education was established in 1867 and the Geological Survey in 1879, but subsequent initiatives represented little more than the recognition that certain functions had outgrown their administrative dependence on other departments. The Weather Bureau (1890), the Bureau of Chemistry (1901), the Bureau of Plant Industry (1901), the Bureau of Soils (1901), the Bureau of the Census (1902) and the Public Health and Marine Hospital Service (1902) were all the natural developments of older civil service functions (the first marine hospital, for example, was being built by the Treasury as early as 1798). Even the one significant initiative, the Hatch Act of 1887 which attached federally funded research units to the agricultural colleges (whose budgets were enlarged by the Adams Act of 1906), represented nothing but a consolidation. Having decided for electoral reasons to subsidise agriculture, the federal government had, therefore, to provide for agricultural research, but such research was highly applied, highly practical and controlled from Washington (the Adams Act brought such research under tight direction). No post-civil war initiative represented anything but the consolidation of existing federal functions. Washington adopted no further responsibilities for science; indeed, so distant had the federal government grown from science that, between 1900 and 1915, it made only four requests for help from the National Academy of Sciences. All this was to change with the re-advent of war.

The parallels between the scientific responses to the Great War and the Civil War were extraordinary. New agencies were created (the Advisory Committee for Aeronautics, for example, which spent over $1 million of federal funds between 1915 and 1919) but, uncannily, old agencies were re-created under new names. The Civil War's Permanent Commission of the Navy Department, for example, reappeared as the Naval Consulting Board, a research coordinating committee under the chairmanship of Thomas Edison, no less. The most far-reaching scientific consequence of the Great War, however, was the creation of the National Research Council, and this owed more to George Ellery Hale than to anyone else. This was uncanny indeed, because Hale behaved like a reincarnation of Bache.

George Ellery Hale was a distinguished astronomer, the editor of the *Astronomical Journal*, the Director of the Mount Wilson Observatory, and the Foreign Secretary of the National Academy. He was a great believer in the importance of government funding for science, and he passionately believed that America should increase its respect for research. He also believed that the National Academy should assume a greater role in the nation's affairs; all these ideas he marshalled in his book *National Academies and the Progress of Research* (1915). In particular, Hale believed that there should be a national research council to coordinate research and to encourage scientific intercourse. Hale felt that if scientists in different disciplines would only converse, their cross-fertilisation would provide unexpected fruit. He also felt that scientists were too isolated in their laboratories, and he wanted more national meetings – indeed, more international meetings – and more journals. He also felt that more discussion would reduce the duplication of experiments.

These aims were doubtless very worthy, but it was only the First World War which enabled Hale to enact them with federal funds. When President Wilson delivered his ultimatum to Germany on 18 April 1916, Hale sprung into action. On the very next day, 19 April 1916, he called an emergency meeting of the National Academy, he persuaded it to resolve to help the President, within a week the Academy had presented its resolution to the President at the White House, and by June the President had agreed to the formation of a new agency of the Academy, the National Research Council. Under the chairmanship of Hale, representatives of government, science and industry met to coordinate the scientific responses to the nation's military needs. Physicists developed techniques for detecting submarines, chemists developed poisonous gas and psychologists developed tests of intelligence and personality.

Throughout this time of excitement, of the mobilisation of science, and of crisis, Hale emphasised to the Government the military value of federally-organised science, and he was rewarded by the adoption of the National Research Council as the Department of Science and Research to the Cabinet's Council of National Defense. But, throughout this period, Hale had his deeper plan. As he wrote in a letter to Isaiah Bowman on 2 May 1933: 'While most of my own experience

.as Chairman was during the war, the plans I always had in mind looked forward to work under peace conditions.'[7]

But the prospect of peace threatened Hale's National Research Council. Could further government patronage be assured? A public change of tack was necessary, and on 27 March, 1918, Hale wrote to President Wilson explaining how a peacetime National Research Council would 'stimulate pure and applied research for the national welfare; the movement to establish councils for the promotion of scientific and industrial research has swept over the whole world since the outbreak of the war'.[8] Hale recruited to his cause distinguished industrialists like Eliku Root who wrote that 'without pure science, the whole system of industrial progress would dry up'.[9] After many such letters, President Wilson cracked, and on 10 May he signed an executive order making the National Research Council permanent.

Hale had once hoped for an Act from Congress, then under the control of the Democratic Party, but as he wrote in a letter to H. H. Turner on 6 March 1916: 'Democratic senators spoke eloquently against all . . . organizations for research and advanced study.'[10] The Democrats, the party of the people, distrusted the public funding of science. Democrats saw science as an elitist activity, for which ordinary people should not be taxed, and they also argued that, if indeed science was of economic benefit, then its proper paymasters were industrialists, not ordinary people.

The National Research Council did not initially obtain as many federal funds as Hale had hoped (at first, Washington's only subvention was to pay the National Research Council's subscription to the International Research Council) but, as Hale wrote to James Garfield on 16 May 1918: 'We now have precisely the connection with the government that we need.'[11] He knew that he had created a channel down which federal funds would one day pour.

But not at first. The 1920s witnessed a reversal to pre-war days, with the government only funding applied research that was integral to its agencies' social or political functions. During the 1920s, that sort of research did grow, and some agencies (in particular the Naval Research Laboratory and the National Advisory Committee for Aeronautics) did very good, even sometimes pure, research. The first half of the 1930s, however, was dreadful, as Government cut back its spending in response to the Great Depression. The Department of Agriculture, for example, cut its research budgets from $21 500 000 in 1932 to $16 500 000 in 1934, and it shed 567 research jobs in 1934 alone.[12] A purer research agency like the Bureau of Standards saw its budgets more than halved from \$3 904 000 (1931) to \$1 755 000 (1934).[13] Cuts in federal budgets for science were probably inevitable during the early 1930s, but they were made even tighter by the replacement of President Hoover by Roosevelt in 1932.

Hoover, an engineer by training and, as a Republican, sensitive to the needs of business, believed that science underpinned economic growth. In a speech during the 1932 presidential campaign he said: 'Progress is due to scientific research, the opening of new invention, new flashes of light from the intelligence of our people.'[14] But Roosevelt and the Democrats believed that improvements in productivity had caused the depression by throwing people out of work. Indeed, they believed that the depression had been caused by overproductivity. If Hoover supported science, they argued, then science had to be a bad thing, particularly as it was a notoriously elite activity, studied by rich college boys and supported by capitalists who were looking for new machines to displace hands. Much of the nation shared that belief and, if there was an understanding that science might improve living standards in the long run, there was also a widespread belief that the time had come for a moratorium on research. In 1935 (vol. 81, p. 46) *Science* magazine collected various statements that people had made: 'there should be a slowing up of research in order that there may be time to discover, not new things, but the meaning of things already discovered'; 'the physicist and the chemist seem to be travelling so fast as not to heed or care where or how or why they are going. Nor do they heed or care what misapplications are made of their discoveries.'[15]

Hoover opposed these ideas. In a 1932 campaign speech he said: 'I challenge the whole idea that we have ended the advance of America . . . What Governor Roosevelt has overlooked is the fact that we are yet but on the frontiers of development of science and of invention.'[16] But Hoover lost, and Roosevelt came to power committed to reducing federal budgets in general, and science budgets in particular – which he did, savagely.

It was the New Deal that caused him to reverse his federal anti-science policies. Initially, the New Deal did nothing for research, but the Works Projects Administration (WPA) one of the agencies that emerged from the New Deal, was committed to providing jobs for skilled people, and by 1939 it was funding science projects in every publicly funded university in the Union. Thus did the Democrats come to endorse the government funding of science as a form of outdoor relief for indigent intellectuals. But the mid-1930s also witnessed a change in Roosevelt's economics, and he came round to the Keynesian view that economic recovery could be assisted by increasing, not decreasing federal budgets. He also adopted the Baconian view that governments should fund science. Science budgets were gradually restored, and by 1938 they again represented 1 per cent of federal expenditure. Many scientists, however, opposed the government's renewed enthusiasm.

In the topsy-turvy world of science politics of those days (topsy-turvy to us now) when Republicans supported publicly funded science and Democrats opposed it, many scientists continued to resist any government intervention in science. Some of the scientists' complaints were unchanged since 1863. Scientists resisted government funding because it led to government control and to a loss of academic freedom.

The politician who most encountered the scientists' complaints was Henry A. Wallace, the Secretary for Agriculture, whose department controlled about half of all federal science funds, since in those days the federal government's major expenditure in research was still largely agricultural. In an article published in *Science* in 1934, he attacked scientists for promoting academic freedom above political control (which he called 'social responsibility'): 'The scientists have turned loose upon the world new productive power without regard to the social implications.'[17] Wallace's greatest problem with the scientists, however, was a new one – they distrusted Keynesian economics. He complained of their attachment to sound money and to balanced budgets: 'In the past, most scientists and engineers were trained in *laissez-faire* classical economics.'[18] Wallace was always pushing for increased federal budgets, but he constantly found scientists 'a handicap rather than a help' in his campaigns.[19] It was indeed a topsy-turvy world when scientists turned down federal money in the national interest. Yet again, however, war was to catalyse change.

If there is one constant in science, it is that at any one time there is at least one powerful person, waiting in the wings, just itching to impose a centrally planned, federally funded science policy on the people of America. Most of these people die unsung, but if they are fortunate, and a war just happens to coincide with their career peaks, then they soar into apotheosis. The Civil War enabled Bache to create the National Academy of Sciences, the Great War enabled Hale to create the National Research Council, and the Second World War provided Vannevar Bush with his opportunity. These science 'czars' [as they were described by the National Academy of Sciences in its *Federal Support of Basic Research in Institutions of Higher Learning,* (1964)] always succeed in linking their own career progression to their commitment to the national interest. Thus Bache became the first president of the National Academy of Science, Hale became the first president of the National Research Council, and Vannevar Bush was to become the first president of the National Defense Research Committee, the first president of the Office of Scientific Research and Development, *and* the first president of the Joint Research and Development Board.

Vannevar Bush was an electrical engineer (PhD in 1916 from the Harvard/MIT joint programme) who rose through MIT (the Massachussetts Institute of Technology) to become Vice-President and Dean of Engineering, but he left in 1938 to become the President of the Carnegie Institution (a private body which was then the USA's largest science funding agency). Bush outlined his views on science policy in his *Science – The Endless Frontier*, which was published in 1945, but which encompassed the views he had tirelessly propagated throughout the 1930s. In that book, Bush outlined the nation's science needs as he saw them. First: 'We have no national policy for science. There is no body within the Government charged with formulating or executing a national science policy. There are no standing committees of the Congress devoted to this important subject.'[20] This, Bush argued, represented a dereliction of federal duty: 'It has been basic United States policy that government should foster the opening of new frontiers. It opened the seas to clipper ships and furnished land for pioneers. Although these frontiers have more or less disappeared, the frontier of science remains. It is in keeping with the American tradition – one which has made the United States great – that new frontiers should be made available for development by all American citizens.'[21]

Advances in science, Bush maintained, 'will bring higher standards of living, will lead to the prevention or cure of diseases'[22] but, and here Bush made an explicit Baconian statement, one of great impact: 'Basic research leads to new knowledge. It provides scientific capital. It creates the fund from which the practical applications of knowledge must be drawn. New products and new processes do not appear full-grown. They are founded on new principles and new conceptions, which in turn are painstakingly developed by research in the purest realm of science.'[23] These, Bush explained, would require vast federal funds, and who better to administer them than Vannevar Bush?

Unfortunately for Bush, the pre-war government ignored him, and he had to content himself with dispensing the Carnegie's funds until, by great good fortune, Japan started to invade its neighbours. By 1940, war between America and Japan looked so likely that Bush was able to lobby the White House, and on 27 June 1940 President Roosevelt created by executive order the National Defense Research Committee (chairman: Vannevar Bush). The National Defense Research Committee promptly moved to integrate the country's research with the needs of war. Subcommittees were formed (ordnance, chemistry and explosives, communications and transportation, instruments and controls, and patents and inventions). Academics, industrialists and service officers sat on these committees, galvanising the USA's civilian science base to support the defence forces.

Curiously, aeronautics was excluded from the jurisdiction of the National Defense Research Committee. In his otherwise authoritative *Science in the Federal Government*, Hunter Dupree innocently speculated that this exclusion might have represented the 'admission that the National Advisory Committee for Aeronautics was the one research outfit in the government already organized for the emergency.'[24] The real reason was more personal: Vannevar Bush was already the Chairman of the National Advisory Committee for Aeronautics, so it was the one body he did not need to bring under the control of his National Defense Research Committee.

In June 1941, the National Defense Research Committee was reorganised as the Office of Scientific Research and Development (chairman: Vannevar Bush) to provide scientists with independent funding through congressional appropriations. This provided the vast funding needed for the huge development of weaponry (the National Defense Research Committee having been limited to Presidential emergency funds). The increase in funding was large indeed. Table 8.1 shows that the total federal expenditure on all research and development in 1940 was $74.1 million. Of this, the largest component was agricultural R&D at $29.1 million. By 1945, federal expenditure on R&D had risen to $1.591 million, the increase being almost entirely accounted for by defence and atomic research (see Figure 8.1) and the Office of Scientific Research and Development (OSRD) was coordinating research in fields as diverse as penicillin and radar as well as nuclear.

But the Office of Scientific Research and Development was only intended as a war-time measure, and with the threat of peace came the age-old threat of dismantling. Bush, of course, had always wanted a federally funded national research foundation through which he could initiate, fund and coordinate a vast peacetime empire of pure science, and it was to help convert his Office of Scientific Research and Development into a national research foundation that he wrote his *Science – The Endless Frontier* in 1945. Initially, this was well received. President Truman informed Congress on 6 September 1945: 'No na-

TABLE 8.1   *The funding of research and development in the USA in 1940 (millions of dollars)*

Federal funding	
Agriculture	29.1
Defense	26.4
Health	2.8
Aeronautics	2.2
Others	13.6
Total	**74.1**
States' funding	
Largely agricultural	7.0
Total government	**81.1**
Private funding	
Industrially funded research and development	234
University- or foundation-funded research	31
Total private	**265**
Total USA funding for research and development	346.1

SOURCES: *Federal Funds for Research, Development and other Scientific Activities* (Washington: National Science Foundation, 1972). Industrial funding figures came from John R. Steelman, *Science and Public Policy*: A *Program for the Nation* (USGPO, 1947). University figures (which are for 1935–6) came from *Research – A National Resource* (Washington, DC: USGPO, 1938).

tion can maintain a position of leadership in the world today unless it develops to the full its scientific and technological resources. No government adequately meets its responsibilities unless it generously and intelligently supports and encourages the work of science in university, industry and . . . its own laboratories.'[25] In July 1945 Senator Warren G. Magnusson had introduced a bill proposing a national research foundation. But Magnusson, an ally of Vannevar Bush, found his bill opposed by Senator Kilgore, who introduced a bill of his own, and for five years Congress, the executive and the scientists argued. It was not until 1950 that the National Science Foundation was created.

Senator Kilgore, Democrat of West Virginia and Chairman of the Senate Subcommittee on War Mobilization, was not opposed to a national research foundation, but he wanted a very different one from Bush. Both men wanted a giant foundation that would coordinate, plan and

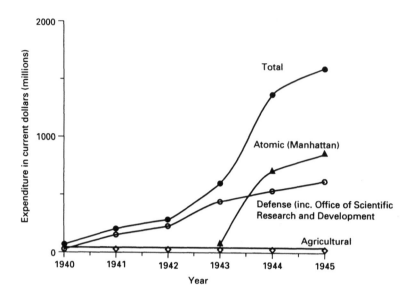

FIGURE 8.1   *US Federal expenditure on R&D (Second World War)*

The data come from Mowery, D. C. and Rosenberg, N., *Technology and the Pursuit of Economic Growth* (Cambridge, 1989).

control the nation's science in accordance with a national policy for research, but they disagreed over that policy. Bush wanted to create a foundation that would be run for the benefit of scientists. He proposed, therefore, that its governing body be made up of academics, industrialists and engineers. He wanted its grants to be distributed according to scientific merit. Bush also wanted scientists to retain any patent rights. But Senator Kilgore believed that a national science policy should be directed by government, and that science should be bent to wider national needs. Kilgore argued that the governing body of a national science foundation should be staffed by civil servants, not independent academics, industrialists and engineers. He argued that grants should be distributed equally across the individual states, rather than be concentrated on the traditional centres of excellence on the East and West Coasts; that the government should retain patent rights; and that government should support the social sciences (Bush denied that the social sciences were sciences).

Congress debated the issues for five years, being largely split on party lines. The Democrats, who distrusted scientists as elitist, and industrialists as capitalists, supported Kilgore, and the Republicans supported Bush. Bush's allies generally won the arguments in Congress, but when they did, President Truman exercised his veto.

To be fair to Kilgore and Truman, there were genuine problems over Bush's plans. It was unprecedented to suggest that federal funds be handed over to an organisation that would, in effect, be answerable only to itself and to its beneficiaries. In explaining his 1947 veto, for example, Truman wrote: 'This bill contains provisions which represent such a marked departure from sound principles for the administration of public affairs that I cannot give it my approval. It would, in effect, vest the determination of vital national policies, the expenditure of large public funds, and the administration of important governmental functions in a group of individuals who would be essentially private citizens. The proposed National Science Foundation would be divorced from control by the people to an extent that implies a distinct lack of faith in democratic processes.'[26] The scientists, moreover, were also divided. Most scientists probably supported Bush, but many shared Kilgore's constitutional concerns over democratic accountability. At least three scientific bodies, the Federation of American Scientists, the Washington Association of Scientists and the Association of Land–Coast Colleges and Universities, supported Kilgore in arguing that democratic accountability was more important than individuals' scientific or academic freedom.[27]

These arguments might have gone on for ever but, as usual, war came to the rescue of the government funding of science. By 1950, the USA was fighting communists in Korea. Indeed the Soviets had been quietly waging a 'cold war' against the West from the day Germany surrendered in 1945. When, in 1946, Winston Churchill made his 'iron curtain' speech in Fulton, USA, warning of Soviet aggression, he was dismissed as a bellicose old fool, just as he had been in Britain during the 1930s, but by 1947 the Soviets had brutally eliminated all opposition in Hungary and Bulgaria, in 1948 they had subjugated Romania and Czechoslovakia, and in 1949 they took over Poland. They had already (1948) tried and failed to starve West Berlin into surrender, and the 1948 Congress of 'Young Communists' in Calcutta had threatened a Soviet expansion into all Asia. By 1949, the Chinese communists had captured Peking, and in 1950 North Korea (a Soviet satellite) invaded South Korea. The USA led the response to these aggressions, organising the Berlin airlift (1948), drafting the NATO treaty (1949) and commanding the United Nations troops in Korea (1950). The chronic conflict against communism finally persuaded Congress, the President, Bush and Kilgore to compromise, and the National Research Foundation was created in 1950, largely along Bush's lines.

Kilgore was prepared to compromise with Bush and the majority of the scientists because of the lessons he believed he had learnt during the Second World War. One remarkable feature of that war had been that Congress's allocations of money for science had always been greater than the amount the scientists could use. There was a simple reason for this: America had run out of scientists. The military's vast demand for new technology could not be fully met by the researchers, whose numbers had been geared to peace-time needs. No one understood this better than Senator Kilgore, the Chairman of the Senate's Subcommittee on War Mobilization, who in 1942, 1943 and 1945 had conducted a series of hearings on America's war-time research shortfalls.

During the war, Kilgore had introduced the legislation which had created the Office of Technological Mobilization, and he had introduced the Science Mobilization Bill in 1943. Faced with Korea, the Warsaw Pact and China, Kilgore sunk his differences with Bush and the scientists, because his aspirations transcended those of science itself. For Kilgore, the major purpose of the National Science Foundation was not the generation of new knowledge (nice though that would doubtless be) but the generation of trained scientists. Kilgore wanted to create a reserve of scientifically trained personnel who could be mobilised for strategic purposes. Since the source of trained scientists was the universities, and since the universities demanded academic freedom, Kilgore was prepared to compromise. As long as they took the money and boosted their production of qualified scientists, Kilgore was satisfied. The National Science Foundation, therefore, was created in 1950, in the same year (and for the same reasons) as the National Security Council.

Later in this book we will address two problems. First, we will ask why the vast US government investment in university science has not yielded any economic benefit and, second, we will ask why the conditions under which scientists work are so often unsatisfactory. The circumstances under which the National Science Foundation (NSF) was created, however, answer many aspects of these questions. First, its progenitors like Kilgore never intended the NSF to be commercially important. Kilgore accepted that the USA's peace-time production of scientists before 1940 was appropriate to the needs of the peace-time economy. For Kilgore, the NSF was always a defence establishment. But this was to cause manpower problems. When the free market prevailed in science, the supply of scientists roughly met the demand,

and the competition of employers for scientists ensured that employers paid proper salaries and that they provided acceptable conditions of work. When the government intervened to push up the supply of trained PhDs beyond the need of the economy, this threatened a flood of surplus PhDs, thus devaluing their economic power. In the USA, this devaluation has not been too extreme, because the government funding of science seems only to have displaced private funding (see Chapter 10) so alleviating a terrible surplus. Moreover, different government agencies and different universities have continued to compete against each other, even on government money, so raising salaries. In Britain, however, where the Government has provided a vast surplus of university places in science, and where it has offered a free university education to even the least qualified schoolchildren (see the postscript on the Robbins report) the consequence has been the flooding of the market with scientists, to drag their salaries down to pitiful levels and to aggravate a horrible competition between them. It has also repelled the most able schoolchildren, who now spurn a profession whose only rewards are relative penury and bitter in-fighting over funding.

(Scientists' salaries in Britain are even further reduced by an unholy national agreement between government science agencies and the universities to fix pay scales as low as possible. Scientists and academics have thus been degraded. A Cambridge University professor during the 1930s, when there was still an academic free market, earned around £80–90 000 a year at current prices (perhaps double that of medical general practitioners) but today he only earns about £37 000, significantly less than a medical general practitioner. At least, in America, the government science agencies and the universities compete for the best scientists in a more open market, which has helped sustain proper salaries.)

These problems, however, were for the future and, particularly, for Britain, (and continental Europe). A more immediate difficulty had arisen in the USA. In 1945 Bush had hoped that the Office of Scientific Research and Development (OSRD) would simply mutate into the National Science Foundation (NSF), but since the NSF did not emerge until 1950, and the OSRD was dismantled in 1945, this left an administrative gap. The three major functions of the OSRD were hived off. Military research, much of it in basic science, devolved to the Joint Research and Development Board, of which Bush retained the chairmanship, but atomic research devolved to the Atomic Energy Commission, and biomedical research was grabbed by the Public Health Service to create the National Institutes of Health. This left the National

Science Foundation, which was created to support academic science, with relatively little to do, and its initial budget in 1950 was only $3.5 million (Bush had hoped for $33.5 million). By 1955 it had only grown to $16 million (Bush had hoped for $122.5 million) but help was on the way.

On 4 October 1957, the USSR launched *Sputnik*, the world's first artificial satellite. The impact on American public opinion was dramatic. Would the Soviets destroy America from space? In the words of Wernher von Braun, '*Sputnik* triggered a period of self-appraisal rarely equalled in modern times. Overnight, it became popular to question the bulwarks of our society; our public educational system, our industrial strength, international policy, defense strategy and forces, the capability of our science and technology. Even the moral fiber of our people came under searching examination.'[28] After intense and bitter criticism of various branches of government, both by other branches and by the media, two important Acts were passed in 1958. The National Aeronautics and Space Administration (NASA) was created, and the National Defense Education Act was passed. The National Defense Education Act poured money into American education at all levels. A student loan programme was created and National Defense fellowships offered to graduates.

NASA's expenditure peaked in 1968 but, the federal government having now accepted a responsibility for science *per se*, new government initiatives conspired to drive up federal expenditure on R&D (see Figure 8.2). During the 1970s, President Nixon declared 'war on cancer' which pushed up health expenditure: In his State of the Union message of 22 January 1971, President Nixon said: 'The time has come in America when the same kind of concentrated effort that split the atom and took man to the moon should be turned towards conquering this dread disease. Let us make a total national commitment to achieve this goal.'[29]

OPEC's rises in oil prices during the 1970s prompted a rise in federal spending on energy research. In 1977, President Carter launched his National Energy Plan, describing the nation's energy problems as the 'moral equivalent of war'. In his State of the Union message he said: 'Scientific research and development is an investment in the nation's future, essential for all fields, from health, agriculture, and environment to energy, space and defense. We are enhancing the search for the causes of disease; we are undertaking research to anticipate and prevent significant environmental hazards; we are increasing research in astronomy; we will maintain our leadership in space science; and

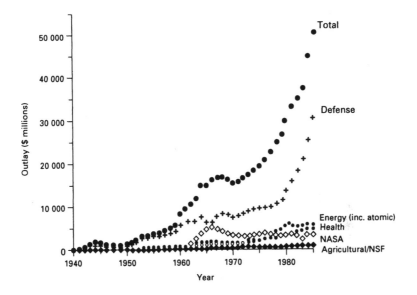

FIGURE 8.2 *US Federal outlay on research and development*

The data come from Mowery, D. C. and Rosenberg, N., *Technology and the Pursuit of Economic Growth* (Cambridge, 1989).

we are pushing back the frontiers in basic research for energy, defense, and other critical national needs.'[30]

President Reagan, in his turn, boosted defence research (the Strategic Defense Initiative or 'Star Wars'), which President Bush cut back as part of the 'peace dividend'. But Bush did, however, expand considerably the federal support of pure science (see Chapter 11). President Clinton took office committed to a further expansion of federally funded research (Robert Solow persuaded him that it would generate wealth) but we do not yet know if he will succeed in raising the budgets (readers of this book will, however, know that a raise will *not* increase wealth). The election of Newt Gringrich to the speakership of the House does, however, threaten the federal science budgets (and readers of this book will know that cuts in those budgets will not harm American science).

As the Russian threat fades, the prospects for the US government's funding of science are now looking poor. In 1993, Congress cancelled the Superconducting Super Collider (see Chapter 12) and President Clinton's proposals for the 1995 fiscal budget, presented on 7 February

1994, effectively cut back on research, once allowance had been made for inflation.[31] But President Clinton did establish a National Science and Technology Council in 1993, and in 1994 his administration did issue a policy statement *Science in the National Interest* which proclaimed a desire to increase the percentage of GDP spent on R&D to 3 per cent. But Congress may cut the science budget as it addresses its obligations to reduce the federal deficit (pork barrelling excepted).

Let us review the consequences of the federal government's post-1940 research policies. In 1940 (see Table 8.1) R&D in the USA was dominated by the private sector, which spent $265 million out of the total of $346.1 million. The government only spent $81.1 million, and this concentrated on two sectors, defence and agriculture (successive administrations having adopted agricultural R&D as a legitimate government expenditure to buy the farmers' votes). The bulk of civil R&D was funded by industry ($234 million in 1940) and basic science, which represented about 10 per cent of the entire civil R&D budget, was dominated by the private sector. Let us review the consequences of government funding post 1940.

Research and Development

The story can be simply told. In 1940, private industry dominated this sector, but after 1940 the federal government moved in massively, largely to fund defence R&D but also to fund space, health, nuclear research and to maintain its agricultural research (Figures 8.1 and 8.2). But, after the war, private industry also continued to expand its budgets, and private and federal expenditure have marched more or less together; the government concentrating on defense R&D and private industry concentrating on civil R&D (Figure 8.3).

Basic Science

In financial terms, this sector is always dwarfed by total R&D (see Figure 8.4, see also in Figure 8.2 how low runs the line for the NSF, the federal agency that funds academic science). None the less, the sums spent on pure science are not trivial (Figure 8.5).

The story of basic science funding since 1940 can be told reasonably simply. In 1940, the private sector dominated, providing about 90 per cent of all the money. There were three major sources of private money: (i) the charitable trusts such as the Carnegie Institution, the Guggenheim Foundation and the Rockefeller Foundation; (ii) the en-

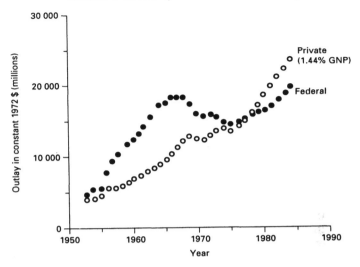

FIGURE 8.3   *US national expenditure on research and development*

The data come from Mowery, D. C. and Rosenberg, N., *Technology and the Pursuit of Economic Growth* (Cambridge, 1989).

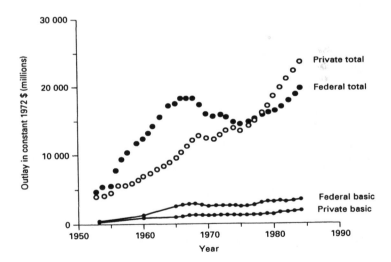

FIGURE 8.4   *US national expenditure on R&D and basic science*

The data come from Mowery, D. C. and Rosenberg, N., *Technology and the Pursuit of Economic Growth* (Cambridge, 1989).

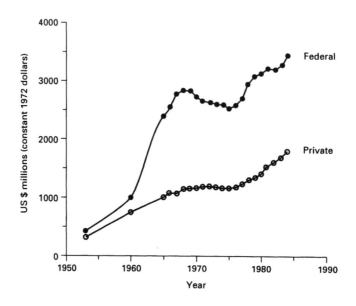

FIGURE 8.5 *US national expenditure on basic science*

The data come from Mowery, D. C. and Rosenberg, N., *Technology and the Pursuit of Economic Growth* (Cambridge, 1989).

dowments of the private universities; and (iii) private industry, which spent approximately 10 per cent of its total R&D expenditure on basic science (though that figure is an estimate). The sums in 1940 amounted to $31 million from the charitable trusts and universities together, and some $20 million from industry. In consequence, basic science in the USA pre-war was very strong. The foundations supported literally hundreds of research fellowships in universities, and the foundations also provided complex and expensive equipment such as major telescopes and state-of-the-art cyclotrons.[32] Industry, too, supported major pure science. The laboratories of American Telephone and Telegraph, Eastern Kodak, Du Pont, Corning Glass Works and Westinghouse were distinguished, and as early as 1900 General Electric dedicated a research laboratory entirely to pure science.

After 1940, however, government funding for basic science started to increase, and by 1953 it had overtaken the private sector's (Figure 8.5). Thereafter it soared, to peak in 1968, but it levelled out and even dipped a bit after that year, only to resume a more gentle expansion. The private sector's basic science, meanwhile, has grown more steadily.

The fluctuations in federal support for basic science are instructive. One reason for its decline after 1968 was the decline in NASA's budgets, but a more important reason intervened – government disenchantment. The post-1940 federal expenditure on civil R&D had not, in itself, represented a fundamental break with past policies. If defence, health, energy and space were to be considered the proper responsibilities of the federal government, then their mission-directed research represented nothing more than an appropriate subvention. But after 1950 the federal government discarded its *laissez faire* policies of nearly two centuries, and took on the responsibility for basic science.

In *Science – The Endless Frontier* (1945) Vannevar Bush had stated: 'Basic research leads to new knowledge. It provides scientific capital. It creates the fund from which the practical applications of knowledge must be drawn.' Government agreed. Administrations since 1940 have been avowedly Baconian. It will be remembered that in his Special Message to Congress of 6 September 1945 President Truman said: 'No government adequately meets its responsibilities unless it generously and intelligently supports and encourages the work of science in university, industry and ... its own laboratories.'[33] In announcing his National Science Board veto of 6 August 1947, President Truman had none the less said: 'Our national security and welfare require that we give direct support to basic scientific research and take steps to increase the number of trained scientists.'

Truman had been influenced, not only by Vannevar Bush and by Senator Kilgore, but also by John R. Steadman, his scientific advisor, who published *Science and Public Policy: A Report to the President* in five volumes between 1946 and 1947. Page 3 of the first volume stated: 'The security and prosperity of the United States depend today, as never before, upon the rapid extension of scientific knowledge. So important, in fact, has this extension become to our country that it may reasonably be said to be a major factor in national survival.'

The huge increases in federal support for basic science seen in Figure 8.5 reflected such beliefs, but during the mid to late 1960s Washington suddenly realised that after two decades of huge government investment, the American economy had failed to take off. Basic science had been sold to government by the scientists as a critical ingredient in economic growth, but where was this growth? Figure 8.6 illustrates the situation. Between 1820 and 1979 the American economy has grown at a linear rate of about 2 per cent GDP *per capita* per

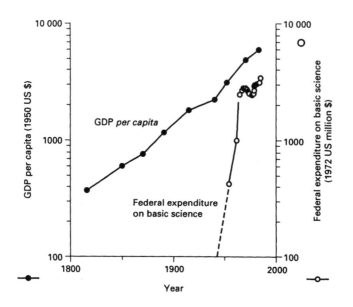

FIGURE 8.6   *USA GDP* per capita *and federal expenditure on basic science*

The federal expenditure on basic science comes from Mowery, D. C. and Rosenberg, N., *Technology and the Pursuit of Economic Growth* (Cambridge, 1989) and the GDP *per capita* data from Maddison, A., *Phases of Capitalist Development* (Oxford, 1982).

year. Neither the sudden intervention of government funding for basic science in 1940, nor its soaring peak in 1968, nor its subsequent small decline or revival, seem to have done anything to the economy.

Moreover, it was not only economic growth that had not been delivered: where were the promised benefits to health? On 15 June 1966, on launching Medicare, President Johnson complained about the irrelevance of all the pure science being carried out at the National Institutes of Health: 'Now actually a great deal of basic research has been done. I have been participating in the appropriations for years in this field. But I think the time has now come to zero in on the targets by trying to get this knowledge fully applied. There are hundreds of millions of dollars that have been spent on laboratory research that may be made useful to human beings here if large-scale trials on patients are initiated in promising areas. Now Presidents, in my judgment, need to show more interest in what the specific results of medical research are during their lifetime, during their administration. I am going to show an interest in the results. Whether we get any or not I am going to show an interest in them.'[34]

But it was the economic argument that galvanised Washington. Officials started to doubt the Baconian model. Did basic science really breed applied science? The Federal Government's largest science funder, the Department of Defense (which had spent nearly $10 billion on research development beween 1945 and 1965) commissioned a vast study, *Project Hindsight*, in which thirteen separate teams of scientists and engineers analysed the factors that had contributed to the development of twenty crucial weapons systems. *Project Hindsight* isolated seven hundred research 'events' that had led to those twenty weapons systems, and of those seven hundred only two had arisen from basic science. Two! *Project Hindsight*, whose final report was published by the Office of the Director of Defense Research and Engineering in 1969, concluded that applied science built on applied science, not on pure science. Indeed, the evidence suggested that, if anything, it was the unexpected discoveries made by technologists or engineers that boosted pure science, rather than the other way round.

The scientists and their pressure groups fought back of course. It was explained that if even more money had been poured into basic science then there would have been an economic boost. Moreover, there would have been no economic growth at all but for the money Washington had spent. President Johnson's attack on pure medical research was dismissed as ill-informed and philistine (no one actually addressed his point that life expectancy was barely rising), and as for *Project Hindsight*, well, the National Science Foundation promptly published *Project TRACES*[35] which proved to the scientists' complete satisfaction that all technology developed from scientific discoveries (but which defeated its case for government funding by having to go back fifty years to find its scientific discoveries, to a time when basic research was privately funded).

The debate continues in the US.

UNITED KINGDOM SCIENCE POLICIES

The evolution of British governments' science policies are remarkably similar to the Americans' – Britain was *laissez faire* until war and a Soviet-inspired conversion to Baconian thinking jolted successive administrations into taxing the nation to fund a government programme for science – to no economic benefit.

We saw in Chapter 6 how Whitehall's bitter experiences with Babbage scared Government from further entanglements with science. Thereafter, Whitehall refused to have anything to do with research. Just like its American counterparts, Whitehall accepted responsibilities for certain functions which, in consequence, involved the support of research, but this was always mission-orientated. A celebrated example was the chronometer. During the early eighteenth century the numbers of shipwrecks rose alarmingly, in Royal Navy as well as merchant navy ships. So in 1714 the Government established the Board of Longitude, to reward with £20 000 the inventor of a really exact chronometer. By 1735, John Harrison, a watchmaker, had developed an excellent one, and in 1762 a ship sailed to Jamaica carrying a Harrison chronometer that only registered a 5 second error for the voyage. (Yet, in a classic example of Shakespeare's 'insolence of office' the Government delayed payment until 1773, and then only paid because it was threatened with unfavourable legal judgements. Ironically, one of the arguments the Government employed for not rewarding Harrison was that the free market was already doing so, as the commercial need for an accurate chronometer was only too obvious to a maritime nation, and he was making a good income from his sales. The Government's intervention into the development of maritime technology had been gratuitous.)

Moving to the nineteenth century, we find that by 1869–70 the Science and Art Department had a budget for the year of over £225 000 and the Civil Estimates for 1869–70 showed that all government departments together spent nearly £400 000 on science, but this money only went on museums, gardens, observatories and other scientific work that supported government functions.[36] Pure science was not recognised as a government responsibility. From the 1820s, men like Charles Babbage and David Brewster continued to campaign for government support, but Whitehall's response was only nominal. Knighthoods and other honours were bestowed on scientists freely, and they were very gratefully received indeed, but of more substantial support came there little.

In 1849, Parliament voted an annual grant-in-aid of £1000 to the Royal Society, but that was just a token of esteem. Even allowing for subsequent inflation, it was a trivial sum of money. Gradually, however, the pressure for greater support from government began to build up. The decisive event was the 1867 Paris Exhibition which, as we saw in Chapter 7, was widely if wrongly believed to have exposed

British scientific weakness. Two of the English jurors returned from Paris determined on political action; Lyon Playfair, as we have seen, wrote his famous open letter to the *Journal of the Society of Arts* demanding the public funding of science and of education, while at the British Association meeting in Norwich in August 1868, another juror, Lieutenant-Colonel Alexander Strange, a little-known, ex-Indian Army officer and amateur astronomer, delivered a paper entitled 'On the Necessity for State Intervention to Secure the Progress of Physical Science'. This aroused immediate support. There already existed an informal association of young scientists, the so-called X-Club, which met regularly to complain of how badly paid they were and how short they were of research money. These men seized on Strange's address to persuade the British Association to create a committee of enquiry into the state of British science.

The British Association obliged and created a committee, and since its membership was largely made up of members of the X-Club (John Tyndall, FRS, physicist; Edward Frankland, FRS, chemist; Thomas Hirst and Thomas Huxley, FRS) its conclusions were hardly surprising; namely that the Government should fund science. The British Association then approached the Government for a Royal Commission to study the problem of science funding. Gladstone, then Prime Minister, obliged, and he appointed William Cavendish, the seventh Duke of Devonshire, to the chair.

The Commission's conclusions were preordained by its composition. The chairman, Cavendish, was a passionate supporter both of science and of the universities, being successively Chancellor of London University (1836–56) and the Chancellor of Cambridge University (1861–91). Of the Commission's eight other members, two, Huxley and Lubbock were already members of the X-Club, two more, William Sharpey FRS, the physiologist, and William Miller FRS, the chemist, were professional scientists who were trying to build up the University of London into a research institution, and of the others, men like Bernhard Samuelson (ironmaster, FRS, 1881) were well-known for their public advocacy of government funding for science. Sure enough, in 1875, the Commission concluded that there should be . . . government funding for science.

Independently, other scientists kept up the pressure. The most powerful advocate was Norman Lockyer FRS, an astronomer who was employed by the Science and Art Department. In 1869 he founded *Nature*, the science journal. Although *Nature* published scientific discoveries, Lockyer primarily forged it as a political platform. Thus he proclaimed in 1872 that 'England, so far as the advancement of knowledge goes, is but a third-rate or fourth-rate power.'[37] Tragically, things got even worse, and in 1873 he announced that: 'It is well known to all the world that science is all but dead in England',[38] which must have come as a shock to Clark Maxwell and William Kelvin, then the world's leading physicists, and to Darwin and Huxley, then the world's leading biologists (*Nature* still publishes this sort of rubbish – on 22 March 1990, John Maddox, the editor, proclaimed that the British Government had 'all but killed off' British research.[39] Maddox wrote this when Britain was second only to the USA in the numbers of scientific papers it published, and in the numbers of citations those papers attracted; see Chapter 11).

The Royal Commission, *Nature* and other advocates of government money for science wanted more salaries for scientists. As a correspondent explained in *Nature*: 'It is the question of scientific careers that is the pressing one, and the one most difficult to settle.'[40] William Crookes, one of England's greatest chemists and physicists, wrote: 'It has become more difficult for a poor man, unaided, to win his way to eminence.'[41] Everyone agreed that science was becoming more middle class, less aristocratic, and that in consequence scientists need salaries, but not everyone agreed that this justified the intervention of the state. No less a giant than Alfred Russell Wallace, the co-discoverer with Darwin of evolution through natural selection, observed that the free market (which he called 'public competition') was providing more and more jobs in science, and at a greater rate and under more honorable conditions, than government could ever provide: 'Experience shows that public competition ensures a greater supply of the materials and a greater demand for the products of science and art, and is thus a greater stimulus to true and healthy progress than any government patronage.'[42] As Wallace noted, someone like Faraday would have been startled to learn that in his day poor men had it easier in science. It was clearly not true.

Gladstone, a Liberal who believed in the free market, had no difficulty in dismissing the scientists' arguments as special pleading. He failed to see why society should be taxed to pay for the interests of an educated elite, particularly when private endowments were continually creating more opportunities for the poor. After all, Gladstone argued, no one had to become a scientist; if people chose to do science, they had to face the consequences. Many scientists agreed. In 1877, William Spottiswoode, the Treasurer of the Royal Society, opposed the introduction of government grants because it would encourage men 'not yet of independent income to interrupt the business of this life' for the sake of science.[43] Gladstone argued that the taxpayer was not a cow to be milked by anyone who wanted to pursue a personal interest. If interested individuals like Cavendish, a Duke, wanted to support science, let them dip into their own pockets (and, to be fair to Cavendish, he did, creating the Cavendish Laboratory at Cambridge, the institution that has subsequently won more Nobel Prizes than any other – a wonderful tribute to the private funding of science).

But in 1874 Gladstone lost the General Election to Disraeli and the Tories. The Tories were less committed than the Liberals to the free market, but much more committed to the nation state. They thought in terms of superpower rivalry, and they saw science as a national resource. The scientists played on their fears. Thus Sir Edward Frankland FRS told the Devonshire Royal Commission that his bibliometric analyses had shown that Germany was publishing six times as many scientific papers as Britain, and that in consequence Germany was bound to overtake Britain both economically and militarily. This horrifying figure was repeated time and time again by science activists though, as Roy Macleod pointed out in his superb 'The Support of Victorian Science'[44] no one has ever checked Frankland's improbable findings.

The most influential advocate of science as a national resource was George Gore, a Birmingham industrial chemist. In 1872, Gore told the Social Science Association in Plymouth: 'Scientific discovery and research is national work and it is the duty of the State to provide and pay for it . . . because nearly the whole benefit of it goes to the nation and scarcely any to the discoverer, and because there exists no other means by which scientific investigators can be paid.'[45]

Little of what Gore said was true. In his day, successful scientists could be well paid. Faraday, even earlier, had been swamped with consultancy work, much of which he had refused, and Gore himself earned his living as an industrial chemist, having set up his Institute for Scientific Research in Birmingham to undertake chemical analyses. Moreover, the expansion in the universities of the later nineteenth century provided an ever increasing number of privately funded positions for scientists (for which Gore was not eligible since he had no formal qualifications, which embittered him). But Gore's blatant nationalism caught the Tory mood, particularly that of two Tories in high office, Salisbury and Derby, who were themselves amateur scientists, and who also believed that the Government should fund science.

Derby paraphrased Bacon. This is what, as Rector of Edinburgh University, he asked the students during his inaugural address of December 1875: 'How many conquests of man over Nature would be secured if, I do not say a numerous body, but even for some 50 or 100 picked men, such modest provisions were made that they might be set apart, free from other cares, for the double duty of advancing and diffusing science. . . . Whatever is done, or whoever does it, I think more liberal assistance in the prosecution of original scientific research is one of the recognized wants of our time.'[46] Derby never explained why 50 or 100 was so special a number. Why not 500 or 5000? Or five or ten? Perhaps it was because there were already 50–100 picked men engaged in the double duty of advancing and diffusing science in British universities at the time, so it seemed a good idea to double their numbers. Whatever the reason, a few months later, in 1876, the Tories voted £4000 a year for the Science and Art Department to fund pure science. The government funding of science had returned to Britain.

Between 1876 and 1913, however, that sum did not rise. What did rise was a tremendous backlash. Public opinion during the later Victorian and Edwardian periods turned against science for religious, aesthetic and ethical reasons. Christians were appalled at the damage that the theory of evolution had perpetrated against their faith. The Metaphysical Society was created in 1869 to 'unite all shades of religious opinion against materialism' and moral reformers like Frances Power Cobbe accused researchers of trying to supplant religion with a 'priestcraft of science'. Even the poets joined in. Tennyson wrote *Lucretius* and *Higher Pantheism* as attacks on reductionism and Lewis Carroll wrote *Fame's Penny Trumpet* to mock the science activists.

The scientists' triumphalism offended the champions of spirituality. In 1881 Gladstone said: 'let the scientific men stick to their science and leave philosophy and religion to poets, philosophers and theologians'. The lack of any spiritual or ethical dimension to science offended those like Ruskin who believed that ethics were more important than materialism (Ruskin opposed the building of a physics research laboratory in Oxford in 1884). Both Gladstone and Ruskin had been upset by the Royal Commission on vivisection of 1876 which had exposed some disgusting and cruel practices within university departments of physiology and anatomy. In his *Physiologist's Wife*, Conan Doyle, himself a doctor, portrayed scientists as cold, heartless and overcontrolled.[47] George Eliot introduced Dr Lydgate, *Middlemarch*'s own research physiologist, with amused contempt.

The popular revulsion against science inhibited further government support, but the scientists themselves had turned against State funding. A typical article by Richard Proctor, the astronomer, was entitled 'The Dignity of Science'[48] and it condemned the 'noisy public begging'. In his *Wages and Wants of Science Workers* (1876) Proctor deplored 'scientific Micawberism'.[49] An 'FRS', writing anonymously as was then often the custom, bemoaned in *The Times* of 16 September 1876 'the wretched whining cry for State funds in aid of research'. Scientists noted that the Government's new money had not actually yielded anything very interesting, and in 1880 a number of distinguished scientists created the Society for Opposing the Endowment of Research (the secretary was William Noble, FRS).

On 25 February 1881 no less a person than Sir George Airy, the Astronomer Royal, outlined the Society's policy in the *English Mechanic*: 'Successful researches have in nearly every instance originated with private persons, or with persons whose positions were so nearly private . . . that the investigators acted under private influence, without the dangers attending conjunction with the State. Certainly I do not consider a Government as justified in endeavouring to force at public expense investigations of undefined character, and at best of doubtful utility; and I think it probable that any such attempts will lead to consequences disreputable to society.' Sir George doubted that the Royal Society's grant had actually yielded anything useful. As *The Times* noted on 23 December 1886, the best research continued to be privately funded, and Government grants had 'not been so conspicuously successful as to create a very strong desire for their extension'.

But unquestionably the strongest revulsion was felt at the behaviour of Lockyer, the editor of *Nature*, the employee of the Science and Art Department, and the leader of the campaign for government money. In 1876, the Government voted the Science and Art Department its £4000, and what did the Department do with it? It created a full-time research post, and a new solar physics laboratory, for . . . Norman Lockyer its employee! Just as in America, it was to be the public advocates of government funding who were to most benefit from it.

Or was it? The transparent croneyism of the award shocked everybody. *English Mechanic*, the science magazine with the highest circulation of the day, condemned both Lockyer and the Department in its issue of 7 January 1876, and it denounced the Devonshire Royal Commission as 'mere tools and cats' paws of a needy and designing confederacy' led by Strange and Lockyer. In his *Wages and Wants of Science Workers* Richard Proctor dissected the individual motivations of each member of Devonshire's Commission, and showed just how biased and self-serving its conclusions were. The public outrage was so fierce that the Government took fright, retrieved its money and transferred the £4000 as a further grant-in-aid to the Royal Society.

The Royal Society accepted the tainted money with reluctance, but thereafter it did try to distribute it fairly. Researchers' grant proposals were judged competitively and efforts were made only to fund the best applications. It was widely noted, however, that many of the grants went to men like William Thomson (Lord Kelvin) and William Crookes who themselves sat on the Royal Society's Grant Committee (to be fair to these men, they did stand down when their applications were considered, but it bred cynicism). Distinguished scientists refused to join the Committee: William Flower, the Director of the Natural History Museum, refused to join. He said: 'The large increase of this method of subsidizing science, accompanied as it is with the (as it appears to me) humiliating necessity of personal application in each case, must do much to lower the dignity of recipients and detract from the independent position which scientific men ought to occupy in this country.'[50]

The *English Mechanic* kept up its criticisms, and on 22 July 1881, noted: 'It is commonly acknowledged that the Royal Society's government grants were awarded in a lax and unbusiness way.' In the end, the Royal Society rued the day it had ever taken the Government's money, and when A. J. Mundella, the education minister (technically the Vice-President of the Privy Council) offered a further £2000 a year to the Royal Society in 1881 to fund research fellowships, the Society refused it with a collective shudder. The Society had come to agree with Richard Proctor: 'Few circumstances have caused true lovers of science in recent times more pain than that outcry for the Endowment of Research' and the money was sent back to the Treasury.

It did not matter; the private endowments for science continued to pour in. When Alfred Mond endowed the Davy–Faraday Laboratory at the Royal Institution, *The Daily Telegraph* noted on 4 July 1894 that the Mond request had done 'more for physical science at one stroke than all the Cabinets of Her Majesty have done since the commencement of that reign, one of the greatest glories of which has been the advance of research in England.' When the Imperial Cancer Research Fund opened in 1901, and the Jenner (now Lister) Institute for Bacteriology soon after, England's privately funded biological research soon matched, if it did not outstrip, its physical.

As ever, it was war that provided the impetus for the government funding of science in Britain. As early as 1868, activists were blaming Britain's difficulties in the Crimean War on the Government's lack of support for higher education. Thus Matthew Arnold wrote 'Our breakdown at the Crimea [1854–6] is distinctly traceable to the ineffectiveness of our superior [i.e. higher] education.'[51] (It was an improbable argument; Lord Cardigan, who led the greatest of Britain's Crimean disasters, the charge of the Light Brigade, was educated at Oxford, Arnold's own adored *alma mater.*)

The triggering event for science was the Boer War (1899–1902) during which the Army had rejected 40 per cent of all potential recruits as unfit. This shocked the War Office, which as an immediate measure lowered the minimum height for infantrymen (1902) but its agitation over the physical deterioration of the British led to the creation of the Inter-departmental Committee of the Home and Education Departments and the Local Government Board. Although the Committee generated little, it provided a forum which helped change official culture. In particular, the Committee looked to Germany.

Ever since 1883, Bismarck had been introducing a Welfare State. Old age pensions, unemployment benefits and medical insurance were all introduced into Germany while Britain remained *laissez faire*. Bismarck, of course, was buying off his people: in a Faustian pact, he offered them economic growth and social security in exchange for their political rights. The British, who enjoyed full political rights, preferred not to pay the taxes. Before 1914, the British state only sequestered about 10 per cent of GDP in taxes (currently it is 40–50 per cent). But before 1914 the British preferred to keep their own money and to arrange their own social security. As Harold Perkin has shown, by 1900 well over half of all working men belonged to the so-called Friendly Societies, voluntary insurance associations which offered medical and unemployment benefits to their members in return for a weekly payment.[52] Nor were the poor neglected: they received free medical care in the voluntary hospitals which local philanthropists had built in all centres of population. But the threat of war with Germany, which hovered over Britain for many years before 1914, lent urgency to the War Office: the German Army was vigorous, the Germans were apparently healthy, the Germans had state health insurance. The equation seemed clear to the military mind.

The British Liberal Party, in the person of David Lloyd George, the Chancellor of the Exchequer, responded sensitively to the War Office. The Liberals were giving up *laissez faire*. The rise of socialism and of the Labour Party was encouraging them into central planning and State control. In his famous, so-called People's Budget, in 1909, Lloyd George proposed to introduce State health insurance, social security and old age pensions. Since these measures were to lead to the creation of the Medical Research Council, we will dedicate the next two paragraphs to describing the politics of their implementation.

The Budget of 1909 was rejected by the House of Lords as unconstitutional, because it purported to be a Finance Bill but was, in fact, a vast social programme that had not been explicitly presented to the electorate at a general election. As F. E. Smith, a leading Tory, explained: 'If the Lords should insist that the people of England pronounce upon the Bill before it becomes law, they will have the most complete Constitutional justification before their action. The rule – imperfectly established and depending only upon resolutions of the Commons – that the Lords shall not interfere with finance has always been qualified by the condition that the Budget shall provide for the finance of the year, and for nothing else. Here again the Government has broken the golden rule that they should all tell the same lie. The Master of Elibank, an influential Minister, has announced that the Budget is a social programme for twenty years.'[53] After vast constitutional crises, however, and no fewer than two general elections in 1910, Lloyd George had his way, and state-run national medical insurance was introduced in the Act of 1911.

Despite the fuss, the Act, in reality, changed little for ordinary working people. First, it only covered working men; women and children received no medical insurance under the Act. Secondly, it only covered men who could afford the weekly levy of 4*d* – the really poor, who could not afford the premium, depended on the voluntary (charity) hospitals as before. But third, and most important, it changed little because the majority of working men already had health insurance. If, by 1900, well over half of all working men were already members of Friendly Societies, the proportion in 1911 was even greater. Between 1900 and 1910, new members continued to join the friendly societies at the rate of 140 000 a year,[54] a rate which progressively increased. By 1910, over 9 million people already belonged to friendly societies, so Lloyd George's Act, which covered a total of 12 million workers, served only to include rather less than 3 million new workers,[55] rather less than a quarter of the workforce – a quarter, which was in any case accelerating the rate at which it was voluntarily joining. Indeed, so complete was the hold of the Friendly Societies, that Lloyd George had to use them to administer his national insurance for him, i.e. all he actually did was, effectively, to nationalise them. He forced them to standardise their weekly levies, and he determined how they were to pay the benefits. From the Act of 1911, however, and from the insurance premiums that now flowed into the State's coffers, Lloyd George created the Medical Research Committee (later Council) which first met in 1913. Its first year's budget was £56 000. It was specifically charged to research into tuberculosis.

Landsborough Thomson[56] and Linda Bryder,[57] the historians of the MRC, have both wondered, innocently, why the MRC was created specifically to research into tuberculosis. The problem that troubled the two historians is that, in 1913, it was clear that much research was already being done on tuberculosis by three voluntary organisations, the National Tuberculosis Association (USA), the National Association for the Prevention of Tuberculosis (UK) and the Tuberculosis Society, later Association (UK). Moreover, the Germans were very active in TB research, Koch having discovered the causative bacterium in 1882. As Simon Flexner, the Director of the Laboratories of the Rockefeller Institute for Medical Research in New York, explained in 1912: 'at the Rockefeller Institute they believe in following leads and at present there are no leads in tuberculosis research'.[58] Indeed, the MRC, once formed, chose to ignore tuberculosis research. So why was it created with that particular mission? The answer, of course, is obvious. David Lloyd George's father had died of the disease. Lloyd George was one of those politicians who treat the organs of the state as their own private fiefdom.

Cruelly, the intervention of the State may even have aggravated the problem. Tuberculosis is a disease of poverty, and its incidence had been steadily falling throughout the nineteenth and twentieth centuries as standards of living rose. Indeed, so crucial are standards of living to the disease, and so curiously irrelevant are advances in clinical medicine to its epidemiology, that by the time effective cures were finally discovered (streptomycin in 1947 and isoniazid in 1953) the incidence of tuberculosis had fallen to a tiny fraction of its former figures and, moreover, the introduction of the drugs seems not to have hastened its eradication.

Tuberculosis is a disease of poverty, yet Lloyd George's 1909 People's Budget threatened the creation of wealth. To pay for his budget, Lloyd George proposed taxing wealth; but when capitalists find their capital threatened, they export it. As F. E. Smith explained at the time: 'Now, gentlemen, I am not on principle opposed to a super tax. I shall have to pay this, and I hope I shall do so cheerfully. But I am arguing one point, and one point only – are we, or are we not, taking all the new burdens upon capital together, piling them to such a height as must aggravate unemployment? And here let me remind you in passing that already before this budget, but under this Government, more English capital was invested abroad in one year than in any single previous year in the whole history of British finance. And remember always you [the working classes] suffer from this – not the capitalist.'[59]

The flight of capital from Britain had long worried contemporaries (though economists these days are more sanguine). Nevertheless, by 1913 Britain's foreign assets were equivalent to 180 per cent of its GDP.[60] When Chancellor of Exchequer in 1906, Asquith had refused to tax wealth to prevent more leaving. Yet, as Prime Minister (1908–16) Asquith found himself too weak to prevent his Chancellor from introducing measures he knew to be disadvantageous. The Tory alternative was even worse – tariff reform. In practice, this meant taxes on food – nothing could have aggravated tuberculosis more than to price the poor out of food. Thus both major parties, while claiming to be helping the poor, actually proposed and implemented measures that aggravated poverty.

In the event, on being founded, the MRC ignored tuberculosis. Instead, it became the instrument of Walter Morley Fletcher, its Secretary from 1914 until his death in 1933. Fletcher, who tried to direct all medical research in Britain, was a monster (even his hagiographer T. Elliot admitted that he was 'too trenchant in his denouncement of others').[61] We have seen how the advocates of the government funding of science, men like Vannevar Bush or Lockyer, invariably aspired to become czars of science. But the politics in which they had to engage, the high profile they attained and the countervailing resistance they aroused, inexorably forced accountability on them, and they frequently saw their pet schemes frustrated and their ambitions dashed. No such separation of powers ever troubled Fletcher; the Government had generated the political will to create the MRC, and Fletcher slipped in with all the anonymity of a civil servant. But he suffered none of the civil servant's constraints, because the organisation to which the MRC, and Fletcher as its Secretary, formally reported, the Committee of the Privy Council for Medical Research, only met once during Fletcher's lifetime. Fletcher, therefore, ruled the MRC with absolute personal authority, and his ego expanded like a monstrous weed, strangling medical research in Britain.

Fletcher, a respectable if not inspired physiologist from Cambridge, had not been an obvious choice for the secretaryship of the MRC, but the first, second and third choices had all turned the job down. Once in post, Fletcher set out to realise his vision of medical research in Britain. He believed, passionately, in a number of principles. First, he believed that he, and only he, should control all medical research in the country. Fletcher loathed the medical charities, whose autonomy outraged him. In 1918 he stated: 'The Medical Research Committee . . . is able to organize and coordinate work at every centre of medical research in the Kingdom. . . . If any fresh private endowment, given generally for medical research, is available, it will be very important either that it should be administered by the Medical Research Committee, or, if it be under a separate trust or management, that this should be in close relation to the Committee's work.'[62]

Unfortunately for Fletcher, some medical charities persisted in their independent research, so Fletcher tried to prevent any more from being founded. When a group of doctors and laymen first proposed creating a Cancer Research Campaign, Fletcher wrote to them forbidding them from proceeding: 'I ought, I think, to remind you that the Medical Research Council is the body specially charged by the Government and Parliament with the duty of supporting and encouraging work in all branches of medicine.'[63] But despite all his threats, Fletcher could not stop them – thank God. By 1991 the Cancer Research Campaign (CRC) was spending £44 million on research, making it the fourth largest funder of medical research in Britain after the MRC (£200 million), Wellcome Trust (£77 million) and The Imperial Cancer Research Fund (£50 million). The CRC's £44 million represented no less than 10 per cent of Britain's entire medical research in 1991 (the 73 member charities of the Association of Medical Research Charities raising £266 million between them) which would have rendered its loss a terrible tragedy had Fletcher prevailed.

Fletcher loathed and despised doctors, which was an unfortunate attitude in the Secretary of the Medical Research Council. When the clinicians who intended to create the CRC refused to desist, Fletcher wrote: 'a committee of eminent clinicians will . . . be perfectly useless, as well as highly embarrassing because clinicians do not possess the requisite education either to add to or even to supervise work which demands highly trained biologists'[64] (there was a personal element to Fletcher's hatred of clinicians; many of the candidates who had been preferred over him for the secretaryship of the MRC had been medical men like Sir George Newman).

Under Fletcher, the Medical Research Council broke off all contact with the Royal Colleges, the professional organisations of doctors, because they were 'unfit' and 'incompetent'. He and his friend Gowland Hopkins used to laugh over the 'idiocy' of the Royal College of Physicians. But these attitudes did terrible damage to medical research in Britain. By deliberately divorcing basic scientists from clinicians, by only funding basic science and by refusing to fund clinical science, Fletcher was destroying the very *raison d'être* of medical laboratory science, namely its interrelationship with clinical practice. George Ellery Hale, for example, was shocked by Fletcher. So were British doctors.

Lord Moynihan, the truly distinguished President of the Royal College of Surgeons, denounced the MRC for its 'deplorable and sinful indifference to medical research' and for its 'disdain for clinical activities' and its threat to 'divorce' basic science from clinical medicine.[65] Lord Dawson, the President of the Royal College of Physicians, accused the MRC of deliberately 'minimising the value of practitioners' work in comparison with that of the research worker by adopting a lofty or superior attitude to clinicians as of "highbrows" looking down upon "lowbrows".'[66]*

Just as bad, Fletcher divorced basic science from public health. When the Ministry of Health initiated research into applied health, Fletcher refused to allow any member of the MRC to communicate with any member of the Ministry of Health. This appalling situation obtained from 1923 until Fletcher died in 1933.

The only organisation in Britain to feel gratitude to Fletcher was the Department of Biochemistry in Cambridge whose Head was Gowland Hopkins, the old friend who had put Fletcher's name forward for the MRC. In an orgy of mutual back-slapping, Fletcher poured money into Hopkins's department, justifying his choice by claiming that only through Cambridge University could British medicine advance. In particular, Fletcher maintained that only Cambridge could successfully research into the biochemistry of vitamins, which were then in the forefront of medical science.

Yet Cambridge biochemistry failed Fletcher. Flushed with MRC money, Hopkins found he could afford to stop researching on vitamins and he

---

* The MRC remains high-handed and haughty. In 1988, when closing some of its units, it did not bother to inform the directors of those units, who first discovered it from the newspapers (see the editorial, *Lancet*, vol. ii, pp. 19–20, 1991). When the directors complained publicly, they were reprimanded (see the editorial in the *Lancet*, vol. ii, p. 200, 1988). The director of one unit was instructed to discipline his own wife for a letter of protest she sent to *Nature* (Concar, D. and Alhouse, P., 'Storm Over Funding Policy', *Nature*, **348**, p. 6, 1990).

   In 1992, to 'counter the belief held by many clinicians that the MRC does not either recognise the problems of, or adequately support, clinical research' the MRC actually devoted a whole issue of *MRC News* (no. 57, December 1992) to clinical research. Big deal. Despite being invited by the issue's editor, Professor C. R. W. Edwards of the Department of Medicine, Edinburgh University 'to be grateful for the role that the MRC has played in promoting clinical research' many clinicians would rather reserve their gratitude for the Wellcome Trust, the medical charities, and the drug companies.

indulged himself in obscure studies on glutathione, his personal hobby-horse, and he appointed to his department characters like Haldane, Needham, Quastel and Whetham. These men made useful, if recondite, contributions to science, but they increasingly devoted their energies to extra-curricular activities. Haldane became the Communist Party's most vocal public advocate, and he wrote more for the *Daily Worker* than for the *Biochemical Journal*. Needham, another Marxist (of the Maoist persuasion) dedicated his life to proving that Chinese science had long been better than that of the West (a thesis he never tried to prove by actually emigrating to China) while Quastel and Whetham dissipated their energies in frivolous activities like the house journal *Brighter Biochemistry*, which was full of 'jokes' like these confessions of Haldane's in the issue for 1929/30:

When are you at your best? My optimum temperature is 0°C, my optimum pH is a question for L. J. Harris, being obtained by saturating a 25 per cent aqueous solution of ethanol with CO_2. (Yes, Mr Editor, you can get iced champagne for 4/- a bottle in Paris).

When are you at your worst? There is no evidence that the depths of my potential iniquity have been plumbed. But probably at the end of an hour's lecture in French, when my notes have only lasted for 45 minutes.

Do you think life worth living? Yes, but I do not think that the majority of resting bugs, dons and bacteriophages are alive. My answer only applies to higher organisms.

These activities were so far removed from what working men believed they were getting from the Medical Research Council for their weekly levies of 4d that even Fletcher became alarmed. Even he knew that basic science had, ultimately, to justify itself through clinical advances, yet Hopkins's abandonment of vitamin research threatened the MRC's strategy. As Fletcher wrote in 1927 to the physiologist A. V. Hill: 'I told Hopkins that, having somehow bagged the credit for inventing vitamins, he spends all his time collecting gold medals on the strength of it, and yet in the past ten years has neither done, nor got others to do, a hand's turn of work on the subject. His place bristles with clever young Jews and talkative women, who are frightfully learned about protein molecules and oxidation–reduction potentials and all that. But they all seem to run away from biology. The vitamin story is clamouring for analysis.... Yet not a soul at Cambridge will look at it.'[67] Fletcher was too arrogant to cut the MRC's funding for Hopkins's department, which would have admitted failure so, in 1927, he persuaded the trustees of the Dunn Estate (a private medical charity!) to endow a Dunn Nutrition Laboratory in Cambridge to do the work on the vitamins that Hopkins so resolutely refused to do.*

These incompetencies lay in the future; let us return to 1913, or rather 1914–18 because, as ever, it was war which catapulted the Government into funding science. In 1916 the Department of Scientific and Industrial Research, the precursor of the Science and Engineering Research Council was created, with a then huge budget of £1 million, the so-called 'Million Fund', to initiate research into poison gas, barbed wire and tanks, the three great needs of Flanders. The Science and Engineering Council is now the largest organ of State funding in Britain. Almost as important is the University Funding Council (now the Higher Education Funding Council), which was created in 1919 as the University Grants Committee (UGC).

The Government had for some time supported the universities in a small way. Indeed, the Scottish Government had long supported the universities north of the border, and after the Union of 1706 these commitments were continued (they amounted to £5077 in 1832, for example). The English universities, however, were self-supporting (although London University received £3320 in 1841 in recognition of its role as the central examination body for the colonies). In 1872, for example, when the Welsh University Colleges asked for support, they were told 'it had never been [government] policy to give financial assistance for the promotion of higher education in England'.[68] Eventually, however, grants had to be made in Wales (£4000 to Aberystwyth in 1882, and similar sums to Cardiff in 1883 and Bangor in 1884) because these colleges were close to bankruptcy.†

Between 1851 and 1892, eleven university colleges were founded, privately, throughout England. Typical was Mason College, later Birmingham University, endowed by Josiah Mason, a successful local

---

* Not surprisingly, advances in vitamins continued to elude the MRC. Vitamin B12, for example, was discovered by Lester Smith of Glaxo.

† The Welsh colleges have always been the weakest institutions of higher education in Britain, they repeatedly come bottom in the UGC's gradings of excellence in research, and during the 1980s University College Cardiff actually went bankrupt, having to amalgamate with another college. For a horrific insight, read Kingsley Amis's *Lucky Jim* (1954), based on his experiences as a university lecturer in Swansea.

industrialist. On laying the foundation stone in 1875 he said: 'I, who have never been blessed with children of my own, may yet, in these students, leave behind me an intelligent, earnest, industrious and truth-loving and truth-seeking progeny for generations to come.'[69]

On being founded, however, the university colleges soon united to press for government money. The academics were jealous of their German counterparts who luxuriated in vast government subventions. On 9 May 1887, representatives of the colleges met in Southampton to coordinate their campaign, and for the rest of the year they bombarded the press. On 21 March 1887, for example, T. H. Huxley wrote in *The Times*: 'We are entering, indeed we have already entered, upon the most serious struggle for existence to which this country has ever been committed. The latter years of the century promise to see us in an industrial war of far more serious import than the military wars of its opening years.' On 1 July 1887, Sir John Lubbock wrote in *The Times*: 'The claims of these colleges were not based alone on their service to learning and study; they were calculated to contribute largely to the material prospect of the country.' The academics' public argument was that unless Britain imitated Germany, its economy would be overtaken. That was a false argument: the American government supported no universities (except for some agricultural colleges) yet the American economy was outstripping both Germany's and Britain's; but Lord Salisbury was persuaded, and in 1889 an *ad hoc* Committee on Grants to University Colleges was given £15 000 to dispense between eleven colleges. By 1903, fourteen colleges were sharing £27 000 between them.

These were small sums, as no college received more than £2000, yet as early as 1872 Birmingham, for example, had a total budget of £13 089, of which about half came from fees, and half from endowments. In 1904, the Committee was given £54 000 to spend annually, £100 000 in 1905 and £150 000 in 1912, but these sums only accounted for some 10–20 per cent of the civic universities' income, and Oxford and Cambridge stayed aloof, glorying in their total independence. The war, however, as ever, changed everything. First, it changed the culture of the country. Before 1914, the universities, particularly Oxford and Cambridge, feared government intervention and despised the Germans for their technocratic, utilitarian, qualification-obsessed universities. A well known nineteenth-century Oxford ditty went:

> Professors we, from over the sea,
> From the land where Professors in plenty be,

And we thrive and flourish, as well as may,
In the land that produced one Kant with a K,
And many Cants with a C.

But by 1918, after four years of national mobilisation, the universities (all of them, Oxford and Cambridge included), could petition the government for money because, in the words of the Universities' Deputation: 'The development of the universities, no less than their maintenance, is a national duty. It is a national duty, and it is a national benefit.'[70]

War re-shapes a nation's culture. During war, a government must intervene to regulate all aspects of a nation's life, and that experience of government control legitimises *dirigisme*, central planning and dependence on the State: 'We all pulled together to win the war, so why can't we now pull together to improve things?' The answer to that question is that the empirical evidence shows that the free market, either in goods or in the philanthropic impulse, actually provides more and better than does the State, but such empirical evidence is ignored in the emotional heat and self pity of a nation recovering from the trauma of war.

A famous *Punch* cartoon that appeared towards the end of the Second World War showed a British soldier in uniform, accompanied by his wife and children, in a grocer's shop, thumping the counter with his fist, and demanding 'What do you mean we cannot have these goods? We've paid for them twice over, once in 1914 and once in 1939.' It was a similar sentiment that fuelled the post-war electoral promise of 'Homes fit for Heroes'. But these sentiments, understandable though they might have been, were violently illogical: war is desperately impoverishing and, during the early years of any subsequent peace, standards of living must fall. Much of the apparent success in 'pulling together to win a war' is based on vast government borrowings, and these national debts need re-paying post-war. Thus the UK national debt rose from £720 million in 1910 to £7828 million in 1920, and from £9083 million in 1940 to £21 366 million in 1945. The US national debt rose from $1191 million in 1915 to $24 299 million in 1920, and from $42 968 million in 1940 to $258 682 million in 1945.

People do not want to hear unhappy truths after a war. In his *History of the Labour Party from 1914* G. D. H. Cole recounts how he and others after both world wars tried to warn that, post-war, standards of living, and of services such as universities or science, had to fall, but these warnings were 'too unpalatable to be given serious attention'.[71]

For the universities in 1918, therefore, the implication of the term 'national duty' had changed. Before 1914, national duties were assumed as the responsibilities of individuals like Josiah Mason, and it was believed that the collective efforts of disparate individuals would meet the national need (the empirical evidence supported that belief); but after 1918, a national duty implied a government responsibility, a requirement for central planning, and taxes.

It is often said that Britain won the wars against Germany during the twentieth century, yet lost the post-1945 peace. In economic terms that is actually doubtful, as we discussed in Chapter 7, but in social terms it is unquestionably true. Before 1914, Britain prided itself as a nation of free individuals. The State only sequestered 10 per cent of GDP. But between 1919 and 1938, the State took around 25 per cent of GDP in taxes, and since 1945 the figure has hovered around 50 per cent. Ironically, the national effort devoted to defeating *dirigiste* Germany converted Britain into an equally *dirigiste* economy.

After 1918, the universities were in difficulties; their staff and students had been largely mobilised, fee income had dried up over four years, and they faced a flood of demobilised students. That flood could have translated into a considerable income from fees, but the universities seemed to have lost faith in their ability to pay their own way. In almost hysterical terms Sir Oliver Lodge, Principal of Birmingham University and a leader of the Universities Deputation to the Treasury, said: 'It is suggested that we might ask for a doubling of the grant now and a doubling soon. But reconstruction is in the air, demobilisation is upon us; we cannot wait; we want these two doublings put together, we want a quadrupling at once. This is what we ask for.'[72]

He got it. In 1919, the Universities Grants Committee was instituted, with a budget of over £1 million. This was not a single gift, designed to help the universities through the post-war period until they could resume an autonomous existence; this was annual, constitutive intervention on the German model. By 1921 the UGC's grant was £1 840 832. In 1936 it was £2 100 000. Since, in broad terms, over half of this money went on the science faculties, this represented a massive government intervention in science. It also changed the nature of the universities. They had effectively been nationalised. In 1914, only some 20 per cent of their income had come from the state, but by 1921, when local education authority grants were included, more than 50 per cent of the universities' income came from government bodies.[73] This proportion continued to rise because, after 1919, private donations to the universities dried up. The universities were increasingly believed to be a government responsibity and, moreover, potential donors were hard hit by the new taxes.

The war had, furthermore, robbed the universities of much of their investment income. The nineteenth century had been a period of stable prices and, as Keynes showed in his 1923 *Tract on Monetary Reform*, the general level of retail prices was the same in 1915 as in 1826. In such a non-inflationary climate, fixed-interest investments were secure and profitable, and the universities, like the other wealthy elements of society, invested heavily in the Government's debt securities called Consols (i.e. Consolidated Annuities). So secure and so profitable were such investments that Kipling could write in his *Education of Otis Yeere*: 'a woman is the only infallible thing in this world, except Government Paper of the '79 issue, bearing interest at four and a half per cent'. When, towards the end of the century, interest rates fell from 3 to 2 per cent, the capital value of Consols rose by 50 per cent, making them even better investments.

The French bourgeoisie and its institutions invested as heavily and as profitably in their own government debt issues, as Balzac's courtesans were only too aware: 'Ah! Shopkeeper, hair-oil seller that you are, you put a price-ticket on everything! Hector told me that the Duc d'Hérouville brought Josépha bonds worth thirty thousand francs a year in a cornet of sugared almonds! And I'm worth six Joséphas! Ah to be loved!, she sighed, twisting her ringlets round her fingers and going to look at herself in the glass' (*Cousin Bette*, 1847). Those who could not obtain government bonds bought railway debentures and other fixed-interest investments.

To repay their vast war-time debts, and to help pay for reconstruction, the post-war governments did not just impose new taxes such as death duties, and they did not not just hike up the old ones such as income tax which rose to 30 per cent in 1920, but they inflated and, as Keynes said in his *Tract on Monetary Reform*, the printing of money: 'is the form of taxation which the public finds hardest to evade and even the weakest governments can enforce'. Inflation destroys the value of fixed-interest investments. By 1920, the after-tax purchasing power of Consols had fallen to only 26 per cent of the value in 1914, while the capital had crashed in real terms to below their value in 1815. Thereafter, the situation deteriorated even further, and Churchill's attempt in 1925 to restore sterling's parity with gold collapsed in 1931, broken by the horrible deflation and a general strike.

The enemies of polite society rejoiced in the collapse of the *rentiers*; Nikolai Bukharin's *Economic Theory of the Leisure Class* published

in Russian in 1919 was little more than a celebration of the *rentiers'*
distress, and even Keynes in his *Treatise on Money* predicted that the
'euthanasia of the *rentier*' was both inevitable and desirable; but the
universities are the quintessential *rentiers*. If academic freedom is to
be maintained, the universities must know that the investments whose
dividends pay the academics' salaries are safe. If governments destroy
those investments through debauching the currency, they destroy the
universities, which must either go bankrupt or take government funding
(the third option, that of soliciting private funds, is impossible when
the entire middle class finds its own income threatened by rising taxes, the inflationary destruction of its own investments and the imposition of death duties which directly threatens bequests). Thus we see
how war, yet again, destroys the universities' autonomy and forces
them into dependence on the State.

Incredibly, in view of the dangers it carried for academic freedom
from government – dangers over which academics had agonised for
centuries – the universities actually wanted to be nationalised. Their
enthusiasm for government funding transcended their immediate financial problems; they were converted to the German model. At any
rate, they wanted vast government funds and, as C. H. Shian acknowledged in the title of her history of UGC, *Paying the Piper*, this effectively amounted to asking the government to pick the tune. In 1918,
for example, Sir Oliver Lodge complained on behalf of the Universities' Deputation to the Treasury that: 'The German universities are
sustained by a State Grant averaging 72 per cent of their total income.'[74]* This was a desperately dangerous argument. As Mark Walker showed in his *Science, Medicine and Cultural Imperialism*[75] the

---

*The bizarre reluctance of the British universities to aspire to fiscal autonomy, and
so academic freedom, was illustrated in 1980, when Mrs Thatcher's government
announced that it would no longer subsidise foreign students' university fees in
Britain. The universities erupted with dire warnings. Vice Chancellors and professors broke into savage print in the newspapers – no foreign student would ever
set foot in Britain again, they would all go to America or France or Germany or
Russia. Masters of Oxbridge colleges were particularly upset that the taxpayer
would no longer subsidise foreigners, which was odd because Oxbridge colleges
have never been forward with their own funds at subsidising anybody or anything.

In the event, numbers of foreign students did fall, from 17 101 in 1979 to 13 623
in 1981, but they promptly recovered (19 315 in 1985 and still climbing) and,
because their fees had risen from around £600 pa to £4750 pa, and because all
this went directly into the universities' treasuries, the universities' income from
foreign students rose from £28 million in 1978/9 to over £200 million by the end
of the 1980s. Yet it took the Government to force this real accession of autonomy
onto the universities – just as it took the Government to persuade the universities
to copy the Americans and try to appeal to alumni and other donors for endowments.

German universities have a shameful record. They supported the nationalist imperialism of Bismarck and the Kaiser, but they actively distanced themselves from the democratic struggles of the Weimar Republic, only to support the nationalism, aggression and racism of Hitler.

The German universities' support for Hitler was extraordinary. As the historian G. W. Craig noted, by 1933 'the great majority of the holders of university chairs' in Germany had joined the Nazi Party, long before they had needed to.[76] In his *Rektoral* address at the University of Freiburg, Martin Heidegger, the philosopher, called on his colleagues to recognise Adolf Hitler as the leader whom destiny had called to save the nation.[77] In another *Rektoral* address in Regensburg, Professor Gotz Freiher von Polnitz proclaimed the accession of Hitler as the 'hour of victory'.[78] In Tübingen the professor of Volksunder (folkloric studies), Gustav Bebermeyer, announced, 'Now the great wonder has occurred. The German people has arisen!'[79] The German universities had been so corrupted by their dependence on government that in 1945 over 4000 academics had to be dismissed for active Nazism; several thousand further members of the party were allowed to retain their positions.

On the return of democratic government in Germany, the universities re-adopted their studiously aloof distance from its problems. The universities, for example, were no bulwark in the fight against the Baader–Meinhoff gang – rather the opposite, in fact. The depths to which German universities can fall can be illustrated by a vignette. One of the first victims of Germany's invasion of Belgium in August 1914 was the library of the University of Louvain (founded in 1425). The burning of the library consumed 300 000 books and over a thousand medieval and original manuscripts. The world was shocked, so Adolf von Harnack (the doyen of German theology) gathered together a further 92 of the most distinguished German academics to distribute a justification of the burning of the library, and the invasion of Belgium, on the grounds that international scholarship should only survive if it supported German foreign policy.[80] In 1915, 352 of the country's most important professors, including Ulrich von Wilamowitz-Moellendorf, Eduard Meyer, Otto von Gierke and Adolf Wagner, signed the Declaration of Intellectuals that announced that it would be reasonable and just for Germany to acquire Belgium, France's channel coast and all of its important mineral areas, Courland, the Ukraine, and extensive colonial territories as the price of peace.

The German universities have rarely, therefore, elevated the search for truth above politics. Because they have been funded by *dirigiste* governments, they have identified with *dirigiste* governments, and they have prostituted themselves in the search for funds. Thus they loved the Kaiser, Bismarck and Hitler because those thugs propagated national ideologies that inflated the importance of universities in the service of the State; but the German universities distrusted the Weimar Republic and the post-war democratic FRG because those governments, in their adherence to democracy, never treated the universities with the semi-religious awe that they considered their due.

German universities have proved extraordinarily homogenous in their ideologies. Although some dissidents have emerged, the vast mass of German academics have thought as one. This illustrates how a monopoly of funding will breed a monopoly of thought. It also shows how academics will, collectively, elevate self-interest over the search for truth.*

Some British academics have understood that only a plurality of funding will feed a plurality of thought. In 1935, when state funding for universities had exceeded £2 million a year, John Murray, the Principal of Exeter University College commented: 'A university is like a man, it may gain the whole world and lose its own soul.'[81] If the universities are to be anything other than an active agent of the state, providing it with its military science and industrial technology, then they must be funded independently. In defence of such precepts, the University Grants Committee was established under the Haldane Principle, named for the distinguished Liberal politician, which theoretically preserved academic freedom. Thus the UGC was largely made up of academics, whose funding came from the State but whose decisions were autonomous. But, in reality, no government will ever surrender its control over its money. Governments will often allow, cunningly, its bodies some initial autonomy, but once they have displaced the truly independent bodies, governments will invariably exploit their monopoly powers to take over. Thus the UGC has now been supplanted by the Higher Education Funding Council (HEFC) which is dominated by industrialists, not academics, and the Science Minister personally

---

*I can illustrate this by a personal example. During the 1980s and 1990s, when I argued that the British Government's cuts in the universities' budgets were neither as extensive nor as damaging as was generally claimed, and that British science was expanding happily, I found that various universities cancelled invitations to me to speak. A paper of mine describing the economics that constitute Chapter 10 of this book was turned down by the journal *Research Policy*, one referee claiming that my work was 'dangerous'.

chairs the meetings of the Council for Science and Technology which supervises the research councils (and he does not hesitate to overrule their decisions, see Chapter 12).

Time and again, however, it is war that drives out these proper concerns. As H. J. Laski commented in his *Reflections on the Revolution of our Time* (1943): 'The two world wars bring us, whatever we wish, to the certainty of a planned society.'[82] The annual UGC grant in 1945 was doubled from £2 149 000 to £3 149 000, with another £1 000 000 added specifically for medical education and its attendant sciences of biochemistry and physiology. In 1947, the annual UGC grant stood at £9 000 000 and in 1952 £15 000 000. Sir Stafford Cripps, the Chancellor of the Exchequer, justified these huge increases with: 'It is on the advances that we make in scientific knowledge and on the energy, initiative, directive capacity and courage of these young graduates that the economic future of the country will largely depend.'[83] That was a pure Baconian statement, and inherently improbable. Major entrepreneurs have rarely been graduates, and even today, when so many people go to university, it is still the non-graduates like Richard Branson who possess the energy, initiative and directive capacity it takes to foster economic growth as entrepreneurs. None the less, the government continued on its Baconian way. The UCG grant in 1953, under the Tories, jumped to £20 000 000, and by 1957 was £25 000 000.

Incredibly, the universities hurried to discard any residual autonomy. When the Government's Barlow Committee on Scientific Manpower demanded, on May 1946, that 'the State should increasingly concern itself with positive University policy'[84] the Committee of Vice Chancellors and Principals replied that 'they will be glad to have a greater measure of guidance from the Government.'[85] This has remained the pattern. When, on 23 March 1987, Sir George Porter, the President of the Royal Society, wrote in the *Independent* asking for more government money for science, he also asked for greater government control. He wrote: 'What Britain needs is a clear and visible long-term policy for the whole of science and technology, determined and accepted at the highest level: a policy which co-ordinates both science education and research, universities and industry. Many of us in this country believe that the Prime Minister should set up, and herself chair, a high-level National Science Council, to determine our overall science policy.'

It almost beggars belief to witness senior professors actually asking for their academic freedom to be removed. Do these people not understand what universities are for? Still, at least they were once grateful.

The expansion in the British universities and their science, funded by government, was 'dazzling' as John Murray of Exeter University wrote in 1947.[86] The Vice Chancellor of Oxford, for example, contemplating the vast funds he was receiving from the State, complained in 1948 that 'we are in danger of being killed by kindness.'[87] The numbers of academics kept on doubling, from 5000 in 1938–9, to 11 000 in 1954–5, to 25 839 in 1967–8. This vast expansion was funded by governments committed to Baconian thinking. On 18 January 1956, for example, the Prime Minister, Sir Anthony Eden, impressed by Russia's so-called economic 'success', said: 'The prizes will not go to the countries with the largest population. Those with the best systems of education will win. Science and technical skill give a dozen men the power to do as much as thousands did fifty years ago. Our scientists are doing brilliant work. But if we are to make full use of what we are learning, we shall need many more scientists, engineers and technicians.'[88]

In an editorial of 15 March 1957, the *Times Education Supplement* declared 'Education is no longer a waif, the neglected child that needs an aunt. Education is a weapon of war, the long, cold war of brains and political subtlety, the war that has to be won, the war that keeps us alive in the very waging.... For education, which for long had only idealists as its supporters, and one knows how much they count in politics, has become a necessary tool of the realists. It is something along with stocks and shares and the production belt which the hard-faced men realise is necessary to the survival of this nation. Education is seen as linked to the atomic energy programme, to the jet engine and the computer; it is all mixed up with the balance of payments and design of the family car. Education is something that pays off. We compete in education as we once competed in battleships.' This is a far cry from Newman's idea of a university as a centre of liberal thought.

Crucially, however, none of this government money for university science was requested by industry. Since the turn of the century, the Board of Education (1909), the Balfour Committee (1929), the Malcolm Committee (1929) and the Deputy Under-Secretary at the Board of Education (1942) had all reported that the supply of scientists and technologists had fully met, and even exceeded, industrial demand.[89] But neither the Government nor the universities were interested in industry's objective needs. Obsessed with Bacon's linear model, and with Germany and with Russia, they believed they knew better than mere industrialists. By 1961, the UGC was spending £39 500 000 a year in recurrent grants, as well as spending £12 000 000 a year on capital expansion. And then the system expanded again!

It is a feature of human philosophical constructs that, when they fail, the protagonists simply redouble their efforts. When General Haig discovered during the First World War that sending 100 000 unprotected men to run against barbed wire, machine guns and trenches only led to them all being killed, his response was to send 200 000 men, then 400 000 and then 800 000. When Freud discovered that psychoanalysing patients twice a week for six months failed to cure them, he increased his prescription to daily for three years (it still fails). And when the protagonists of Francis Bacon failed in Britain, they simply increased their government funding for the universities and science.

By the late 1950s and early 1960s, it was obvious that decades of vast government-funded expansion of the universities and their science had failed, economically. Germany and France, the old bugbears, were fast overtaking, and new competitors loomed. The Government reacted by creating yet more universities. A UGC minute of 1958 initiated the universities of East Anglia (1963), Essex (1964), Kent (1965), Lancaster (1964), Sussex (1961), Warwick (1965) and York (1963). Then, in 1961, the famous Robbins Report on Higher Education was commissioned. It was published in 1963,[90] and it had such an impact that it promptly spawned a further fourteen universities: Aston (1966), Bath (1966), Bradford (1966), Brunel (1966), City (1966), Cranfield (1969), Dundee (1967), Heriot-Watt (1966), Loughborough (1966), Salford (1967), Stirling (1967), Strathclyde (1964), Surrey (1966) and Ulster (1965). Consequently, the number of university academics doubled, from 15 682 in 1962/63 to 32 738 in 1976/77, and the numbers of students also doubled. A huge expansion – but relative economic growth came there none.

Robbins had identified a new competitor, the USSR (he ignored Japan and completely failed to anticipate its growth). The success of *Sputnik* in 1957, and the Soviet Union's central planning, persuaded Robbins and his colleagues that its Baconian model for science would enable it soon to overtake Britain. Every one apparently believed this. In his famous 1959 essay on the so-called *Two Cultures* C. P. Snow wrote that: 'The Russians have a deeper insight into the scientific revolution than anyone else'.[91] In paragraph 194 of the Report, Robbins quotes with approval the Russian official who explained that the USSR's economic supremacy, based on vast education in science, was inevitable because 'the Soviet Union would always have use for people who had been trained to the limit of their potential ability.' But Robbins also argued that Britain had to expand its universities because 'almost everywhere we have travelled, we have been impressed by an urge to educational development.'[92] This sort of argument would have been found very persuasive by the Gadarene swine. The politicians were certainly convinced, and in 1964 they launched Britain on a fascinating economic experiment – wealth creation that was to be primarily achieved through the government funding of science. The politician who powered the programme was Harold Wilson, and his model was the USSR.

The USSR was Marxist, and Marx (as we shall explore further in Chapter 12) was a passionate Baconian. Wilson, in his turn, was obsessed by the 'success' of the USSR which he attributed to its Baconian science policies. In 1957, for example, while still in opposition, he explained on the eve of one of his repeated trips to the 'scientific' Soviet Union: 'This is an age of Sputniks and space travel and of scientific achievement proceeding at a staggering rate. All this produces a new challenge to any Government, and that is why it is a tragedy that this country, which seemed to be leading the world, is still governed by a group of obsolete Edwardians.'[93] Actually, that was unfair to the 'group of obsolete Edwardians', many of whom were as transfixed by the myth of the USSR as Wilson (see Eden's 1956 quote above) but it took Wilson's governments of 1964–70 and 1974–76 to turn the myth into the central economic policy of the nation.

Wilson really believed. On 10 January 1956, he wrote in the Daily Mirror: 'In the next generation Russia's industrial challenge may well dominate the world economic scene.' Later that year he published with Douglas Jay and Hugh Gaitskell his influential *We Accuse: Labour's Indictment of Tory Economic Policy*[94] in which he explained that Britain was growing slower than Russia because the Tory government had failed to imitate Moscow's centralised state funding of science and technology.

Wilson led a party of believers. It was Nye Bevan, the left's darling, who delighted the Labour Conference of 1959 with the prediction that the world's economic leader would soon be the USSR: 'The challenge is going to come from Russia. The challenge is not going to come from the United States. The challenge is not going to come from West Germany nor from France. The challenge is going to come from those countries who . . . are able to reap the material fruits of economic planning. In a modern complex society it is impossible to get national order by leaving things to private economic adventure.'[95] Other Labour leaders echoed these sentiments. Dick Crossman claimed to detect a 'terrifying contrast between the drive and missionary energy displayed by the Communist bloc and the lethargic, comfortable indolence of the West'.[96] In 1960 he wrote that 'the Communists are overtaking us'.[97] Thomas Balogh, the Balliol economist who closely advised Wilson as Prime Minister, wrote: 'Russian output per head will surpass that of Britain in the early 1960s and that of the US in the mid-1970s.'[98]*

Faced with this so-called Russian threat, Labour prepared by copying. Throughout the 1950s, Wilson, Gaitskell and Patrick Blackett, the socialist physicist, chaired meetings of a 'scientific and technological policy group' at the Reform Club. The group was very distinguished, including great scientists, prominent thinkers, senior politicians, and a certain publisher: Annan, Bernal, Bowden, C. P. Snow, Florey, Wynne-Jones, Lockspeiser, Robert Maxwell, B. R. Williams, C. F. Carter, Peart, Bronowski, D. M. Newitt, Ritchie Calder, Callaghan and Robens.[99] This group produced some very important policies, including the decision that the Minister of Science should be third in the order of Cabinet precedence, following on the Prime Minister and the Minister of Planning, but senior to the Chancellor of the Exchequer, the Foreign Secretary and other ministerial riff-raff (when Harold Wilson announced this at a scientists' dinner, he caused them almost to swoon with delight, as Richard Crossman solemnly noted in his diary).[100]

The group helped produce Labour's 1961 pamphlet *Science and the Future in Britain*, in which Harold Wilson wrote that the modern world required 'the full planning and mobilization of scientific resources',[101] and the group helped write the most influential of Wilson's speeches, his celebrated 'White Heat of Technological Revolution' speech delivered to the Party Conference at Scarborough in 1963. In that speech he promised to implement Robbins's proposals in full, regardless of expense, but he went much further. He wanted whole new industries: 'This means mobilizing scientific research in this country in producing a new technological breakthrough.... If we were now to use the technique of R & D contracts in civil industry, I believe we could within a measurable period of time establish new industries that would make us once again one of the foremost industrial countries of the world.' Government investment in R & D would 'provide the answer to the problem of Britain's decaying industries.... We would train and we would mobilize the chemical engineers to design the plants

---

*The Master of Balliol during Balogh's tenure was the Marxist, Christopher Hill, who visited Stalin's Russia at the height of the terror and of the show trials. Balliol is proud of its input into *The History Workshop*, which is subtitled *The Journal of Socialist and Feminist History*, which title bears as much relevance to the search for truth as would a *Journal of Socialist and Feminist Physiology* (though I can think of at least one Balliol scientist who is probably itching to edit it).

that the world needs . . . those of us who have studied the formidable
Soviet challenge in the education of scientists and technologists, and
above all, in the ruthless application of scientific techniques in Soviet
industry, know that our future lies.'[102]

The speech was a triumph. It caught the imagination of the country,
it created a new *cliché*, and the new exciting 'scientific' policy it ad-
umbrated tipped the balance for Labour at the subsequent general election.
So profoundly did Britain during the 1960s adopt science, technology
and higher education as its totems, that it sometimes seemed that the
whole nation was mesmerised. The collective creed was perhaps best
expressed by Bill Maitland, the hero of John Osborne's *Inadmissible
Evidence* (1964), whose only faith was 'in the technological revolution, the pressing, growing pressing, urgent need for more and more
scientists, and more scientists, for more and more schools and universities and universities and schools, the theme of change, realistic decisions based on a highly developed and professional study of society
by people who really know their subject, the overdue need for us to
adapt ourselves to different conditions.'

The preface to Labour's 1964 election manifesto read: 'A New Britain – mobilizing the resources of technology under a national plan;
harnessing our national wealth in brains, our genius for scientific invention and medical discovery.' It worked, and in 1964, Harold Wilson
became Prime Minister. In power, he was as good as his word. Frank
Cousins was brought into the Cabinet to be the Minister of Technology,
with C. P. Snow as his deputy. Michael Stewart, another Cabinet heavy-
weight, was appointed as Secretary of State for Education and Science. Moreover, George Brown, the Deputy Leader was given a new
ministry, the Department of Economic Affairs (DEA) to implement
the National Plan and to overrule any Treasury objection. The Plan,
which was based on the USSR's five-year plans, set a target of 25 per
cent growth in GDP by 1970 (an annual average of just below 4 per
cent), which was to be achieved by 39 separate actions including, of
course, a vast expansion in science, R & D and higher education.

Labour honoured its commitment, and it expanded science, R & D
and higher education as comprehensively as it had promised. Politically, the Government stuck to its guns. When Cousins resigned from the
Ministry of Technology, he was replaced by Wilson's then-trusted rising star, Anthony Wedgwood Benn, who admitted that he found technology 'spiritual'. Wilson also boosted the ministry with a further junior
minister, the gifted Richard Marsh. When George Brown resigned from
the DEA, Wilson replaced him with Michael Stewart. When he failed

to make a great success of it, Wilson decided to take it over himself, so seriously did he view the ministry: 'the idea came to me – why shouldn't I take over Michael's job myself?'[103] Peter Shore, a Wilson ally, was formally installed as Secretary of State at the DEA, but he was very much the Prime Minister's agent.

What was the result of all this science, technology and higher education? Economic disaster. So comprehensively did the policy fail that, in September 1976, the Labour Government had to hand over the management of the economy to the International Monetary Fund in return for a vast loan. Britain had joined the ranks of the Third World – which was not, perhaps, so surprising because the country on whose economic and scientific policies it had modelled itself, the USSR, was about to join the Third World even more comprehensively. So bad was Britain's economic performance after 1964, and so disillusioned with science did the politicians become, that in 1971 Mrs Shirley Williams, who had been Labour's Secretary of State for Education and Science, warned that 'for the scientists the party is over' (and Anthony Wedgwood Benn abandoned technology for socialism, dropping one and half names on the way and becoming Tony Benn). The Baconian experiment had failed.

It even failed in scientific terms. During the 1970 General Election campaign, trying to justify his vast expenditure on science, Wilson claimed three great advances: 'In nuclear energy, our prototype fast reactor now going ahead at Dounreay is 3 or 4 years ahead of anything else in operation or planned in any part of the Western world.' The 'Cephalosporin antibiotic' was earning vast royalties, and carbon fibres were 'a product of work in a Government research laboratory developed by public enterprise under Labour Government legislation in collaboration with industry'.[104] A meagre list indeed, particularly when one considers the utter commercial failure of the British civil nuclear industry, the fact that cephalosporin research had enjoyed much private funding, and the fact that Labour's policies introduced, *inter alia*, an inflation rate of 29 per cent.

Wilson's Baconian experiment on science funding failed – a lesson the politicians never forgot (even if the scientists never understood), and since 1971 successive British governments have only increased science funding in line with inflation – to the rage of scientists who had grown accustomed during the previous 50 years of government science budgets which had, in real terms, grown much faster than national wealth.

SCIENCE POLICIES ELSEWHERE

The history of government science policies worldwide can be easily summarised. First Britain, and then the USA, grew rich through *laissez faire*. Governments all over the rest of the world, trying to catch up, decided to imitate the private British and American universities and research laboratories by funding their own. All other countries' science, therefore, is predominantly government-funded (this applies even to Switzerland, which is often mistakenly assumed to be *laissez faire* in science; only Japan is, relatively, see the next chapter). Britain and the USA, observing with alarm the speed at which competitors were catching up economically, wrongly attributed it to their government funding for science, so their own governments started to copy the governments that were copying their own *laissez faire* science, and they started to fund science. The British and American governments were encouraged to do this because their competitors (Germany, Japan, the USSR and, on occasion, France) could be so hostile, and were systematically subjugating science to military ends. It is no coincidence that it was a Prussian cavalry general who wrote: 'War is the father of all things.'[105]

The conversion of the USA and the UK to government funding of science did absolutely nothing for their economic growth rates, which disproved the suggestion that Germany *et al.* were growing so fast because of government science funding, but by the time the American and British governments had discovered their mistake, it had become too difficult to withdraw. The universities and the scientists make a powerful and very persuasive lobby, and it is desperately hard for governments to divest vested interests – as Adam Smith pointed out repeatedly in his *Wealth of Nations*.

POSTSCRIPT ON GERMANY, USSR, JAPAN (THE 'OTHER COUNTRIES') AND THE STATE

During the nineteenth century, and the early part of the twentieth, it was fashionable to admire Germany's system of higher education and state-funded science. After Germany's collapse in 1945, and particularly after Sputnik's orbit in 1957, the USSR's state-funded science became the cynosure. In its issue of 15 January, 1987, *Nature* acknowledged the USSR as the world's largest relative investor in R&D at 3.73 per cent of GDP. Although much of that was for defence, it still dwarfed West Germany's 2.84 per cent, Japan's 2.77 per cent, the USA's 2.72 per cent, the UK's 2.18–2.38 per cent and France's 2.10 per cent. After the collapse of the USSR, it became fashionable to admire Japan and MITI's State direction of industrial technology. Each fashionable wave has inspired government-funded programmes in imitation elsehere, yet curiously these are never dismantled after the collapse of the original model. But this does raise germane questions on the role of the state.

The early *laissez faire* successes of the UK and the USA inspired active copying elsewhere, and generally that was state-directed. Thus it was the German, French, Russian and Japanese governments that, during their countries' early periods of catching up, often initiated industrialisation and scientific education. The institutions those governments created were necessarily different from the *laissez faire* models; so, for example, a German nineteenth century university was different from Mushet's private industrial research laboratory, because it was doing a different thing. The cost, rigidity and distance from industry of a German university made it inferior to Mushet's laboratory as a vehicle for technological progress, but it did provide the State with an instrument for changing the culture of its people. (Though for the lead *laissez faire* countries to then copy the institutions of the catching up countries was folly, as is shown by the current economic malaise of the UK and the USA).

Those catching-up governments of the nineteenth century that then retreated into *laissez faire* after having provided the initial capitalist impulse (as did the Japanese government, for example, after 1884) were rewarded with flourishing economies. Those governments that retained their direction of the economy, like that of the USSR, created poverty. We can conclude, therefore, that the role of the State is thus: the State must provide the framework for capitalism, as without the rule of law, anarchy will descend (witness much of Africa today). But once a capitalist culture is rooted, the government should adopt *laissez faire*.

The irony of Japan as the model for today is that it has succeeded because of *laissez faire*, not because of MITI, yet it has only spawned government intervention amongst its imitators. Now that Japan grows ever more interventionist, so does its economy suffer.

POSTSCRIPT ON INFLATION AND THE UNIVERSITIES

Let us not shed too may tears over the loss of the government debt issues. There were many who argued that they were immoral as they provided easy profits for the fundholders out of the taxpayers who had to service the national debts (indeed, radicals like Sir Francis Burdett and William Cobbett claimed that the rich deliberately lobbied their political friends to raise the national debt). But the abrupt destruction of the fundholders' investments after 1918 was damaging, largely because it was so abrupt. Universities are desperately immobile. Unlike companies, universities cannot just shed staff, or easily change activities. The nature of teaching and of research is highly conservative, and long-term stability is crucial to the development of teaching departments and research groups. Even relatively small drops in income, therefore, are dreadfully disruptive.

Curiously, inflation *per se* is not damaging to economic growth. It is generally believed it is, but that is to confuse anecdote with systematic study. Admittedly, hyperinflation, such as that seen in Germany between the wars is destructive, but more gentle inflation is not. In his *Phases of Capitalist Development* (Oxford, 1982) Maddison not only provides GDP *per capita* data for the 16 leading capitalist countries since 1870, but their inflation data as well. When economic growth rates since 1870 are compared with inflation rates, no correlation emerges, i.e. low inflation economies such as Germany's or Japan's have not, in fact, done any better than high inflation ones, after account has been made for their different GDPs *per capita* in 1870. This is because, of course, long-term economic growth is a consequence largely of technology and of industrial training, and industry will adapt to different economic climates to promote those, such are the pressures of competition.

Inflation does, however, change societies. A low-inflation society will breed *rentiers*, and a high-inflation society will breed speculators. A low-inflation society protects the old, the young, the sick and the vulnerable, a high-inflation society fosters the strong, whether they be the trades unions of skilled workers or predatory entrepreneurs.

The precise nature of their financial environment is not intrinsically important to universities – the private American colleges now thrive under very different conditions than those of before 1914, as do the contemporary Oxbridge colleges – but economic predictability is essential. Thus it was not unreasonable for British governments to support the universities temporarily post-war, especially as it was the governments' own policies that were undermining them, but long-term, constitutive financial support should have been anathema.

POSTSCRIPT ON HAROLD WILSON

Perhaps the decision of Wilson's that was the most fateful – it was also his very first – speaks of his entire philosophy. Less than 24 hours of taking office as Prime Minister in 1964, having only appointed two ministers to his government, Wilson broke Britain's commitments under EFTA (the European Free Trade Association) to impose import surcharges. Free trade was anathema to Wilson. It is often now forgotten that his long-standing opposition to the then-Common Market, now European Community or Union, was that 'the whole conception of the treaty of Rome is anti-planning'.[106] A man who finds the rule of Brussels unacceptably *laissez faire* really is *dirigiste*. (Jacques Delors originally opposed the Common Market in 1957 because he also found it too *laissez faire*, saying: 'one cannot simply hand over the destiny of the French economy to the blind mechanisms of the market.')[107]

Wilson's EFTA decision, although now largely forgotten, may be considered his most fateful, because it not only enraged Britain's EFTA partners, it also convinced them that a mere treaty of free trade between sovereign nations could never survive. There had to be a loss of national sovereignty to a central body to impose nations' adherence of treaties, i.e. EFTA had to be superseded by the EEC. This was gratuitous, of course, and it is shameful that Britain under Harold Wilson discarded its treaty obligations so lightly.

Wilson's Baconian experiment on science funding failed because he did not believe in the free market. A long-time collaborator of Beveridge and a war-time civil servant, Wilson actually did believe in the dictum of his colleague Douglas Jay that 'the gentleman in Whitehall really does know better'. As Prime Minister, Wilson was *dirigiste* to a degree that now seems fantastic. He tried to fix the salaries of every employee in the country, he tried to regulate the investment decisions of every company, and he even tried to regulate where people took their holidays. It could be argued that Wilson's White Heat of the Technological Revolution failed because of everything else, i.e. that the government was right to be *dirigiste* in science, but that it should have been *laissez faire* in everything else. (That is analogous to the apologists' argument that 'planning' under the DEA would have worked but for the government's failure to devalue earlier, as if government decisions on devaluation did not represent the purest form of planning.) That argument over selective *dirigisme* in science funding will be demolished in Chapters 9 and 10.

Let me add a personal speculation. No one really knows why Wilson resigned so unnecessarily in 1976, although he may have anticipated his decline into ill health, but I suspect that, despite his reputation as an unprincipled operator, Harold Wilson really did believe in the central planning of the economy – and, crucially, in the government funding of science – and when he realised that his policies had failed, he lost heart.

POSTSCRIPT ON THE ROBBINS REPORT

In 1963 Lord Robbins and his committee published their famous Report on Higher Education. The Committee comprised ten men and two women, and it was the very stuff of the educational establishment. Between them, its members could muster a peerage, a KCMG, a DBE, three knights bachelor, a CB, four CBEs, an OBE, an FRS and two professorships. But those gongs were but seed corn; after the publication of their report, the members of the committee were to be showered in honours.

The Robbins Committee advocated the government funding of the universities and their science for two reasons: (i) economic growth and (ii) the promotion of higher cultural standards. Let us examine its achievement.

Of economic growth came there none – which was not surprising because, as we have seen, the Committee modelled itself on that economic exemplar, the Soviet Union.

Of the promotion of higher cultural standards came there little either – mainly because the Committee appeared to loathe proper universities. Robbins reported in 1963, his recommendations were anticipated and immediately implemented, and by 1968 the universities were awash with drugs, were germinating an epidemic of venereal disease, and were ablaze with riots, sit-ins and direct action. For several years during the late 1960s and early 1970s higher education was in thrall to violence, promiscuity and Marxist *chic*: 'property is theft'. There is every reason to believe that the universities' chaos was caused by the

Robbins expansion, and that their recovery was effected by the halt of that expansion; there is also every reason to believe that the universities have not only failed to transmit 'a common culture and common standards of citizenship' but that they have helped brutalise their own society. The intolerance, violence and drug abuse of the students' unions will have legitimised, for many a young thug, direct action and brutality. The anti-police sentiments, sexual licence and drug dependence of many offenders still echo the fashionable *mores* of the 1968 university intellectual.

How did the universities fall so low? Robbins's defining axiom, the famous rationale that underlay his entire report, was that 'courses of higher education should be available to all those who are qualified by ability and attainment to pursue them and who wish to'. This might appear to be an unexceptionable, even obvious axiom, except that Robbins's definition of those 'qualified by ability and attainment' encompassed all those who passed two 'A'-levels at grade 'E'. Robbins produced statistics which showed that, whereas 90 per cent of all schoolchildren who had attained at least three 'A'-levels were embarking upon full-time education, only 62 per cent of schoolchildren with two 'A'-levels were entering it; but with most of that 62 per cent training to be teachers. Only 22 per cent of schoolchildren with 2 'A'-levels were actually going on to university. Robbins believed that these statistics justified immediate expansion. Robbins complained that 'there has been increasing competition for entry to university' and that it was 'most undesirable that this should increase in future . . . in fact it should be reduced' (paragraph 156). Robbins actually believed that even two 'Es' were too demanding an entrance requirement. At his instigation the ten Colleges of Advanced Technology, places such as Bradford, Loughborough and Salford, were promptly converted into universities, yet at the time 15 per cent of their students had only attained one 'A'-level at admission – and a further 25 per cent did not even have that: they were admitted on ONCs.

Most dons would agree that students with two grade Es at 'A'-level are simply insufficiently prepared to benefit from a British university education as it is generally understood. Three grade Cs might represent the minimal acceptable preparation, although few dons would really want to teach an undergraduate who had not attained at least a grade 'B' in the 'A'-level that approximated most closely to the degree subject. Robbins knew all that: he was in fact trying to destroy the British tradition.

His report advocated the German *Abitur* or the French *Baccalaureat* systems as ideal models. Contintental universities are different from the British: whereas the British only admit undergraduates competitively for a limited number of places, continental universities admit all applicants who have passed the government *Baccalaureat* or its equivalent. A continental university, in consequence, is a huge impersonal diploma factory while a British one, traditionally, is an *alma mater*.

Some continental universities, frankly, are slums. The French, for example, regard their own with such horror that they direct their best pupils to the *Grandes Ecoles*. These have stiff entrance requirements, high academic standards, close tuition, low failure rates and no riots. Yet Robbins wanted to convert the British high quality *Grandes-Ecoles*-type universities into continental-style ones. and for a time he nearly succeeded.

Robbins has proved to be the greatest purveyor of anomie in British intellectual history. He took an entire generation of youngsters and pushed them into an environment where they simply could not survive. Two grade Es just do not equip an eighteen-year-old to flourish at university. Quite properly, university life is unstructured: students have hours, even days, to negotiate alone. Well-stocked, robust minds love it, of course, but ill-educated, ill-informed minds find it frightening and purposeless. Inevitably such minds will leap at the crude certainties of international revolution or drug-induced mind expansion. To use the then-fashionable terminology, the sit-ins, marches and demos of the 1960s and 1970s were cries for help from young people bewildered by an environment they could not master.

British students were, in fact, alienated. All adolescents have to fight insecurity, but the adolescent at university with two Es knows that he is academically inadequate in the one place which values academic achievement most highly. That insecurity, coupled to the loneliness of the long, empty campus hours, will provide precisely those conditions that Durkheim predicted would breed alienation. He, at least, would not have been surprised by the mess, silliness and sometimes pure wickedness of the average students' union.

Were the Robbins universities actually designed to breed alienation? Why do so many of them look so grim? Their grey windswept concrete towers shelter, only too often, low, grey, ill-lit cemented corridors. And they are isolated. The concrete University of Bath, for example, occupies an exposed hill top some five hundred feet above the lovely mellow-stoned city that nestles in the cosy valley of the River Avon.

Robbins anticipated that his expansion would increase the wastage rate (the percentage of students who fail to complete their courses).

The French or American rates are about 50 per cent, but Britain's are only 14 per cent. Incredibly, Robbins was contemptuous of Britain's low rate, dismissing it with 'an average wastage rate of 14 per cent in universities as selective as ours is nothing to boast of' (paragraph 576). Robbins did not think that wastage mattered. He believed that even some exposure to higher education was better than none. But wastage is synonymous with failure, and there are few worse things with which to burden a young person launching into adult life than a sense of failure. An inadequate exposure to philosophical ideas such as Marxism, moreover, may do more harm than no exposure at all. It is curious how glibly Robbins ignored the wisdom of Pope's *Essay on Criticism*: 'A little learning is a dang'rous thing; Drink deep, or taste not the Pierian spring: There shallow draughts intoxicate the brain, And drinking largely sobers us again.'

As well as stuffing British universities with poorly prepared students, Robbins then gorged them on second-class dons. The 1960s expansion was so precipitate that the universities simply could not find good people to appoint. Britain, in 1939, possessed just 5000 university dons. Several centuries of natural university development had flowered in just 5000 scholars; but of such quality! One Cambridge college alone, Trinity, could boast of the overlapping fellowships of Bertrand Russell, Wittgenstein, G. E. Moore, G. M. Trevalyan, F. A. Simpson, Littlewood, Hardy, Ramanujan, A. E. Housman, Gow and Adrian. By 1962/3, following some dramatic post-war expansion, Britain had 15 682 dons. The reservoir of talent must have been fairly exhausted by then, but within five years Britain had increase that number to 25 839. Whence were 10 000 university dons to come? In five years Britain was somehow to generate twice as many academics as the universities had managed to produce after six centuries of development.

It could not, of course, be done, and the universities found themselves appointing as lecturers and professors men and women who were little better educated than the students they were expected to teach and guide. No wonder the universities collapsed into chaos.

The British universities are better places now, but only because their expansion was halted around 1970. Following the arrest of their expansion, the universities' standards have risen. Only 15 per cent of undergraduates are now admitted with less then three grade Cs at 'A'-level. Even this remains too high, although some unreconstructed Robbinsonian dons will still frighten television viewers with their persistent claims that there remains a shortage of university places.

As academic standards rise, so have standards of behaviour. It is instructive that the remaining troublesome universities are the least good ones: Aston, for example, has barred Conservatives from holding meetings on campus because it cannot afford the costs of security; and Swansea recently had to provide no fewer than five security guards to protect Ian Grist, MP, Under-Secretary to State for Wales.

The worst excesses are now reserved for the polytechnics, now the new universities, another legacy of Robbins. Although few were ever as degraded as the North London Polytechnic, whose horrors were chronicled by Cox and Marks in *The Rape of Reason*, they can still be unpleasant places. A recent lecture by the South African diplomat Louis Mullinder at the then Wolverhampton Polytechnic was aborted by a mob that stormed the hall.

It is not fair to blame Robbins for everything that has gone wrong in British higher education. The student riots of the 1960s and 1970s were international – not only because they were *chic*, but also because everybody expanded their higher education after the war. The Continental and American universities also expanded – the French named their's *l'explosion scolaire* – and if their violence exceeded Britain's, it was because they started from a higher base.

But Robbins can be blamed for joining that particular merry-go-round. One of his major justifications for a British university expansion was the Continental and American one – 'Almost everywhere we have travelled we have been impressed by an urge to educational development' (paragraph 129). But could he not read the statistics showing the thousands of places – particularly in science – going unfilled? (There are 6000–7000 unfilled science places today.)

Curiously, the most public warning against Robbins did not come from an academic, but from a novelist, the vagaries of whose career had driven him to teach temporarily in one of Britain's more mediocre university colleges – Swansea. It was Kingsley Amis who warned that 'more means worse'. It is strange how often the novelists have warned, unsuccessfully alas, of the academics' ambitions. Amis's *Lucky Jim* (1954) exposed only too well the inadequacies of some provincial universities and their dons, while Tom Sharpe's *Wilt*, though set in a university when it was still a technical college, beautifully chronicles the sheer irrelevance of educating students beyond their abilities. Malcolm Bradbury's *History Man* portrays horribly the consequences of radical *chic* propagated at the taxpayer's expense.

Violence, drug abuse and promiscuity have always dogged the universities – for obvious reasons. Medieval Oxford, according to some chroniclers, was little better at times than a drunken brawl spread across its brothels. Young men will always be naughty, and the excesses of the 1960s embodied much tradition. But over the centuries the universities developed defences: Oxford, for example, threw up proctors, selective admission and rustication. But the Robbins Committee subscribed to a naive, fashionable view of the young, and the new universities smothered their students with tolerance.

Those students were remarkably privileged. Predominantly middle class, paying no fees, and in receipt of mandatory grants, those students could hardly have believed their luck. In return for very modest academic attainment, they were being invited by the taxpayer to have a riot – as of right. So they did. Oddly, Robbins could have anticipated that had he studied one particular Russian example. During Tsarist times, the Russian universities were free to the poor – but that bred no gratitude nor civic loyalty. It bred the revolutionaries and assassins (like Lenin's brother) that were to drag the nation into socialist hell.

Brideshead Revisited is the universities' least popular novel because it describes only too well the antics of over-privileged, upper-class undergraduates. Now, thanks to Robbins, we have all heard the sound of the newly-privileged middle-class baying for the sound of broken glass. And now the government has so expanded higher education in that one-third of all school leavers enjoy it, free. Yet no one has actually demonstrated that mass 'free', higher education benefits anyone, economically, culturally or psychologically.

9 The Economics of Research: Why the Linear Model Fails

During the early 1980s, the Japanese Government initiated a major project: the fifth-generation supercomputer. It was to be a massive but 'user friendly' parallel computer of 1000 processors, with a software based upon logic rather than on conventional structured programming. This caused a terrible panic in Europe. The Japanese were going to take over the world's electronics industry!

Careful observers might have noticed that the Japanese Government's real initiative was to spend taxpayers' money on research into technology – $400 million at least. Japan's previous technological successes had been privately funded, and its failures (such as its space programme or its nuclear power programme) had been funded by the State – but the early 1980s witnessed the apotheosis of the MITI myth, and European governments were alarmed. They scrambled to compete. Britain launched the Alvey programme, the EEC launched ESPRIT, and over the decade 1982–92 the European taxpayer matched the Japanese taxpayers' generosity to computer research. What was the result? Complete failure. The wonderful Japanese fifth-generation computer makes a good doorstop, but little else.[1]

Despite the subsidies, and despite the government programmes (which, in addition to Alvey and ESPRIT, include BRITE/EURAM, COMETT, COST, EUREKA, MONITOR, RACE, SPRINT, Telematics and VALVE) the European electronics industry is in a desperate condition – Philips of Eindhoven and Germany's Nixdorf, two of Europe's supposedly strongest companies, are struggling, while most of Britain's ICL has been sold to Fujitsu of Japan. The successes have been American where companies such as IBM, Microsoft, Intel, Texas Instruments, Oracle and Apple have produced a world-beating range of pocket calculators, microcomputers, programs and personal computers (though even the Americans have engaged in some panic and unnecessary subsidising, see the postscript on Sematech).

The European and Japanese taxpayers' venture of $800 million into computer science failed for the same reason that the British taxpayer lost all the money invested in Babbage's computer one and a half centuries earlier – governments are dreadful judges of commercial opportunities. MITI had had to fund the fifth-generation supercomputer itself because it had failed to persuade industrialists, bankers or venture capitalists to do so. Believing it knew best, MITI proceeded anyway and, in a chain reaction of hubris, the European and British governments imitated the error. Had that been an isolated error, this book may never have been written, but such error is institutionalised, and governments continue to spend money on uneconomical technology. Moreover, governments increasingly fund academic science. Why do governments fund research?

Government funding for research is based on the linear model, which proposes that academic science breeds technology which breeds wealth:

academic science → technology → wealth.

The model, as proposed by Francis Bacon, also supposes that government money is required to fund academic science, but most supporters of the model further propose that governments should fund technological research:

academic science → technology → wealth.
　　　↑　　　　　　　↑
government　　　government
money　　　　　　money

Let us review the model, taking the second step first (does technology breed economic growth?) and then considering the first step (does academic science breed new technology?). At each step we will ask a subsidiary question: Is government funding required to optimise the step?

To simplify the review of the linear model, and the elaboration of the economics of research, this chapter will be broken into three headings, each of which has sub-headings:

Does New Technology Breed Economic Growth?

Reasons for the support of industry by government:

(i) Social reasons
(ii) Defence
(iii) The support of manufacture
(iv) Countersubsidies
(v) The example of the civil aircraft industry
(vi) The empirical evidence
(vii) Conclusion to this section

Does Academic Science Breed Technology?

Whence does new technology come?
Cross-fertilisation
The separate autonomy of science and technology
Can government funding help industry develop new technology from old?

Would the Free Market Supply Enough Basic Science?

The theories of Nelson and Arrow . . .
. . . are disproved by the observations of Mansfield and Griliches
First-mover advantages
Second-mover advantages
The costs of second-mover research
The dangers of not understanding the costs of second-mover research
Is the government funding of academic science a good economic investment?

DOES NEW TECHNOLOGY BREED ECONOMIC GROWTH?

We have already addressed this question in Chapter 7, and we concluded that of course it does; essentially, technology *is* wealth. But there is a qualification. Technology appears to be material, a matter of physical objects, yet it is only the physical expression of a human culture, and only capitalism has, thus far, provided the appropriate culture. Thus non-capitalists will not create commercially-useful technology – witness the old USSR. In the absence of the appropriate culture, technological objects will assume no greater economical importance than *objects d'art*; when non-capitalists are offered other people's technology, they fail to exploit it – witness the rusting tractors that litter vast tracts of the Third World. Yet, even under capitalism, technology is not precisely synonymous with wealth; technology is a tool, to be employed by people and directed by the market. Thus other aspects of culture such as education, training, and the effective exploitation of capital are as integral as technology to the creation of wealth even though, as we noted in Chapter 7, we cannot calculate exactly their relative importance.

Knowing, however, that technology crucially underpins wealth, let us ask the secondary question: can government funding help industry to transform technology into wealth? Since it is the function of industry to do just that, this question really asks: should governments subsidise industry? This is a wide question that transcends this book, and which is addressed daily by thousands, if not tens of thousands, of economists, politicians, journalists and industrialists. This book is, therefore, unlikely to shed any new light, but let us review a few basic principles.

Wealth is created by industry, not by governments. Governments can only raise money through taxation, and taxation is an unmitigated evil for industry, either because industry is taxed directly (which reduces its capital, profits and economic freedom) or because its customers are taxed (which reduces demand for industry's products).* To tax industry, therefore, to enable governments to subsidise it must, in general terms, be absurd. There are those who argue that politicians and civil servants could run industry better than professional industrialists, but even if they could, they would not need to subsidise to control: governments could use their coercive power to direct industry without subsidising it. But after the experiences of the twentieth-century, not many people outside North Korea or Cuba do believe that governments could do a better job than industrialists. Everyone agrees that governments have the crucial role of maintaining the legal infrastructure within which capitalism prospers, but capitalism should be entrusted to the experts – capitalists.† Britain's recent history confirms

---

*Governments could, alternatively, just print money, but the debauching of the currency (which is politely called inflation) is a social horror. Some economists believe, following Keynes, that in the short term governments can stimulate growth by monetary and fiscal measures to stimulate demand. In this book we are considering long-term economic growth, and there is no evidence that governments can do anything whatsoever to promote that other than supporting the rule of law, the free market and free trade.

One reason governments cannot translate short-term stimuli into long-term growth is that short-term stimuli also translate into inflation or deficits in the balance of payments, which require a reversal of the monetary and fiscal stimuli.

† At the risk of repeating some points already made in a footnote in Chapter 6, let

the poor judgement of politicians. Thus the electronics and car industries, for example, have received massive government funding, yet they languish, whereas Britain's pharmaceutical or popular music industries have been neglected by government, yet they flourish. Indeed, governments often only support industries because they are failing, and then they thrash around for justifications for their actions.

Governments offer at least four justifications for their subsidy of industry with taxpayers' money: (i) social reasons; (ii) defence; (iii) the support of manufacture; and (iv) to counter competitors' subsidies. Let us consider these justifications in turn.

(i) Social Reasons

It is not unreasonable for governments to facilitate the gradual closure of redundant technologies to minimise local unemployment; but no one should assume that such subsidies are economic investments. For example, only 2.2 per cent of Britons are employed in agriculture, but vast subsidies in France and Germany keep 6.4 and 3.9 per cent respectively on the farms.[2] Since the actual production of food in Britain, France and Germany is comparable, Britain is enriched, relative to France and Germany, by its greater agricultural productivity, which liberates labour for other areas of the economy. Indeed, the French and German economies are now wobbling, in part because of their enormous subsidies to agriculture. The social benefits of artificially supporting the farmers do not come cheap.

Very occasionally, governments will support failing companies when

---

us review that in *The Road to Serfdom* Hayek gave three reasons why governments fail to run industry better than industrialists. (i) Governments are divorced from the workplace, and their information depends on a long, vulnerable chain of communication; but industrialists are close to the markets, and their decisions are based on immediate knowledge (proximity really matters, even in relatively small and highly educated organisations; the engineers who built the Hubble telescope knew it was flawed, but their knowledge never reached up the chain to the decision makers; see *Nature*, vol. 348, p. 471, 1990). (ii) Governments are relatively small, and decisions have to be taken by a handful of people, but markets express the collective wisdom of thousands, so they are more likely to make correct decisions. (iii) Finally, governments are free of immediate commercial pressures, so they can afford to make uneconomic decisions (the ones that impoverish society) while private investors need to make a profit, so they will strive for one. Thus bureaucrats can fund fifth-generation supercomputers or baroque schemes like Alvey or ESPRIT with impunity; if the managing director of an electronics company lost $400 million on an ill-conceived idea, he would lose his job, but it is a matter of record that Alvey and ESPRIT actually boosted the careers of the politicians and civil servants who conceived and fostered them.

the market will not, and those companies will go on to become profitable again. Witness Chrysler in the USA or Rolls Royce in the UK. These examples are sometimes cited as 'market failures' (i.e. government funding was justified on economic grounds, yet the free market failed to invest). In practice, however, such government investment is made primarily for social or military reasons, and even though companies can occasionally re-emerge into profitability, a full balance sheet of costs and profits rarely reveals government funding to have been a good investment. The return would have been higher had the money been invested in the bank. De Lorean is the usual outcome. Indeed, the list of companies which, on receiving government assistance, have then gone on to require more and more, is depressingly long.

(ii) Defence

Governments will finance military technology. Had, for example, the British government subsidised the production of khaki dye in 1914, the country would have been better prepared for war – but, again, no one should confuse the subsidy of military technology with the creation of wealth – look at the old USSR; it was a military giant and yet became an economic pigmy.

(iii) The Support of Manufacture

Manufacturing industry lobbies for government money, because manufacture is now in dramatic relative decline. Service industries now dominate the lead economies. Thus 72 per cent of the USA's GDP in 1991 came from service, not manufacture, and this percentage has been increasing rapidly (it was 59 per cent in 1970). Manufacture now only accounts for 23 per cent of jobs in the USA. All advanced economies are now dominated by the service sector.* Consider the UK. In 1971, 11 627 000 people were employed in services in Britain, and 8 065 000 in manufacturing, yet by 1990, 15 849 000 were employed in services and only 5 113 000 in manufacturing.[3]

---

*Sixty-six per cent of Canada's GDP in 1991 came from services (up from 59 per cent in 1970) and similar figures can be seen across the OECD: France 65 per cent in 1990, up from 47 per cent in 1970; Italy 64 per cent in 1990, up from 49 per cent in 1970; Britain 63 per cent in 1990, up from 48 per cent in 1970; Japan 62 per cent in 1990, up from 52 per cent in 1970; West Germany 57 per cent in 1990, up from 42 per cent in 1970; figures from *Industrial Policy in OECD Countries: Annual Review 1992* (Paris: OECD, 1992), reviewed in the *Economist*, 20 February 1993.

These figures provoke alarm in some people. Surely manufacturing is more 'real' than service? Is service not merely parasitic on manufacture? If manufacture is in decline, should it not receive government support? This is an echo of the nineteenth-century alarm over the rise of manufacturing. Everybody in 1800 thought they knew that agriculture was the only real activity, and if people migrated from the fields to the factories, how could they eat? We now really do know that it was the increased productivity of the Agricultural Revolution that spared labour for the factories, just as we should now know that it has been the improving efficiency of manufacturers that has spared labour for the service industries (which include the health services, schools and university lecturers in clinical biochemistry). Manufacture needs service. The old USSR, for example, glorified manufacture, and it made lots of tractors and machines, but because it despised transport and retailing as service industries, food rotted and people now go hungry. So crucial is service to manufacture that, in lead capitalist countries, about half of all jobs even in so-called manufacture are actually service (jobs such as advertising, financial management and delivery).

Service is the purpose of production – manufacture is but a means to an end. It should be the aim of all economies to render the agricultural and manufacturing sectors so efficient that they will be dwarfed by a huge service sector – thus providing people with service, which is what they want. The more nurses or teachers or writers a country supports, and the fewer miners or production line workers or blast furnace operatives, the healthier, better educated and more interesting it will be. Just as, currently, in lead economies, only about 2 per cent of the population need be employed in agriculture, so manufacture will decline comparably. By the year 2100, only a few per cent of the population will actually make anything, yet productivity will be so high we will have machine mountains just as we now have food mountains.

There does not need to be any manufacture, or agriculture, in any particular national economy. Services can be used to buy goods. Consider the Baltic Exchange in London, the major international market for the sale and hire of ships. The British merchant fleet has been shrinking for decades, to general cries of alarm, but Britain earns over £1.5 billion a year in foreign exchange from the Baltic Exchange. The British fleet may have been undercut by cheaper foreign labour, but Britain has capitalised on its standards of education and training to provide an international service, whose profits finance national imports. The free market, therefore, has allocated resources rationally.

Highly paid, highly educated labour in Britain runs complex operations like the Baltic Exchange, while low paid, poorly educated labour in developing countries performs the simple repetitive tasks required to make and transport consumer durables.

For a Western government to raise taxes (which is synonymous with taxing industry, now dominated by the service sector) to subsidise the manufacturing sector, is that government's way of saying that it is a better judge of commercial opportunities than the markets – a discredited proposition. Yet it is one that influences too many governments, including the German, whose industrial policies direct vast sums at manufacture to the detriment of service (the figures in the footnote above show that West Germany not only has the smallest service sector of the advanced countries, but that its service sector is smaller than the USA's was in 1970). German subsidies have been directed at the large manufacturers, whose output is falling, and this has not only harmed service, it has also damaged the small and medium-sized manufacturers which have historically been the engine of German economic growth. The German economy will continue to wobble over the next two decades, not because of the cost of unification (which will receive the blame) but because German *dirigisme* will do for industry in the twenty-first century what it did for agriculture in the nineteenth.*

(iv) Countersubsidies

If one government subsidises a product, should other governments counter with subsidies for their own domestic industry? A comprehensive answer to this question lies outside the competence of this book but, on first principles, we can argue that the first government to subsidise is taking a decision that, by definition, is uneconomical because, had it been commercially reasonable, private investors would have done it. If private investors baulk, that implies that the first subsidising government is giving foreigners a free gift of cheap goods.

The free marketeer might argue that foreigners should happily accept the gift, and that they should then follow the market forces that would direct domestic labour into other areas, to the general enrichment of the country that imports subsidised goods. Some countries do indeed pursue aspects of that policy very successfully. Consider motor

*One small reflection of the German backwardness over service is provided by money. Electronic transactions with cards have become usual in Britain or the USA, but the Germans still carry around vast wads of bank notes. German high street banks are so inefficient that they routinely take 12 weeks to transfer money from America (see A. Gimson, *Sunday Telegraph*, 23 October 1994).

cars. Two countries in Europe have no domestic car industry, Belgium and Denmark, but this benefits the Belgian and Danes hugely. The Belgians impose 100 per cent duties on cars (so foreign manufacturers have to price them very cheaply) while the Danish government simply fixes the prices of cars artificially low. Thus the Belgians and the Danes benefit from the subsidies spent by the British, French, German and Swedish taxpayers.

But governments may be pressured to react to subsidised imports with their own domestic subsidies for at least two good reasons, since (i) the social costs of closing domestic industries are high and (ii) military needs might override economic judgements. We have discussed those above. But science policy activists often suggest a third reason. This is less valid, and should be viewed cautiously: science policy activists argue that governments should countersubsidise to prevent their domestic industry being driven into permanent extinction. They claim that, once a particular country has acquired technological dominance in a particular industry, it will be able to sustain that in perpetuity. That claim grows out of a famous academic study of Arrow's. Entitled 'The Economic Implications of Learning by Doing',[4] the study showed that relatively few technological advances in industry derive from 'break-throughs'; rather, most advances are small (indeed, many are never published or recorded, but simply emerge from the growing expertise and experiments of the workforce). In time, the accumulation of such small advances may lead to a vast increase in productivity, to give an edge to those who 'do', since they are the only ones who can 'learn'.

But 'learning by doing' is a poor argument for government subsidies. First, many of the most important advances are in fact 'break-throughs'. Consider the watch. For years, based on mechanical expertise, the Swiss dominated that industry, but their dominance crumpled over-night when the Japanese invented an electronic technology. Market dominance must always be transitory because all technology is transitory, and technological monopolies can rarely be sustained because technological advance is so unpredictable. Thus the German dominance of the chemicals industry during the early twentieth century has been superseded by America's and Britain's current successes, while America's and Britain's former hegemony over the motor car, motorcycle and ship has been acquired by Japan (yet Britain is now manufacturing motorcycles again, and Japan is losing orders for ships to the other emerging economies of the Pacific Rim).

Moreover, if 'learning by doing' really is so important, then that emphasises the importance of the industrial workplace to technological advance; if, therefore, governments were to raise taxes (i.e. tax all industries) to support a particular industry, then governments would be impoverishing the successful centres of technological development in the land to support one that is failing. That is a doubtful strategy. Furthermore, 'learning by doing' does not preclude the poaching of skilled foreigners should the need arise. Skills and expertise, being attached to people, are highly mobile. Contrary to the protectionist myth, imports (even subsidised imports) stimulate domestic invention (see the study of Sokoloff, already referenced in Chapter 7) and governments which subsidise are generally sustaining outdated technology.

(v) The Example of the Civil Aircraft Industry

The folly of countersubsidies, and of subsidies, is illustrated by this industry. The Americans currently dominate the civil aircraft industry, and this has often been adduced as a triumph of government funding.[5] Since the Second World War, Washington has poured some $23 billion into military aeronautical research, which American manufacturing companies have transferred, cost-free, into their civil divisions.[6] In consequence, companies like Boeing and McDonnell Douglas now dominate the civil aircraft industry worldwide. But, for the American taxpayer, the sums do not add up. The manufacturing companies may appear to have made good profits, but only when those profits are calculated as the return on the $3.5 billion of their own money that the manufacturers have invested, on average, on each new airliner design – it has certainly not represented a good return on the $23 billion the American taxpayer has invested in aviation R&D since 1945. The American taxpayer has been stung twice: not only by his own domestic manufacturers, but also by the foreign carriers who have bought Boeing and McDonnell Douglas aircraft at, effectively, discounted prices.

Actually, as David Edgerton showed in his *England and the Aeroplane*,[7] Mowery and Rosenberg were wrong to adduce the civil aircraft industry as an example of government subsidies enabling a country to dominate a free market, because there has never been a free market in aircraft manufacture. We cannot easily assess the true costs of any civil aeroplane, because aeronautical R&D has always been dominated by defence: 'it is hard to think of an aeronautical advance that was not funded by a government defence department'. The military implications of flight were recognised as early as the Napoleonic wars, when armies spied on each other from balloons. After the Wright brothers first flew in 1903, soldiers everywhere were alarmed, and by 1914, Britain, France and Germany all possessed air forces. The military investment in aviation research has always dwarfed the private, civil component, and the world's major civil aircraft manufacturers have always been the world's leading defence researchers (hence Germany's JU 52 between the wars, hence Britain's Viscount, Comet and VC10 after the war, hence France's post-war Caravelle, and hence the USA's current dominance). Contrary to myth, these countries did not dominate civil aviation because they possessed the largest internal market (or Japan would now be a much more significant producer), they dominated as a military spin-off (which is why Concorde was Anglo-French, and which is why Germany and Spain have had to cooperate with Britain and France to build Airbus).

Certainly, the decisions of the European governments to subsidise Airbus in competition with the USA have been remarkably foolish. We do not know the exact sums that Airbus Industries receives in direct subsidies because, in true Continental fashion, the consortium has been relieved of the burden of publishing proper accounts as it is a *groupement d'interet economique,* but as the European governments have so far admitted to direct subsidies of $6 billion ($8.5 million per aircraft built!) we can be sure that the figures are even higher. Yet these frightening sums cannot even be justified as a challenge to an American monopoly, because Boeing and McDonnell Douglas have always competed desperately. Damningly, the introduction of the Airbus has driven McDonnell Douglas to the wall, which has merely replaced the competition between Boeing and McDonnell Douglas with one between Boeing and Airbus. Even worse, Boeing and Airbus will no longer compete, because they have now agreed to cooperate on a new super jumbo jet with a capacity for 800 passengers. Thus, the British, Spanish, French and German taxpayers have spent $6 billion to destroy a healthy competition between Boeing and McDonnell Douglas (one the American taxpayer was kindly subsidising) to replace it with an unhealthy cartel between Boeing and Airbus (one which will continue to be a drain on the four European countries' taxpayers). The only beneficiaries will be the politicians whose sense of self-importance will be vastly inflated by the huge empowerment they will receive from disbursing these gargantuan sums (the work force will barely benefit because, their business depending on political whim, they will never be able to enjoy the security or the satisfaction that flows from genuine achievement in a free market).*

* Airbus has also inflated industrial corruption. Airbus's salesmen, who have close links to the EC Commission, can blackmail potential customers. Thailand's second largest export commodity, for example, is tapioca, and thousands of poor

(vi) The Empirical Evidence

The empirical evidence confirms that, overall, subsidies or counter-subsidies fail. Thus Switzerland and Japan are the least two taxed countries in the world, and the least industrially subsidised – and the richest. The Swiss are particularly naughty; they import subsidised goods in large quantities from their neighbours in Germany, Austria, France and Italy, yet Switzerland is actually richer than they. But the low Swiss taxes have enabled private investors to fund the R&D of Roche, Sandoz and Ciba-Geigy in Basle so generously that these companies' products sell abroad effortlessly. Indeed, so damaging are government subsidies, it is suspected that the EC is anxious for Switzerland to join so that the Commission in Brussels can burden the Swiss with EC levels of subsidy, and so reduce its competitive advantage.

(vii) Conclusion to this Section

We have been considering the final stage of the linear model, the conversion of technology into wealth. We have concluded that, at that stage, the linear model is correct; to a large extent the creation of wealth is synonymous with the introduction of new technology.

Since this book is concerned with government funding, we have touched on the government funding of industry to illustrate two general principles: (i) government money comes, ultimately, from industry, and (ii) markets make better commercial judgements than governments. We have noted that governments may have legitimate, non-economic justification for intervening in the economy (especially for social or military reasons), but that too often politicians invoke those to intervene to bolster their own interests as a political caste. However, we must concede a major difficulty. Our arguments have assumed that everyone is working in a free market, but they are not. If we define a free market as one where (i) there are a large number of competing producers, (ii) there are a large number of customers and (iii) both producers and customers play on a level field (to use contemporary fashionable lan-

---

Thais depend desperately on tapioca exports to Europe for their livelihood, yet Ian McIntyre has reported how Airbus salesmen have threatened Thai officials with an EC ban on tapioca imports unless they bought Airbus [Ian McIntyre, *Dogfight* (Praeger, 1993)]. There have been other such threats across the globe.

guage) then many markets are not free. The cost of entry (to use economists' jargon) into civil aviation, for example, is huge, and the world economy could not support a large number of competing producers.

Politicians and economists* invoke the innate lack of freedom in the market for major technology as a justification for subsidies. Yet politicians have their own vested interest in subsidies as these create a dependent, client sector within the electorate. They also confer great power on the politicians. True statesmen would seek to overcome the innate lack of freedom of the market in big technology by supporting GATT (now the World Trade Organization) or by negotiating bilateral arrangements that precluded unilateral subsidies, but national politicians will use other countries' subsidies, or social or strategic arguments, or the evidence of the economists that the market is not entirely free, to justify their own self-interested subsidy of industry or the erection of tariffs. Consider Switzerland. A major anomaly in the Swiss commitment to *laissez faire* is its vast subsidy for agriculture, which outstrips even that of the EC's Common Agricultural Policy (CAP). The economic futility of such a waste of taxpayers' money recently prompted a referendum, which voted for a marked reduction in subsidies. But the politicians bypassed the vote. The representatives of agricultural constituencies persuaded their fellow politicians, in exchange for mutual favours, to help them in defusing the referendum. Thus even in Switzerland, the most democratic country in the world, the people's desire for rational economic policies has been frustrated by the politicians' needs for power and influence (and, sadly, corruption; subsidies, controls and licences offer a pot of corruption for politicians and, all over the EC, bureaucratic structures are being created in imitation of those of Italy). Thus we see that the German, French and Central European tradition of government-sponsored cartels owes as much to politicians' self interest as to any philosophical construct of the State or economists' distrust of the market.

DOES ACADEMIC SCIENCE BREED TECHNOLOGY?

To answer this question, the first link in the linear model, we have to ask a wider one; from whence does new technology come?

---

*In my experience, economists rarely believe passionately in, or care passionately for, the free market. They are generally more concerned to reveal the market's imperfections, to further their own professional importance.

Whence does New Technology Come?

Here is a central question in Man's millennial search for a better life: whence does new technology come? Fortunately, a number of careful empirical studies have now answered this question. The results are startling.

The most comprehensive answer to this question emerges from the study of Edwin Mansfield.[8] Mansfield surveyed 76 major American firms which, collectively, accounted for one-third of all sales in seven crucial manufacturing industries – information processing, electrical equipment, chemicals, instruments, drugs, metals and oil. He discovered that, for the period 1975–85: 'about 11% of new products and about 9% of new processes could not have been developed, without substantial delay, in the absence of recent academic research'. An earlier survey of American industry by Gellman Associates[9] also showed that around 10 per cent of industrial advances were attributable to recent academic research. These findings represent, therefore, a gaping flaw in the linear model: only 10 per cent of new technology emerges from academic research. Ninety per cent of new technology arises from the industrial development of pre-existing technology, not from academic science.

A British study by Langrish *et al.* spelt this out: reporting on the origins of 84 technical innovations in British industry (so important that they won the Queen's Award for Industry) they found that 'although scientific discoveries occasionally lead to new technology, this is rare'.[10] Generally, 'technology builds on technology' and that building of technology occurs within the R&D departments of industry. And we have already noted the US Defense Department's *Project Hindsight* which found that, of 700 research 'events' which had led to the development of 20 weapons systems between 1945 and 1965, only two had arisen from basic research.

Indeed, the implications for the linear model are even grimmer than these statistics show. Mansfield may have found that around 10 per cent of new products or processes arose from the industrial development of academic research, but these innovations tended to be economically marginal: they only accounted for 3 per cent of sales and 1 per cent of the savings that industry made through innovation. These studies, therefore, show that 90 per cent of industrial innovation, and well over 95 per cent of industry's profits through innovation, arise in-house from the industrial development of pre-existing technology.

The linear model, therefore, needs a fork and lots of arrows:

academic science → new technology → economic growth.

↑↑↑↑ ↑↑↑↑

old technology

(We can now see immediately why *dirigiste* France failed to create wealth. By pouring money into academic science, the French followed the wrong model. But *laissez-faire* Britain, freed from centralised planning, could follow the market, which correctly directed investment into technology, not science.)

The so-called linear model not only requires a fork, it also requires a reverse arrow. Technology can lead as vigorously to advances in basic science as vice versa. We have already noted such earlier examples as Carnot's *Reflexions sur la Puissance du Feu* published in 1824, which transformed thermodynamics, but which was inspired by the failure of contemporary physics to understand Newcomen's steam engine, then over a century old. This pattern continues to be repeated. Consider radioastronomy. This appears to be one of the purest of sciences, yet it was born and fostered by industry. During the late 1920s, Bell Labs inaugurated an overseas radiotelephone service, but it was plagued by static. Jansky, an engineer, was employed to discover the source of this noise, and in 1932 he identified it as coming from the stars. Thus was born the science of radioastronomy. Two other Bell Lab employees, Penzias and Wilson, studying the problems of microwave transmission, discovered the cosmic microwave background radiation, and made the experimental observations that underpin the Big Bang theory (they won Nobel Prizes in 1978). It was only after these industrial discoveries that the universities took up the study of radioastronomy.*

Consider the current academic ferment over ceramic superconductivity. Departments are mushrooming in universities all over the globe

---

* It is astonishing how many nineteenth- and twentieth-century discoveries in astronomy, that purest of sciences, have been made privately. In Chapter 6 we noted how the British hobby scientist William Parsons, the third Earl of Rosse, built the world's then-largest telescope in 1845. Later, rich Americans competed to outstrip him, with first Lick, then Yerkes and then Lowell (1855–1916) claiming the palm. The Lowell Observatory still stands at Flagstaff, Arizona, having, *inter alia*, fostered Vesto Slipher's discovery of the large Doppler shifts that reveal that the galaxies are expanding radially. Most famously, it was the Lowell Observatory that discovered the ninth planet, Pluto, in 1930, following Lowell's own prediction of a 'Planet X' to account for discrepancies in the orbit of Uranus [for fuller history, see William L. Putnam, *The Explosion of Mars Hill* (Phoenix, 1994)].

(the one in Cambridge was founded in 1988) but ceramic supercon-ductivity did not exist until two IBM engineers, Bednorz and Muller, discovered it in 1986. The industrial technology came first, the aca-demic science followed.

The same pattern can be seen in biology. Even mighty molecular biology arose from applied science. How do we know that DNA is the molecule of inheritance? Because O. T. Avery, a doctor, was studying pneumonia at the (privately-funded) Rockefeller Institute. During the 1940s, pneumonia was still a killer and Avery was studying *Pneu-mococcus*, one of the bacteria that causes it. *Pneumococcus* comes in two forms, encapsulated and non-encapsulated, and Avery discovered that if he injected DNA from the encapsulated into the non-capsulated, he transformed it permanently into the encapsulated. Thus did Avery discover that DNA regulated inheritance (DNA had been described decades before, but no one knew what it did). Once Avery had discov-ered the importance of DNA, the subsequent development of molecu-lar biology was inevitable – thus the discovery of the structure of DNA in 1953 became a race between three teams, Linus Pauling's in Caltech, Watson and Crick in Cambridge, and Franklin and Wilkins in London. Molecular biology has continued to be a race ever since, as teams across the globe have chased each other towards predictable goals.

Cross-Fertilisation

To counter the popular misconception that technology is dependent on academic science, we have highlighted the dependence of academic science on technology but, of course, the two can fertilise each other. Consider, as one out of innumerable advances, solid state physics. During the 1930s, only two universities in the world taught it, MIT and Princeton. It was a Cinderella subject, of little interest to anyone. But in 1948 one of MIT's graduates, W. Shockley, while researching at Bell Labs, developed the transistor. Thereupon, academic physicists scrambled to research this new discipline. J. A. Morton, the head of transistor re-search at Bell Lab, wrote during the late 1940s that it was impossible to hire people 'because solid state physics was not in the curriculum of the universities'.[11] Shockley had to run a course at Bell Labs for academic physicists, and the standard university text book was Shock-ley's famous *Electrons and Holes in Semiconductors*. Industrial re-search continued to drive the academics: 'Even many years later, in centres of semiconductor activity such as Silicon Valley, it has been far from unusual for university courses in the solid state to be taught by part-time professors from local industry.'[12] Thus the flow of theoretical knowledge in solid state physics has alternated; in the early phase it was a small, university-based discipline, but it took off in the industrial labs, and the flow of knowledge was reversed. And the Nobel Prize in Physics, of course, went to Shockley and his industrial colleagues.

The Separate Autonomy of Science and Technology

To counter popular misconception, we have highlighted the mutual dependence of science and technology but, in daily practice, these two interact relatively rarely. In daily practice, academic science is as self-contained as technology. Just as 90 per cent of new technology arises from old technology so, for most of the time, new academic science builds on old. The two disciplines largely grow separately and I, as a practising biochemist, rarely observe my colleagues reading journals that are not biochemical. The so-called linear model, therefore, requires yet a further modification to reflect that both science and technology are largely self-contained, growing on themselves:

basic science $\leftrightarrow$ technology $\rightarrow$ wealth.
$\qquad\uparrow\qquad\qquad\quad\uparrow$
old science   old technology

Science and technology remind me, in their dissimilarity but distant recognition, of the Lion-faced Boy and the Heir Apparent in the second stanza of Robert Graves's poem *Grotesques*:

> The Lion-faced Boy at the Fair
> And the Heir Apparent
> Were equally slow at remembering people's faces.
> But whenever they met, incognito, in the Brazilian
> Pavilion, the Row and such-like places,
> They exchanged, it is said, their sternest nods –
> Like gods of dissimilar races.

Can Government Funding Help Industry Develop New Technology from Old?

Now that we know whence comes new technology, we can ask this question. The answer has to be no. Governments can only raise money through taxing industry, either directly or indirectly, and for governments to tax industry, to fund and direct its R&D, would suppose that government knows better than the directors of each individual enterprise how best to optimise that enterprise's R&D. That is an absurd supposition. As we saw above, it would also be absurd for governments to tax industry generally to specifically boost manufacturing industry, because that would suppose that governments were better commercial judges than the market.

It is often argued that long-term research projects are neglected by the free market because bankers or the City are obsessed by short-term profits. Only the Government, it is claimed, will fund such projects, yet governments' records across the globe in funding long-term projects are terrible. Such projects are rarely, if ever, profitable, witness Concorde, the British nuclear power programme or the Alvey and ESPRIT programmes that introduced this chapter. One must conclude, therefore, that free markets are shrewder than governments at judging the prospects of profit.

Bankers are right to be wary of long-term projects, because science and technology advance so rapidly, and so unexpectedly, that long-term projects are often rendered redundant before they are complete. Long-termism can, therefore, be uneconomic. Consider the Leeds and Liverpool Canal, a dramatic engineering enterprise which still survives (the staircase of five locks at Bingley, the so-called 'Bingley Five-Rise' is glorious). Started in 1770, this challenging canal (it crosses the Pennines) was not finished until 1816, at a cost of £1 200 000 – a very long-term project. But in 1841, the opening of the Leeds and Liverpool railway made the canal redundant. The canal's investors would have done better to have eschewed the long term and to have put their money into the short. So, after 1945, would the British taxpayer.

It is often forgotten that Labour's major justification for the nationalisation of the 'commanding heights of the economy' between 1945 and 1951 was that major industries would perform better if they were removed from the short-term pressure of the market, and run according to long-term goals. By 1975, those monuments to long term planning were collectively losing £2 750 000 000 (i.e. £2750 million) a year – and in 1975 £2750 million was an enormous sum of money. On privatisation between 1981 and 1989, the loss makers, on subjection to 'short-term' market forces, became as profitable as they had once been liabilities.

That new profitability developed because, it emerges, politicians are the shortest short-termers of all – but their short-termism is electorallydetermined rather than profit-based. In his detailed *History of the British Coal Industry*, William Ashworth showed that time and again it was the politicians who overruled the managers' long-term rational planning to court short-term electoral popularity.[13] One remembers as well the appalling loss-making Humber Bridge that the Labour government built just to help win a by-election, and the famous Cabinet meeting which decided to close a loss-making railway in Wales until the Chief Whip intervened with 'but that runs through seven marginal constituencies'. The private sector makes much better commercial judgement than politicians because the private sector is focused on commerce – politicians buy votes with taxpayers' money.

But it is specifically argued that British and American financial institutions are more obsessed with the short term than their German or Japanese counterparts, hence the greater economic success of Germany and Japan. As suggested in Chapter 7, it is doubtful if the German or Japanese economies have performed better than might have been predicted, and even if they have, it is almost certainly not due to their far-sighted financiers. Bankers are not altruistic, they will always prefer the security of a short-term risk, and if German or Japanese bankers lock themselves into long-term projects, they will have commercial reasons. Japanese and German bankers certainly lack the opportunities for short-term profits offered by the flexible, international money markets of London or New York and, indeed, the situation in Japan is more dire than is generally realised outside the country. Japanese investors are often stuck with their investments because Japanese business accountancy is so primitive that it can actually be impossible to know if any particular company is profitable or not. When, for example, Showa Shell Sekiyu (a partially-owned Japanese subsidiary of Royal Dutch/ Shell) admitted in 1993 to losses of $1050 million on foreign exchange futures, it was also revealed that the company had disguised these vast losses for over four years. In the words of the *Financial Times* for 23 February 1993: 'It is no tribute to Japanese accounting standards that the irregularities seem to have gone undetected for several years.'

This failure of Japanese business practice only surfaced in the Western press because of its effects on the share price of the parent Western company. The *Financial Times* commented that 'the [Japanese] ministries of both trade and finance would like to portray the development as an isolated case, but the extent of the losses, the fact that they did not surface for so long [and] the fact that under Japanese accounting rules companies are not required to disclose unrealised losses' led the paper to suppose that the Showa Shell Sekiyu débâcle was only the revealed tip of a very large submerged problem. Japanese banks are as incompetent as Japanese accountants, and they often do not know if their investments or clients are sound since their own practices and technology are so primitive (readers who find this hard to believe should read Tim Jackson's *Turning Japanese*).[14] And now the banks are collapsing under a flood of bad debts. The Tokyo Government had to give (yes give) $140 billion of taxpayers' money to the banks in 1994, just to stop them from going bankrupt, and it is anticipated that this will have to be ten times greater in 1995 at $1.4 trillion.[15] The incompetence of Japanese bankers is truly breathtaking.*

One aspect of the German banks' long-termism derives from a Continental culture that is alien to Britain and the US – government interference. European governments, with their *dirigiste* traditions, simply do not allow large companies (the ones who tend to do research) to go bust. This was illustrated in 1993 when the Anglo-Dutch truck manufacturer Daf went bankrupt. The British half was dissolved and the work force dismissed, but the Dutch part was rescued by the Dutch Government (at a huge cost to the Dutch taxpayer). Continental banks, therefore, can lend with insouciance, knowing that the taxpayer will always bale them out, while in Britain and the US, banks have to achieve a return on their investment (witness, also, the refusal of the British government to rescue Barings. No continental government would have failed to save an equivalent domestic bank).

British and American bankers are rational, and when circumstances change, so do they. Consider the Channel Tunnel, a huge long-term project. The British half was built, and privately financed, by the Eurotunnel Company. Sir Alistair Morton, Eurotunnel's chairman, in a speech to Birmingham Polytechnic on 9 June 1992, explained that he had succeeded in raising the money by selling shares because they

---

*The incompetence of the Japanese belies their image abroad, though the aftermath of the Kobe earthquake has disillusioned some of the more naïve observers. The only international conference I have attended where my hotel booking was mislaid was in Tokyo. The only industrial country whose shop assistants will still use the abacus is Japan. None of which, of course, alters the fact that the Japanese are, individually, charming.

The problem with Japan is that government policy has been to foster manufacture, to the detriment of service (which is not only too low at 62 per cent of GDP, but which is also growing too slowly, having been 52 per cent in 1970). Japanese manufacture, therefore, is suffering from inadequate service. Fortunately, the recent huge falls in Japanese manufacturing output (down 7.6 per cent in 1992) are being coupled with a rise in services (overall GDP rose by 0.9 per cent in 1992) and if government attempts to restore manufacture fail – as it is hoped they will – the Japanese economic growth into the twenty-first century will be assured.

were issued soon after the crash of 1987. Robbed of short-term prospects, investors were forced into the long-term. As Sir Alistair said: 'all investments are long term after the market collapses'.[16] Much of British and American technological short-termism has been no more than investors' recognition that the American and British finance markets' internationalism and efficiency offered surer profits than the relatively inefficient, parochial financial markets of Germany, France or Japan. The London and New York markets have been more rational and more effective and they have, therefore, served their countries well.

High Definition Television (HDTV) provides a graphic illustration of the dangers of long-termism in research. HDTV is known as the 'wrinklies' enemy', because it will produce sharp images that will portray warts and all, but the history of the funding of HDTV has produced a telling lesson. In 1989 Mr Pandolfi, the EC industry commissioner, proposed an Ecu 850 million (£688 million) R&D programme for HDTV. It was to be based on analogue transmission. It took three years to consolidate the proposal but in 1992, just as it was about to be ratified by the EC industry ministers, it was vetoed by Britain – to the fury of the other eleven EC nations. Britain vetoed because its minister at the time, Edward Leigh, believed in the free market, and he believed that if research into HDTV were a good investment, industry should fund it (Leigh was supported by the British Treasury, which was then short of money). Leigh's doubts were well-founded. By early 1993, the analogue transmission of the EC's proposed HDTV was already falling behind new digital technology that was being developed by American companies, and it was emerging that any long-term R&D in analogue systems would have been wasted.[17]

Were Britain's EC partners grateful for the veto which saved £688 millions of taxpayers' money? Were they hell. Commissioner Andriessen actually complained to Mr Major, the British Prime Minister, over Mr Leigh's impertinence, and this contributed to Mr Major's sacking of Mr Leigh in May 1993 (Mr Major sets much store by appeasing the EC, see Mr Leigh's description of his sacking in the *Spectator*[18]). The EC's concern with HDTV reflected no passionate desire to improve technology for its own sake but, rather, a typical Continental attempt to rig the market. The EC hoped that if it could develop analogue quickly, it could establish technical standards that would have to be accepted internationally. This would lock the rest of the world into inferior European analogue technology at the expense of the superior American digital technology.

An earlier EC scheme for HDTV in 1986 had already foundered on rapid American advances in satellite technology that had enabled broadcasters to circumvent the EC's so-called 'Mac' standards, and ever since the Commission had been looking for ways to circumvent superior American technology through European wiles and tricks. How silly! As Michael Porter explained in his *Competitive Advantage of Nations*[19] 'national differences are at the heart of competitive success; nations prosper in industries that draw most heavily on unique elements in their histories and characters'. Or, as David Ricardo noted during the nineteenth century, some nations enjoy 'comparative advantage'. It would be much better for the Commission to accept that, in electronics, America enjoys unique elements that predispose to success and, rather than waste taxpayers' money, the Europeans should gratefully import new American technology while fostering their own comparative advantages (see the postscript on technical standards and competitive national advantage).

WILL THE FREE MARKET SUPPLY ENOUGH BASIC SCIENCE?

Although we have established that most technology arises from preexisting technology, some does arise from basic science. Will the free market supply enough? The overwhelming academic consensus is that it will not; no serious scientist, economist or politician currently advocates *laissez faire* in the funding of basic science (this book is the exception). Even a passionate champion of the free market like Arthur Seldon, the founder of the Institute for Economic Affairs, a follower of Hayek and of Milton Friedman, and an advocate of the privatisation of the schools, hospitals and of social security, conceded in his *Capitalism*[20] that governments had to fund science.

The Theories of Nelson and Arrow...

The two seminal academic studies that persuaded everyone were the papers of Nelson[21] and Arrow.[22] Nelson and Arrow argued that governments should fund basic science because industry will not. They claim that industry will not for two reasons: first, the outcome of basic research is unlikely to be commercially applicable directly, so why should industrialists fund it? Second, the discoveries of basic science invariably become widely known, so why should industrialists fund discoveries from which competitors could benefit? (In jargon, private funders cannot 'capture' all the benefits of the basic research for which they pay, so they will not invest in it.) Because, therefore, they claim that companies will not fund basic science, Nelson and Arrow argue that governments should do it instead.*

... are Disproved by the Observations of Mansfield and Griliches

Unfortunately for the theories of Nelson and Arrow, companies do, in practice, fund basic science comprehensively. Moreover, they find it highly profitable. Two major surveys by Mansfield and Griliches have shown that the more basic science a company performs, the more likely it is to grow and to outperform its competitors. Mansfield studied 16 major American oil and chemical companies for the years 1960–76, and he showed that all those firms invested in pure science; crucially, the more a firm invested in basic science, the greater its productivity grew.[23] Griliches, in a comprehensive study of 911 large American companies for the years 1966–77, showed that the companies that engaged in basic research consistently outperformed those that neglected it; the more basic research a company performed, the greater its profits.[24] Basic research, therefore, turns out to be commercially highly gainful, and the companies that fund it ultimately grow to command the market. Nelson and Arrow, therefore, are wrong; the market positively rewards companies who perform pure science. Why?

First-Mover Advantages

Basic science produces two types of commercial benefit, 'first-' and 'second-mover advantages'. First-mover advantages (as defined by M. Leiberman and D. Montgomery in their 'First Mover Advantages'[25]) are exactly what they are called, they are the advantages that accrue from discovering something first. They are discoveries.† The commercial fruits of first mover basic research can be gargantuan, and individual companies can expand to dominate an entire industry through

---

*Discoveries can, of course, be patented, but Arrow disapproves of patents on the grounds that they inhibit economic growth by denying other companies free access to the patented research. (Conventionally, of course, patents are believed to stimulate economic growth by providing an incentive for the private funding of science). Arrow would prefer to see governments perform research to make it freely available. However, governments these days patent furiously, so his arguments do not apply.

† Even if a discovery cannot be patented because it is such basic science, the prior

the exploitation of an advance in fundamental knowledge; but basic science is unpredictable and if, as Arrow and Nelson assumed, it only led to first-mover advantages, it would indeed be too risky for generous industrial support. Second-mover advantages are much more important, and more certain. What are they?

Second-Mover Advantages

Consider a company that decides on an R&D strategy based on basic science. It plans to explore the basic science in a particular area, and then create a product based upon it. How to explore the basic science? It could employ a team of scientists, lock them away in a laboratory, and tell them not to emerge until they have made some discoveries. The scientists will love it, they will seize on interesting problems, write papers, present their results at international meetings, and generally have fun. But the company's chances of their making a commercially useful discovery are slim, and restricted to the particular experiments performed in-house. Yet, across the globe, there are hundreds if not thousands of other scientists, in universities, research institutes and company laboratories, publishing their findings on related problems in related fields, and if the company directed most of its scientists' efforts at following those activities, the company would be much more likely to uncover a potentially useful advance. The time a scientist spends in the library or in the seminar room is more valuable to the company that employs him than the time he spends at the bench. Successful companies have long understood this, and in Chapter 7 we presented the research prospectus of Mitsui Pharmaceuticals of Japan which stated that its first priority in R&D was: 'The exhaustive survey of the [scientific] literature, collection and analysis of various information, establishment of R&D targets, and coordination between various research groups.' Only secondly did Mitsui prioritise: 'The chemical synthesis of new compounds ... and the exploration of new biologically active substances from natural sources.'

Let us consider a specific example of second-mover research, Glaxo's Zantac. The treatment of stomach ulcers has been revolutionised by drugs known as histamine-2 blockers and Zantac, the brand leader,

---

knowledge of it enables the discoverer to establish the technology, acquire the appropriate assets (such as raw materials, machines or transport) and establish a brand before his competitors. To use jargon, prior knowledge enables an industrialist to progress first along a learning curve and to establish himself as a market leader.

makes Glaxo, annually, hundreds of millions of pounds of profit. But histamine-2 blockers were first discovered by James Black who worked at SmithKline, not Glaxo. Black spent years in 'first-mover' science. He was studying stomach ulcers, and he knew that such ulcers were caused by stomach acid, and he knew that such acid was regulated by a chemical called histamine. But no anti-histamine then available could block acidity, so he hypothesised that there had to be a second class of histamine receptors, histamine-2, that needed to be specifically blocked. He spent years, and a lot of SmithKline money, before discovering the first commercial histamine-2 blocker, Tagamet. That helped him win his Nobel Prize in 1988, but Tagamet achieved only modest profits for SmithKline, because Glaxo made a 'second-mover' killing on Zantac.

The head of Glaxo's research team was David Jack, and M. Lynn has described in his *The Billion Dollar Battle: Merck v Glaxo*[26] how David Jack attended a lecture of Black's one evening, during which Black described his new histamine-2 blocker, and Jack returned to Glaxo the next morning determined to copy Black, but with a more potent derivative. Copying is much easier than original research, and Jack soon came up with Zantac. In his own words, Jack acknowledged that: 'It was a straight piece of medicinal chemistry because the original thinking had been done by Jim Black. It does however show you something very important. The second prize in this business can be bigger than the first.'[27]

This was a classic piece of second-mover research; Black had made a major first-mover discovery, and Jack moved in to make an intellectually small, but commercially lucrative, second-mover improvement. David Jack, moreover, was correct in guessing that second prize can be bigger than the first. Indeed, this has been systematically studied. In a recent survey of the ten major Japanese pharmaceutical companies (which together enjoyed $13 000 million sales in 1981) Odagiri and Murakimi found that each individual company could expect, on average, an annual return on its investment in R&D of 19 per cent.[28] But its competitors benefited more. By exploiting the advances made by any one company, the other nine companies could create new drugs which, together, provided the equivalent of a 33 per cent annual return on the first company's research. Second-mover advantages were worth nearly twice as much as first-mover advantages.

This was a beautiful study by Odagiri and Murakimi, but they ruined it by their conclusions. The arguments of Arrow and Nelson are so ingrained within the souls of academics who investigate the economics of research that Odagiri and Murakimi, like everyone else in the field, just parrot them. Arrow and Nelson argued that, because companies can only sequester a small proportion of the benefit of their research, they therefore underinvest in it – hence their call for government funding. Odagiri and Murakimi drew the same conclusion in their paper: 'Thus the private returns are argued to be smaller than the social returns, and government support is called for to make up for the reduced private incentive to research.' But this is nonsense. Second-mover advantages are not free; they are very expensive, and that expense is dedicated to research, much of which is first-mover or basic. While Zantac was being discovered, Glaxo's profits fell from £87 million (1977) to £60 million (1980) as research costs rose from £17 million (1976) to £40 million (1981). Let us examine the costs of second-mover advantages.

The Costs of Second-Mover Research

The biggest myth in science funding is that published science is freely available. It is not. Access to it is extremely expensive. Consider an analogy with the law. No one assumes that legal knowledge is freely available. Anyone could, if they wished, consult the law books and journals to defend themselves in court, but it would take years to master the law and the court-room lore that is never published, so people employ lawyers and pay them for their accumulated expertise. So it is with science. It takes years of training before a scientist can read research papers properly and understand their implication for the future. It takes hours, every week, to read all the new research papers, to assimilate them and to integrate them into a future research strategy. As discussed above, so important is the collection and integration of scientific data, that successful companies see it as the prime role of their scientists. And the accumulated expertise that enables scientists to collect and interpret data does not come cheap. (Oddly, the artists seem to have understood this where the economists of science have not. When Whistler was asked how he could justify charging two hundred guineas for a painting that only took two days to paint, he replied that he was not charging that fee for: 'two days' labour. No, I ask it for the knowledge of a lifetime.')

But the true cost of second-mover research presents companies with a problem. Scientists are activists and egotists – they like to do research – they do not wish to spend all their time in libraries. And the better the scientist, the more active and the more egotistical he is. To retain good scientists, therefore, companies – essentially – bribe them with laboratories, money and the freedom to publish, much as a company pays lawyers' fees. Good scientists have to be bribed with considerable liberty. It was no coincidence that David Jack was employed with: 'final control of everything the research department did'.[29] In effect, scientists are in-house consultants, and their first-mover research costs could be discounted as consultancy fees. Of course companies hope to exploit any first-mover advantages that might accrue, but scientists' second-mover advantages are the more important.

Let us summarise and repeat this point, since it is so important. Companies are primarily interested in second-mover research; they need to know what everyone else is doing so that they can exploit advances from all over the world. The only people who can monitor other people's research are the scientists. Companies, therefore, have to employ scientists. But scientists themselves are only really interested in first-mover research, and the best scientists are obsessed by their own, first-mover research. Yet even scientists need a salary, so companies and scientists agree a *modus vivendi*. Companies pay scientists to do the first-mover research that they, the scientists, enjoy; while in return the scientists, through their reading and attendance at conferences, keep the company informed of developments worldwide. No irreconcilable conflicts of interest arise, because scientists cannot do good first-mover research without keeping abreast of developments, and the only people with the ability to keep abreast of developments in an area of science are the first-mover researchers in that area. But the financial consequences are that companies have to invest very heavily indeed in their researchers' first-mover science to retain them as second-mover consultants.

We can see, now, that the argument that 'thus the private returns are argued to be smaller than the social returns, and government support is called for to make up for the reduced private incentive to research' is false; first- and second-mover advantages are indissolubly linked, and one cannot be performed without the other. And second-mover advantages enforce a vast expenditure on basic science. David Jack had been formally employed by Glaxo to research anti-asthmatic drugs, and indeed he led the team that produced Ventolin, a lucrative anti-asthmatic. When Glaxo was funding David Jack's research, therefore, it thought it was funding first-mover science, and was probably grumbling all the time about its competitors' second-mover advantages – while actually proving itself the most accomplished second-mover in the business.

Basic scientists do even more for companies. In his 'Why Do Firms

Do Basic Research with their Own Money?'[30] Nathan Rosenberg of Stanford University showed how scientists possess analytical, creative and problem-solving skills that are employed at all levels within a company. Engineers and technologists consult them over their problems, as will the marketing, accountancy and sales departments; in return, scientists can integrate ideas from these other divisions with their own research. Within a successful company, the linear model is completely irrelevant; instead, the various divisions interact and cross-fertilise in a matrix mode.

Within a country, too, the linear model is also irrelevant, and scientists largely boost their national economic performances by means of the foreign research they capture. A series of recent reports (summarised in the *Economist*, 18 March 1995, p. 112) have shown that much of the benefits of national R&D 'spills over' to foreigners. Moreover, the smaller the country, the more its economy depends on research performed elsewhere. The *Economist* worried that this would discourage politicians from funding domestic R&D, but that is silly. Without the first mover domestic R&D, second-mover R&D will never be captured.

The Dangers of not Understanding the Costs of Second-Mover Research

Not everyone understands the cost of second-mover research. The most famous advocate of the government funding of science today is Paul Romer. The *Economist* wrote, breathlessly, on 4 January 1992, p. 18: 'With hindsight, intellectual historians will probably date the revival of growth theory to 1983 and a University of Chicago doctoral thesis entitled "Dynamic Competitive Equilibria with Externalities, Increasing Returns and Unbounded Growth". Its author was Paul Romer, now a professor at the University of California at Berkeley and a fellow of the Canadian Institute for Advanced Research.' Paul Romer believes that Government has to fund science because the free market will not. In a recent paper, he gives his two reasons; only one of these is original, and it is wrong.[31]

Let us quote him exactly (readers are warned that the next few paragraphs are heavy; economists write badly and use horrible jargon): 'There are two reasons to expect that too little human capital is devoted to research. The most obvious reason is that research has positive external effects. An additional design raises the productivity of all future individuals who do research, but because this benefit is nonexcludable, it is not reflected at all in the market price for designs.'

This is the unoriginal reason, derived from the work of Nelson and Arrow. The term 'external effects' means second-mover advantages, i.e. other people can build on any individual's research. The term 'nonexcludable' means that individuals cannot prevent others from building on their research. Romer is saying, therefore, that individuals will not invest in research, because others will accrue most of the benefits as second movers. As we have seen above, this argument is false because it ignores the very heavy investment on research that is required to be a second mover, and which integrally funds considerable first-mover science.

Romer proceeds to his second, original point: 'The second and an equally important reason why too little human capital is devoted to research is that research produces an input that is purchased by a sector that engages in monopoly pricing. The markup of price over marginal cost forces a wedge between the marginal social product of an input used in this sector and its market compensation . . . the producer of a design captures only a fraction [of its] net benefit to society.' This is Romer's original point. He claims that there cannot be a free market in research; instead, a middle-man exploits a monopoly to extract unfair profits from both the researcher and the consumer. Let us examine Romer's papers, because they reveal how a distinguished economist can get everything wrong.

Here is Romer's claim that any piece of research (he calls it a design) is freely available (which he calls 'nonrival' as there can be no rivalry over something that is freely available): 'In this sense, a design differs in a crucial way from a piece of human capital such as the ability to add. The design is nonrival but the ability to add is not. The difference arises because the ability to add is inherently tied to a physical object (a human body) whereas the design is not. The ability to add is rivalrous because the person who possesses this ability cannot be in more than one place at the same time; nor can this person solve many problems at once. As noted above, rivalry leads to a presumption that human capital is also excludable. Thus human capital can be privately provided and traded in competitive markets. In contrast, the design is nonrival because it is independent of any physical object. It can be copied and used in as many different activities as desired.'

The reader will at once detect the irrelevance of this argument. A design obviously differs in a crucial way from a piece of human capital such as the ability to add, but a design cannot be 'used in as many different activities as desired' without the intervention of people. The use of a design demands, obligatorily, the exploitation of human capital; therefore any economic arguments about designs must incorporate the economics of rivalry (to use Romer's terminology). Let us restate this point, and represent arithmetic and a design as two separate entities. They are nonrivalous. Each exists as a corpus of knowledge, but neither can be exploited without a human body which is, respectively, able to add or to read a design. Therefore, the use of a design becomes as rivalrous as the use of arithmetic.

But Romer, having established (wrongly) that science is freely available, discovers that under the free market no company would fund it (a logical deduction from a false premise). Romer summarises his argument thus: 'The distinguishing feature of technology is that it is neither a conventional good nor a public good; it is a nonrival, partially excludable good. Because of the nonconvexity introduced by a nonrival good, price-taking competition cannot be supported. Instead, the equilibrium is one with monopolistic competition.'

Romer means that because advances in technology are freely available ('nonrival, partially excludable'), no one will fund research ('nonconvexity'). Since, clearly, in the real world companies do fund science, yet Romer believes he has proved that they would not under a free market, Romer concludes that there is no free market in technology (another logical deduction from another false premise) and he further concludes (darkly) that some one must be rigging the market (because, in a free market, protagonists 'take prices', but a monopolist can 'set prices'). Some one must be exploiting a monopoly. Who?

Unfortunately, Romer cannot blame the likely suspects. He cannot accuse the producers of rigging prices because, as we will explain in the postscript on 'who profits from research?' a number of academic studies have shown that consumers benefit disproportionately from producers' research. Nor can Romer accuse the inventors of exploiting their monopolies, because his whole thesis is that they are undercompensated. So Romer creates a middle-man, whom he calls an intermediate producer, who buys the invention from the inventor at a cheap rate, and then exploits his monopoly to sell to the final producer at an inflated rate. This is all fiction, and as powerful an argument for the government funding of science as Aesop's tales.

Is the Government Funding of Academic Science a Good Economic Investment?

The final academic study that is used to justify the government funding of academic science is the one of Edwin Mansfield's discussed above, the one where he showed that around 10 per cent of new industrial products or processes arose from recent academic research.[32] Mansfield may have found that only 10 per cent of such processes arose from academic research, but he then performed a remarkable calculation: he calculated the profits that accrued from those 10 per cent, he calculated the investment in academic science internationally, and then he calculated the annual return on that investment as 28 per cent. Mansfield himself did not use his calculation directly as a justification for further government support, but his friend and sponsor at the National Science Foundation, Leonard Lederman, certainly did, briefing the press and politicians on the excellent return on investment, and demanding more.[33] Mansfield himself, moreover, did assume that politicians (code-named policy-makers) would be the major funders of any expansion in academic science, and would be influenced by his findings: 'for policy-makers who must decide how much to invest next year in academic research, this incremental rate of return is of primary significance'.

There are, however, a number of assumptions in Mansfield's work. First, he only actually showed a 5 per cent annual return to producers, and he created a figure of 28 per cent annual return to society at large by assuming that consumers benefit more than producers in a ratio of 2.8 : 1 (the details of this calculation are explained in the postscript on 'who profits from research?') Mansfield also added in a figure for new processes developed with some help from academic research (as opposed to those that arose entirely through recent academic research). Second, there is an average of seven years of delay between academic research and its exploitation as a commercial product, and it is not clear from Mansfield's paper that he corrected for that (i.e. if governments have to wait seven years for a return on their investment, has Mansfield corrected for the seven years' lost interest that might have accrued had the government simply put the money in the bank?). But even if Mansfield's assumptions and calculations are appropriate, there is one major assumption for which he has not corrected. In his own words, he is asking: 'What would happen if the resources devoted to academic research were withdrawn – and not allowed to do the same or similar work elsewhere.' But that is an impossible question. If the Government funding of academic science ceased, and taxes were cut accordingly, companies would certainly increase their expenditure on research for all the reasons given above. But it is reasonable to ask if companies would increase their research by as much as their tax cuts, or would some of that extra money go elsewhere? In the next chapter we will show that the money spent on academic research would be as great under *laissez faire*, if not greater.

POSTSCRIPT ON TECHNICAL STANDARDS AND COMPETITIVE NATIONAL ADVANTAGES

Struggles over technical standards persist and, fascinatingly, they follow the nineteenth-century European/American cultural divide. But, like a Tom and Jerry cartoon, the EC Commission, which embodies the French/German tradition of government-led cartels, is forever frustrated in its attempts to cheat the free market by the vigour of American capitalism. The Japanese Government's investment in HDTV has been similarly frustrated – no less than 30 years (yes, 30) of MITI-sponsored HDTV research was ended on 22 February 1994 when Akimasa Egawra of the Tokyo Ministry of Posts and Telecommunication announced its abandonment. After 30 years and billions of taxpayers' yens, a total of only 20 000 analogue HDTV sets had been sold in Japan, with the cheapest costing $6000 (see the report in the *Economist*, 26 February 1994, p. 83).

Technical standards in electronics are invariably established by American companies, generally IBM, which simply do a better job than the flat-footed conspirators of Brussels, Bonn, Paris and Tokyo. Currently, as IBM fades, the American electronic companies are beginning to agree on common, open standards, but each still tries to impose its own standards on the whole industry. Each company claims that its own standards are freely accessible or 'open' but, in the words of Tim Brespahan, an economist at Stanford University, 'The eskimos have 21 words for snow, these guys need 21 words for "open"' (quoted in the *Economist*'s review of the computer industry, 27 February 1993, p. 20).

POSTSCRIPT ON WHO PROFITS FROM RESEARCH?

It appears that most of the profit from research is not sequestered by the funder, but by the consumer. A survey by Mansfield *et al.* of 17 inventions showed that the average annual return to the inventor on his investment was 25 per cent, but the gain to the consumer (in terms of cheaper or better products) amounted to 56 per cent of the investor's investment (E. Mansfield, J. Rapoport, A. Romeo, S. Wagner and G. Beardsley, 'Social and Private Rates of Return from Industrial Innovations', *Quarterly Journal of Economics*, vol. 91, pp. 221–40, 1977). Repeated academic studies have confirmed that the market assigns most of the economic benefit of innovation to the consumer, not the inventor (see, for example, Z. Griliches, 'Research Cost and Social Returns: Hybrid Corn and Related Innovations', *Journal of Political Economy*, vol. 66, pp. 419–31, 1958; W. Peterson. 'Returns to Poultry Research in the United States', *Journal of Farm Economics*, vol. 49, pp. 656–69, 1967) but this is not an argument for government compensation to the inventor. That balance of social benefits is common to all transactions in the market. When, for example, the Ford motor company sells a family car for, say, £10 000 (let us assume for the sake of the argument that it is a car based on old engineering, into which no recent research has gone) the profit to Ford is only, perhaps £2000. But if the family ran the car for, say seven years, averaging 12 000 miles a year, the benefit to that family (in discounted car-hire costs) is much greater. Fortunately, producers are also consumers, so the social costs balance out, just as they do for the inventor. No government compensation is required, as a 20 per cent return on one's investment is good in anybody's terms, even if others do better.

POSTSCRIPT ON SEMATECH

The fifth-generation supercomputer caused such distress in America, which believed that the Japanese model of MITI-coordinated inter-company collaborations would blow the Americans out of the computer business, that the 1890 Sherman Anti-Trust Act was rewritten in 1984 as the National Co-operative Research Act, specifically to encourage joint research projects. The failure of the fifth-generation supercomputer has not, however, precipitated the repeal of this unfortunate breach of America's historic commitment to competition.

In 1987, the American government put up $100 million for the Sematech collaborative research venture with 14 electronic companies. This was in response to Caspar Weinberger's discovery in 1986 that half of the chips in the F-16 fighter's fire-control radar came from Japan, as some of America's chip makers, particularly the bigger ones, were falling behind. America's 200 successful chip makers, generally the smaller, niche ones, were sceptical of Sematech, echoing the words of T. J. Rodgers, the president of Silicon Valley's Cypress Semiconductor: 'Consortia are formed by people who have lost.'

Some of the companies involved in Sematech have done well. Others like Semi-Gas were sold to the Japanese, or, like SVGL, have gone into partnership with the Japanese (the Japanese are very grateful for this gift of American taxpayers' money) while other companies like GCA have gone bust because, despite the myths, their failures were not caused by poor research but by poor quality control (see below).

Because some of the Sematech partners are now flourishing, Vice-President Al Gore has demanded more government/industrial collaborations. But, as the *Economist* showed in a thoughtful study (2 April 1994, pp. 91–3) the Sematech partners have done no better than similar companies that did not join. Indeed, in 1991 two companies, Micron Technology and LSI Logic left Sematech as they found it damaging. The *Economist* found that America's chip makers had been failing because of the American recession, high American interest rates, and a cavalier disregard of quality control. Research was not the major problem, and Sematech solved nothing but the administration's need to be seen to be doing something.

Nor, of course, has anyone calculated the cost to the rest of the economy of extracting Sematech's subsidy out of it.

10 The Real Economics of Research

In Chapter 9 we tackled error. We showed that the linear model, which arose in the pre-industrial era as an idea of Francis Bacon's, was wrong – for reasons that Bacon would have appreciated, since the linear model was an example of Aristotelian thinking: it arose *de novo*, and silly science policy-makers have ever since been trying to deduce truth from it. Here we shall be good Baconians, we shall collect data first, and only then shall we attempt to induce truth therefrom. We have already, in Chapter 7, introduced the principles of the economics of science, but here we shall elaborate them (there will be, therefore, a degree of overlap with Chapter 7, but it will be very small). We shall follow the structure of the previous chapter, and first discuss the economics of R&D, and then the economics of academic science. For each section, we shall ask the same subsidiary question: is government funding necessary to optimise the step?

DATA

We shall draw our raw data from a number of different sources.

Population and GDP Per Capita

These figures come from the data published by the OECD, the grouping of the 24 richest capitalist countries (for analysis, we generally exclude the 25th country, Yugoslavia, as communist, and we generally exclude Iceland and Luxembourg as too small, so most figures are made from 22 countries).

Expenditure on Civil R&D

This too comes from the OECD. Officials within the OECD collect data on expenditure on civil R&D from companies, research associations, and all the other bodies that perform R&D. (Statistics for civil

R&D encompass the relatively small national expenditures on academic science. Figure 8.4 shows that national budgets for civil R&D are about ten times larger than for academic science.)

Academic Science

The most reliable data on the amount of academic science performed by different countries comes not from measuring funding, which is always hard to completely catalogue, but from measuring output, since academic scientists write up their work as scientific papers (which explains the old joke, that the definition of a laboratory rat is a creature which, on being injected with a new chemical, produces a scientific paper). The numbers, and national origins, of scientific papers can be chronicled. The most assiduous chronicler is Professor Braun's team at the University of Budapest. With the help of a powerful computer, Braun *et al.* catalogue all the papers published in the 4000 major international scientific journals, to assign them to their countries of origin.*

Having explained whence comes the raw data, let us start by exploring the economics of civil R&D.

THE ECONOMICS OF CIVIL RESEARCH AND DEVELOPMENT

When national statistics for civil R&D are collected by the OECD, no immediate pattern emerges. But when the statistics are related to national GDP *per capita*, a strong correlation emerges: the higher the national GDP *per capita*, the higher the percentage of GDP devoted to civil R&D (see Figure 7.13). This, of course, is very similar to Figure 7.11, which showed a strong correlation between the numbers of scientific papers published *per capita* and national wealth, and very similar to the patent data which shows a strong correlation between the numbers

---

*The terms 'pure science' and 'academic science' do not refer to exactly the same things, since universities do all sorts of science, not always pure or basic, while much pure science takes place outside the universities; yet the two terms 'pure' and 'academic' are similar enough to be used interchangeably here. Some science policy experts refer to 'academic and related' science to encompass that pure science which is prosecuted in university-like institutions such as the NIH in the USA, the research council units in the UK, the Centre Nationale pour la Recherche Scientifique (CNRS) in France, the Max-Planck institutes in Germany, the Netherlands Organization for Scientific Research (NWO) laboratories in Holland, or Moribusho's intramural institutes in Japan.

of patents filed *per capita* and GNP *per capita*.[1] These similarities allow us to formulate a general observation: the richer the country, the more research it does (below, we shall elevate this into the First Law of Research Funding).

We saw in Chapter 7 that if we go back to early OECD statistics, to 1965, we find that similar correlations between national wealth and the percentage of GDP or GNP devoted to civil R&D continue to hold (see Figure 7.14). A comparison between Figures 7.13 and 7.14 also provides longitudinal confirmation of the relationship, because each of the OECD countries was much richer in 1985 than in 1965, allowing for inflation, and each of them also spent a higher percentage of GDP or GNP on civil R&D in 1985 than in 1965.

Recent data from the OECD provides further longitudinal confirmation of the First Law. Because this is so important, we shall describe it meticulously: the average GDP *per capita* of the OECD countries in 1975 was $7191 (in 1985 prices).[2] By 1985, this had risen to $8989 (1985 prices).[2] If the percentage of GDP devoted to civil R&D is proportionate to GDP *per capita*, then it can be seen from Figure 7.13 that, between 1975 and 1985, the percentage of GDP devoted to civil R&D, averaged across the OECD, should have risen from approximately 1.2 per cent to approximately 1.5 per cent. Gratifyingly, that is exactly what happened (1.26 per cent in 1975 to 1.55 per cent in 1985, to be precise).[3]

Further, the recession of the late 1980s witnessed falls in expenditure on R&D in many OECD countries. Thus between 1990 and 1991, preliminary OECD figures – which unfortunately have not been analysed as civil R&D but only as total R&D – show that total R&D fell in the USA by 0.3 per cent; it fell from 2.88 to 2.86 per cent GNP in Japan; it fell by 7 per cent in Britain; and it fell from 2.41 to 2.40 per cent GNP in France.[4]

We have already, in Chapter 7, touched on the underlying economics of this relationship between increasing wealth and increasing research, and we concluded that it showed that research obeyed the Law of Diminishing Returns, i.e. one has to do more and more research to sustain the same rate of economic growth. But we also concluded, happily, that increasing wealth afforded greater opportunities for research – a virtuous cycle. Yet, in Chapter 7, we did not speculate if the statistics were artefactual. We assumed that they reflected the workings of a fundamental economic law; but might they just not reflect the workings of politicians? How do we know that Figures 7.13 and 7.14 do not merely tell us that politicians, as their countries get richer, grab ever increasing resources to squander on prestige projects like research? To answer that, we have formally to ask the question: what is the effect of the government funding of civil R&D?

What is the Effect of the Government Funding of Civil R&D?

We can address this question with relative ease because the different governments of the OECD have, bless them, followed very different policies. Some, such as the governments of New Zealand, Australia, Greece and Portugal, have effectively nationalised civil R&D, and the state may pay up to 80 per cent of it; other governments, such as those of Japan and Switzerland, have neglected it, and only pay for some 20 per cent (and at least half of that 20 per cent is spent on academic science rather than on industrial R&D). Other OECD governments have adopted compromise positions: in the UK, France, Germany and the USA, for example, about half of civil R&D is funded by the state, and about half privately (almost entirely by industry). These different policies enable us to answer the question: what difference do they make?

A superficial glance at Figure 7.13 reveals that New Zealand, Australia, Greece and Portugal lie below the linear regression line, but Japan and Switzerland above it. This suggests that those countries whose governments spend the most on civil R&D actually end up with less spent, overall, nationally, i.e. Figure 7.13 suggests that the nationalisation of civil R&D has damaged it by lowering national expenditure on it. That is such an important observation that we need to explore it systematically. We need to compare the degree of nationalisation of each OECD country's civil R&D with that country's expenditure on it. Fortunately, that can be easily done (it is extraordinary that no one has before). There are only two major sources of funding for civil R&D, industry and the state (the private funding of academic science is the third significant source, but it is small compared with the first two sources of funding for civil R&D). The OECD regularly publishes data for different countries' ratios of business : government funding for civil R&D, which provides indices of nationalisation (i.e. a country with a business : government ratio of 5 : 1 for the funding of civil R&D has a largely privatised civil R&D, while one with a ratio of 1 : 5 has been largely nationalised).

Figure 10.1 plots the ratio of business : government funding for civil R&D for the OECD countries against the percentage of GDP they spend on civil R&D, and it shows that those countries whose civil

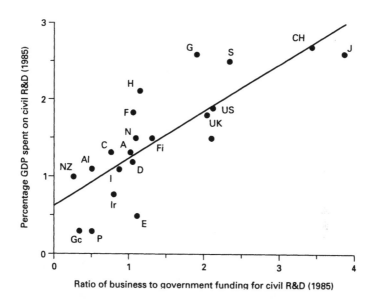

FIGURE 10.1   *Business: government funding (part 1)*

The data have been calculated from the figures in *OECD Science and Technology Indicators No. 3* (Paris: OECD, 1989). The equation is $y = 0.624 + 0.616x$, $r = 0.81$ ($P < 0.001$). See Figures 7.5 and 7.11 for a key to symbols.

R&D is predominantly funded by industry spend more than those whose is predominantly funded by the state. This really does indicate that nationalisation reduces civil R&D budgets. Below, we shall elevate this conclusion into the Second and Third Laws of Research Funding, namely that the state's funding of research displaces private funding (the Second Law) and that the funding is disproportionate, since the state displaces more private funding than it does itself provide (the Third Law). But we cannot just propagate these laws on the correlations of Figure 10.1. Conclusions based on correlations should be treated with caution, as they can mislead; for example, there are positive correlations worldwide between the national incidences of colour television sets and coronary heart disease, but no one believes that coronary heart disease begets colour television sets. Let us, therefore, play devil's advocate. What alternative explanations, other than displacement, could we produce for Figure 10.1? There are at least two, as follows.

Alternative explanation no. 1

The countries that spend most on civil R&D are the rich countries, and the countries that spend most on civil R&D are also those whose R&D is privatised, so perhaps we only have a correlation that demonstrates that rich countries tend to have a privatised R&D. That would be an important finding. It might, for example, indicate that private civil R&D is 'better' than state civil R&D, and that rich countries became so because they left civil R&D to the private sector, which spends money more wisely than politicians. That may, indeed, be partly true (see below; see also the old Soviet bloc whose vast funding of R&D was entirely nationalised, only to produce poverty) but to confirm the Second and Third Laws of Research Funding we need to show that, at every level of national wealth, the intrusion of the state into funding civil R&D disproportionately displaces private funding. We need, therefore, to establish for each country, for its particular level of wealth, an 'expected' expenditure on civil R&D. We then need to compare that expected level with the one actually observed, and we then need to compare that ratio of expected : observed funding for civil R&D with each country's ratio of business : government funding.

We can do all this: from Figure 7.13 we can calculate for each country the expected level of funding for civil R&D by extrapolating up to the line of linear regression; we can then observe the actual national funding of civil R&D, to create for each country, a ratio of observed : expected funding appropriate to that country's wealth. Thus from Figure 7.13 we can see that Switzerland and Norway, say, have similar GDPs *per capita*, but Switzerland's percentage of GDP spent on civil R&D is much bigger than Norway's, and that we can say, therefore, that Switzerland has a ratio of observed : expected funding of about 1.2 to Norway's 0.8. When each OECD country's ratio of observed : expected funding for civil R&D is compared with its ratio of business : government funding, a positive correlation emerges (Figure 10.2). Figure 10.2 shows, therefore, that regardless of national wealth, the nationalisation of civil R&D displaces private money. Moreover, the displacement really is not equal: for every pound or dollar the government spends, it displaces more than the private sector would have spent. The Second and Third Laws of Research Funding appear to be true. But there is yet another alternative explanation, no. 2 (divided into 2a and 2b).

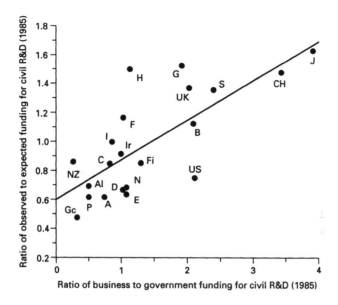

FIGURE 10.2   *Business: government funding (part 2)*

The data for the *x*-axis come from the figure in *OECD Science and Technology Indicators No. 3* (Paris: OECD, 1989). The *y*-axis comes from calculating the ratio between the observed and expected percentage of GDP spent on civil R&D from Figure 7.13. The equation is $y = 0.597 + 0.276x$, $r = 0.74$ ($P < 0.001$). See Figures 7.5 and 7.11 for a key to symbols.

Alternative explanation no. 2a

Perhaps some governments never nationalised anything. Perhaps some countries' industrialists never have done much civil R&D, relative to their national wealth. Thus the leaders of industry in countries such as Australia, New Zealand, Portugal and Greece might simply have neglected R&D. In compensation, their governments might have poured money into civil R&D, but if government largesse failed to compensate for private neglect, it would generate a constellation of countries such as Greece, Portugal, Australia and New Zealand which have low overall spending on civil R&D, low private spending on civil R&D, yet high government spending on civil R&D. Does this explain Figures 10.1 and 10.2?

This alternative explanation no. 2a is, politically, very different from the first. In the first explanation, Government emerges as the bad guy, whose funding for civil R&D displaces private money, but here Government is the good guy, trying desperately to compensate for industrialists' errors. But such an argument rests on an improbable proposition. If a nation's industrialists are collectively stupid, it is unlikely that its government, which in OECD countries will be highly influenced by them, will be more enlightened (remember that we are discussing R&D here, not academic science, so we are being asked to believe that governments will forcibly fund technical advances within the factories of reactionary capitalists).

Alternative explanation no. 2b

Alternatively, it could be argued that a nation's industrialists are right to underinvest in civil R&D, because that particular nation's portfolio of industries happens not to include many industries that require much R&D. This, too, is not a very probable argument, because advanced nations now tend to possess similar economies. (With the major exception being that aircraft R&D is so expensive that countries with aircraft industries do demonstrate significantly raised total R&D budgets for that reason alone – but most of those figures are subsumed within defence, not civil R&D. Moreover, some countries like Britain or Holland benefit from being the research headquarters of huge multinational companies like Shell or Unilever.) None the less, if the argument is correct, and some nations' economies are skewed towards non-R&D industries, this raises a further question: why should governments artificially spend extra on R&D? Is that not a waste of money?

We do not need to explore these alternative arguments any further, because we can produce strong empirical evidence that destroys both alternative explanations, and which directly supports the Second and Third Laws. Moreover, it is not correlative data, but longitudinal and predictive, and therefore powerful.

The Longitudinal Evidence

The mutual displacement of state and business funding for civil R&D can be demonstrated by a longitudinal study. Because this is so important, we shall describe it meticulously. We have already shown above that, as the average GDP *per capita* of the OECD countries rose from $7191 in 1975 to $8989 in 1985, so the percentage of GDP devoted to civil R&D rose across the OECD from 1.2 to 1.5 per cent. But (and this is intriguing) governments barely increased their funding: between 1975 and 1985, average OECD government funding for civil R&D only rose from 0.66 per cent GDP to 0.69 per cent GDP, far short of the extra funding that was both predicted from Figure 10.1 and which was actually seen in practice. But the average OECD business funding for civil R&D rose from 0.60 per cent GDP in 1975 to 0.85 per cent GDP in 1985.[5] This met the shortfall and confirmed the prediction based on Figure 7.13. This establishes, therefore, mutual displacement. Government is the bad guy.

The Economic Laws of Funding Civil R&D

This longitudinal study thus confirms the following economic laws of civil R&D funding, as follows.

The First Law of Funding for Civil R&D states that the percentage of national GDP spent increases with national GDP *per capita*.

The Second Law of Funding for Civil R&D states the public and private funding displace each other.

The Third Law of Funding for Civil R&D states that the public and private displacements are not equal: public funds displace more than they do themselves provide.

The consequences of the First Law: diminishing returns

The first law confirms the suggestion made in Chapter 7: R&D is subject to diminishing returns. We know that lead countries have enjoyed steady economic growth rates for nearly two centuries, yet we also know that, as they have enriched themselves, they have had to invest a greater and greater proportion of their wealth in R&D to sustain that steady rate. Thus we conclude that civil R&D is subject to diminishing returns; as technology becomes more complex, so companies have to invest even more on R&D to make advances. Fortunately, their increasing wealth affords them the increased resources. The free market, therefore, both enforces virtue and also finances it.

Further consequences of the First Law: future economic growth

One major implication flows from the First Law of Civil R&D Funding; we can predict future economic growth. The Law of Diminishing Returns places an ultimate limit on economic growth. Lead countries will continue to grow, economically, at 2–3 per cent a year, doubling their GDPs *per capita* every 25 years, until the percentage of GDP spent on R&D approaches some 10 per cent of GDP. Further growth in that budget will probably be intolerable for two reasons. First, any further expansion will compete seriously with other crucial needs such as health, education and welfare, and second, the manufacturing base will by then have been reduced to some 10 per cent of GDP, and as R&D is performed largely in support of manufacturing, it is unlikely that its budget will actually swamp that of its *raison d'être*.

On present trends, it will only take till the end of the twenty-first century for manufacturing and civil R&D to each account for 10 per cent of GDP; thereafter, each will plateau and, with the plateauing of civil R&D growth, will come a plateauing of GDP growth. But, by then, we will have experienced between four and nine successive doublings of GDP *per capita*, so we will be much, much richer than we are now. Sadly, we will probably be no wiser, and as absolute wealth does not seem to make people happy, we may not even be more cheerful (the evidence suggests that once they have crossed the breadline, people are only made happy by relative wealth – i.e. they need to feel richer than others) and now that the Third World is rapidly catching up economically, the West will grow more and more hysterical as it contemplates another 20, or even 100, countries like Japan. Since the West has an almost infinite capacity for self pity (look at what the EC has done to Third World agriculture) we can expect more trade bullying in the future.

Consequences of the Second and Third Laws

The discovery that the government funding of civil R&D disproportionately displaces private funding is desperately important. Governments, when they started to fund civil R&D, hoped that their funding would be additive (i.e. merely add to the amounts that industry already spent) but if that funding is displacive and, even worse, if governments actually displace more than they themselves provide, then the government funding of civil R&D damages the enterprise. This conclusion will dismay all those with a vested interest in the government funding of civil R&D (politicians, bureaucrats and self-appointed lobbyists) yet, as will be shown below, their lamentations will not avail them: governments are inexorably being forced out of the funding of civil R&D.

What is the Mechanism of Displacement?

Private money is displaced because industry is taxed, so that government might spend its money; but because companies have been taxed, their ability to fund their own R&D has been reduced. Moreover, the assumption by the state of responsibility for R&D creates an expectation in private enterprises that, should they wish to prosecute a research programme, they should lobby the government for support rather than provide it themselves. But why are the displacements unequal? How do governments actually displace more private money than they do themselves provide?

R&D is not unique: nationalisation always lowers budgets, whatever the enterprise. This has long been obvious, anecdotally, to anyone who travelled in a British Rail train, or who tried to persuade the GPO in Britain to install a telephone in the days before privatisation, but it has been formally established by Galal *et al.* in a World Bank report.[6] Galal *et al.* studied the consequences of privatising 12 previously nationalised industries worldwide, ranging from airlines to lotteries. When those industries were privatised, Galal *et al.* showed that increased investment, and more appropriate management decisions, inexorably ensued. Employees' morale improved, as did customer satisfaction. To the extent, therefore, that civil R&D has been nationalised, one would have expected it to have been under-resourced and poorly managed.

Why are nationalised enterprises under-resourced? In theory, nationalisation is meant to confer disinterested, benign, long-term management on an enterprise, freeing it from the horrors of the marketplace; so why in practice is it so damaging?

Since the Second World War, Western governments have committed themselves to providing vast services; social security, education, health and defence amount to an enormous share of GDP. Yet democratic countries will rarely submit to taxes above 40–50 per cent GDP. Governments have, therefore, overcommitted themselves, and the control of budget deficits, or of inflation, has been a recurring, universal problem. Worse, as countries enrich themselves, people demand ever higher shares of GDP to be devoted to social security, education, health and other benefits, just as they do on R&D, so the economic pressures on government grow ineluctably. Governments respond, therefore, by economising, and they will spend less on the services they provide than the free market would have. This might not matter, despite the expense of raising taxes and employing civil servants, if government-induced shortfalls could be made up privately, but nationalisation raises the costs of entry (to use economists' jargon).

Raising the Costs of Entry

Consider a company that needs a laboratory costing, say, £10 million a year. Now consider that the government taxes that company and, in return, creates a laboratory – but only funds it at £8 million a year. To enable the laboratory to function properly, the company will need to buy in with a further £2 million. But the company will want exclusivity on the whole laboratory's findings. The government will almost certainly not grant that, because the taxpayers' money is general. Therefore the company, if it wants a proper laboratory, will still have to find £10 million on top of the millions it has already been taxed. Many companies will simply not have the extra money, and so the national civil R&D enterprise gets underfunded.

It can be demonstrated that bureaucracy raises the barriers to entry; consider the money disbursed by the Department of Trade and Industry (DTI) the British Government's largest agency for funding civil R&D. A survey in 1992 by the consulting firm KPMG Peat Marwick showed that, of 324 companies, 79 per cent of them avoided the DTI's schemes, finding them too cumbersome, complicated or bureaucratic. Eighty-six per cent of companies favoured tax credits for research.[7] Thus 79 per cent of British firms have, through their taxes, contributed to central research funds, to which their access is then blocked by inappropriate bureaucracy. Thus does the partial nationalisation of civil R&D displace more private money than governments themselves provide.

Bureaucratic Nonsenses

Nationalisation also damages civil R&D by substituting the judgement of governments for that of industrialists. Government schemes can be remarkably asinine. Consider the Alvey and ESPRIT schemes which introduced the previous chapter. The bureaucrats in London and Brussels decided that their national electronics companies were not collaborating sufficiently with each other, or across national borders, or with the universities, so they decided to force greater collaboration. To obtain government money, companies had to find partners in different countries, then they had to fill in long, complicated forms explaining exactly what they wanted to do. These forms were then read by committees and working parties, and after innumerable delays, a decision was taken to fund, or not to fund, on a 50 : 50 basis (i.e. the taxpayer met 50 per cent of the R&D costs).

Anyone who has done research will know that it is horrible to collaborate with strangers who are foreigners, who speak a different language and who work hundreds of miles away. Edwin Mansfield has surveyed this issue. He examined the collaborations that 66 of the USA's major manufacturing companies had forged, spontaneously, with university research departments, and he found that proximity was a major determinant of the success of collaboration. Indeed, he found that 'holding research quality constant [of a collaborating university department], the probability that a firm will support research at a college or university 5 miles away tends to be 50% higher than that it will support this research at a college or university 100 miles away'.[8]

If 95 miles are so damaging to collaboration, how much more damaging must a language or national barrier be! Moreover, anyone who had done any research in a competitive field will know that it is futile to pretend that one can draw up detailed plans outlining experiments two or three years hence, and then to wait around for months while a group of bureaucrats weigh their merits. Consequently, no company would endanger any important research by exposing it to bureaucratic delay or outside interference. One reason Alvey and ESPRIT were unsuccessful was that companies only submitted projects of marginal importance (when Alvey money ran out, for example, GEC in Britain discontinued much of the research and sacked the researchers). Of course, there are times when collaboration with other companies or with universities is required, but it is absurd to believe that companies cannot forge those links for themselves. Indeed, one of the great pathologies of capitalism is that companies will forge cartels if they can, and only bureaucrats could believe that taxpayers' money has actually to be spent on bringing competing companies together.

Alvey and ESPRIT even failed academically. Those schemes poured money into university, as well as industrial computer science, yet the number of British and European academic papers in computer science actually fell, relatively, during the course of the schemes. Even Irvine and Martin, those uncritical advocates of more and more and more government money for more and more and more science, concluded that Alvey and ESPRIT, by forcing scientists to work so hard to sustain artificial collaborations across national, language, bureaucratic and institutional barriers, left them with insufficient time for the actual business of research.[9]

The problem, of course, is that the European Union is abusing research by treating it as a vehicle for fostering ever-closer links between its disparate people rather than by supporting it for its own sake or for the sake of the economy. Ultimately, therefore, both science and economic growth will suffer. Much good that will do for European harmony.

Good industrial research is confidential; let us re-visit L. T. C. Rolt's description of Robert Mushet's pioneering metallurgical research in *laissez faire* nineteenth-century Britain:

> 'Robert Mushet became a pioneer worker in the field of alloy steels.
> In 1862 he formed the Titanic Steel and Iron Company . . . and the
> Titanic Steel Works was established in Dean Forest. This was a small
> crucible steel-making plant where, in great secrecy, Robert Mushet
> carried out his alloy steel experiments. Mushet's greatest success
> was a special self-hardening tool steel of immeasurably superior cutting
> power to the carbon steel used previously . . . and R. Mushet's Special
> Steel, or 'RMS' soon became famous in machine shops on both sides
> of the Atlantic. Mushet took the most extraordinary cloak-and-dag-
> ger precautions to keep his RMS formula secret. The ingredients
> were always referred to by cyphers and were ordered through inter-
> mediaries. The mixing of the ingredients was carried out in the se-
> clusion of the Forest by Mushet himself and a few trusted men'.[10]

This does not sound much like ESPRIT or Alvey, but it would be
familiar to Kilby, Merryman and van Tassell of Texas Instruments,
Steve Jobs and Stephen Wozniak of Apple, Bill Gates of Microsoft or
Gary Kildall of Intergalactic Digital Research who developed the pocket
calculator, the Apple Macintosh computer, MS-DOS and the CP/M
operating system respectively. They retired to the seclusion of their
laboratories to work with a few trusted collaborators. Their finance
came from corporate investors or venture capitalists, not the State (which
would have found it very difficult to deal with impatient, non-corporate
characters like Gary Kiddall, a man who often does not wear a tie).

Civil R&D Is Being Privatised Everywhere

However, for civil R&D, the situation is self-correcting; it is being
privatised by default. Prosperous countries are now devoting 2 per cent
or more of GDP to civil R&D, and with tax limited to some 40 per
cent of GDP, this forces a government intent on funding all civil R&D
to allocating it 4 per cent of tax revenue. This is too much in the face
of competing demands. Governments are not actually cutting budgets,
but they are fixing them, and as the requirement for R&D rises with
rising wealth, so effectively it is being privatised. Thus we saw that,
between 1975 and 1985, average OECD funding for civil R&D rose
from 1.26 per cent GDP to 1.55 per cent GDP, yet governments could
only afford to raise their contributions from 0.66 per cent GDP to 0.69
per cent GDP; the rise in the business contribution from 0.60 per cent

GDP to 0.85 per cent GDP signalled the beginning of the effective privatisation of civil R&D across OECD countries – and from Figures 10.1 and 10.2 this is very much to be welcomed.

Conclusion to the Section on the Economics of Civil R&D

The First Law of Research Funding, is not, perhaps, very surprising: as countries enrich themselves, they need to spend more and more on civil R&D because of diminishing returns. But the Second and Third Laws reveal that, contrary to myth, the government funding of civil R&D is not only not additive, and not only displacive, it is actually disproportionately displacive of the private funding of civil R & D. In blunt terms, the Australian, New Zealand, Portuguese and Greek Governments are wasting their money, as are those of the USA, Britain, France, Germany, Canada, Belgium, Holland, Italy, Spain and Scandinavia. Only Switzerland and Japan get it roughly right, and even their Governments are probably overgenerous.

It would be nice to think that this analysis might prompt a re-assessment of government policy, but that would be naive. The myths are too deeply ingrained. Most governments started to fund civil R&D in imitation of the Germans, and few governments seem to understand that (i) the rise in the German economy was unremarkable, and largely an exercise in catching up, and (ii) throughout the period of German militarism that so scared other governments into imitation, Germany was actually a poorish country – certainly poorer than the *laissez faire* USA or the *laissez faire* UK of the day. Fortunately, however, governments will be powerless in the future to prevent the inexorable privatisation of civil R&D; and its private budgets will rise until the human race will finally settle to standards of living about 20–100 times greater than they are now. Not bad.

THE ECONOMICS OF ACADEMIC SCIENCE

The Quantity of Academic Science

Academic science has, everywhere, largely been nationalised. Because it accounts for a much smaller share of GDP than civil R&D it will, moreover, tend to remain nationalised, unless governments lose their desire to keep it under control. Since politicians, everywhere, ache to control as much as they can, especially if it is prestigious, they will hold on tenaciously to the universities and to pure science generally. But, nationalised or not, the economics of academic science are remarkable. Figure 7.11 shows that the amount of science a country performs correlates closely with national wealth, i.e. the governments of rich countries spend more on academic science, both absolutely and as a share of GDP, than the governments of poor countries.

The Quality of Academic Science

Not only is the quantity of academic science determined economically, so is its quality. It is notoriously difficult to measure the quality of science, but the best criterion is the so-called citation index. All scientific papers build on preceding achievements: 'If I have seen further, it is because I have stood on the shoulders of giants' (Isaac Newton) and in their papers, scientists cite the earlier studies on which they have built. Thus major papers such as Einstein's on relativity are cited more frequently than minor ones such as Augustus Threepgood's survey of the distribution of the Lesser Yellow Billed Tit in Eastern Java. Some years ago, Eugene Garfield, the publisher of *Current Contents*, noted that the numbers of citations to any particular paper by all the other papers in the world could be correlated to generate a hierarchy of scientific excellence.

There are problems with citations. Some papers are cited only to criticise them (I am always citing the papers of John Irvine and Ben Martin of the Science Policy Research Unit, Sussex University, England, to expose their flaws) and some scientists cite their own papers as much as they can (see, for example, the excellent study by T. Kealey, 'The Growth of British Science', *Nature*, vol. 350, p. 370, 1990). Papers describing new methods tend to get cited more than most (although that should not be a problem: advances in science depend on new methods). Finally, some fields of science, by their nature, tend to generate more citations than others: advances in pure science tend to be cited by applied scientists, but the opposite rarely holds. Some fields of science are bigger than others, and so generate more citations.

When comparing different countries' citations, however, most of these problems cancel out. Negative or critical citations are, in practice, rare, and are common to all countries, as are the problems of self-citation and of over-citation to methods. Moreover, Braun and his colleagues[11] have constructed a Relative Citation Index, to compensate for the different rates of citation in different areas of science. This index determines the average number of times a paper is cited for each of the 4000 major science journals (the ones catalogued by the Institute for Scientific Information), and then averages the relative citation rate for each country's papers across all 4000 or so journals. Since each journal tends to serve a narrow area of science, this corrects for the different citation rates of different sciences. When national relative citation rates are correlated against national GDPs *per capita*, a very tight correlation emerges. The richer the country, the better the science (Figure 7.12).

Consequences for Scientists

These economics are not surprising. Good science requires resources, facilities and time, which rich countries can more easily afford. But the tightness of the correlation in Figure 7.12, however, does convey a depressing message to researchers. Scientists see themselves as pioneers, individuals whose courage, intelligence and penetration enables them to uncover Nature's secrets. Historically, this heroic view of scientists has never carried conviction, because scientists, however eminent, have always competed. Newton raced Leibniz, Wallace published before Darwin, Mendel's work was independently re-discovered, and Einstein was racing Lorentz, Minkowski, Hilbert and others. No one in science has ever appeared, anecdotally, to be indispensable. The economics in Figure 7.12 confirms the anecdotal history: the quality of science is economically determined. The limiting factor is not scientists' intelligence or their courage but their resources. Had Newton, Darwin or Einstein been born in eighth-century Ukraine, say, or modern-day Somalia, they would have lived and died anonymously – a fact that Thomas Gray appreciated during the eighteenth century. Of poor villagers he wrote:

> But knowledge to their eyes her ample page
> Rich with the spoils of time did ne'er unroll;
> Chill penury repres'd their noble rage,
> And froze the genial current of the soul.
> *Elegy Written in a Country Churchyard* (1751)

The Politics of Economics

Yet the tightness of the correlation in Figures 7.11 and 7.12 is astonishing: those figures confirm that the quantity and quality of academic science are determined economically (i.e. the funding of academic sci-

TABLE 10.1   *Research statistics, OECD countries, 1985*

	Percentage of GDP spent on all R&D	Percentage of GDP spent on civil R&D	Percentage of GDP spent on military R&D	Percentage of GDP spent on academic science	GDP per capita (US $ in 1985 current prices)
USA	2.8	1.9	0.82	0.37	16494
Sweden	2.8	2.5	0.23	0.68	12006
West Germany	2.7	2.6	0.12	0.40	10243
Japan	2.6	2.6	0.02	0.37	10977
Switzerland	2.4	2.7(1986)	0.05	0.40	14195
UK	2.3	1.8	0.49	0.32	7943
France	2.3	1.8	0.41	0.34	9251
Holland	2.1	2.1	0.02	0.49	8628
Norway	1.6	1.5	0.08	0.36	13960
Finland	1.5	1.5	0.01	0.29	11024
Belgium	1.5	1.5	0	0.28(1979)	8022
Canada	1.4	1.3	0.04	0.32	13635
Austria	1.3	1.3	0	0.44	8743

Denmark	1.2	1.2	0	0.30	11319
Australia	1.1	1.1	0.02	0.29	9847
Italy	1.1	1.1	0.06	0.22	6278
New Zealand	1.0	1.0	0.01	0.20	6722
Ireland	0.8	0.8	0	0.14	5123
Spain	0.5	0.5	0.02	0.09	4255
Portugal	0.4	0.3	0	0.10	2032
Greece	0.3	0.3	0.01	0.07	3294
Turkey	0.2	N/A	N/A	0.13	1057

NOTES: This table shows that national expenditures on civil R&D correlate with national wealth, as do expenditures on academic science (which accounts for around 10–20 per cent of civil R&D). Expenditure on military R&D, however, is historically determined, the major spenders being printed in bold.

SOURCES: The economic data on R&D and academic science came from *OECD Science and Technology Indicators No. 3* (Paris: OECD, 1989). Strictly, the data for 'Percentage GDP spent on military R&D' are only for that spent by the government, but in practice that covers almost all military R&D. The GDP *per capita* data came from *OECD Economic Surveys 1986/1987, United Kingdom* (Paris: OECD, 1987).

ence is determined economically) yet we know that the actual decisions over funding are taken by mere politicians. How are the political decisions to fund academic science made so economically sensitive?

Some political decisions are not determined economically. The funding of military R&D, for example, is a political decision that has been determined historically, not economically. Those OECD countries that won the Second World War (USA, UK and France) and Sweden (which believes that its policy of armed neutrality was justified) continue to fund military R&D, while the losing countries (Germany, Japan, Italy and Austria) are revolted by their failed policies, and they abjure military R&D (although Germany does actually spend a little, as indeed do Norway, Switzerland and Australia). The other OECD countries feel marginalised, historically or economically, from military R&D, and they spend very little. The figures are provided in Table 10.1.

Since only four OECD countries spend significantly on defence R&D, and even their expenditure does not swamp their civil expenditure, Figure 10.3, which examines the OECD countries' *total* expenditure on R&D is not too different from Figure 7.13 for *civil* expenditure.

Lobbies

But the political decision over funding academic science is clearly one that has been determined economically. What is the mechanism? It appears to emerge from the balancing of different lobbies, of which there are at least five: (i) the scientists; (ii) the politicians, (iii) industrialists; (iv) the general public; and (v) the national treasury. Let us see how they interact.

(i) The scientists

Researchers always have, and always will, lobby for money. In 1726, in his *Gulliver's Travels*, Jonathan Swift satirised the Royal Society thus: 'The Academy is not an entire single building, but a continuation of several houses of both sides of a street. . . . The first man I saw . . . had been eight years upon a project for extracting sunbeams out of cucumbers, which were to be put into vials hermetically sealed, and let out to warm the air in raw inclement summers. He told me he did not doubt in eight years more that he should be able to supply the Governor's gardens with sunshine at a reasonable rate; but he complained that his stock was low, and entreated me to give him something as an encouragement to ingenuity, especially since this had been

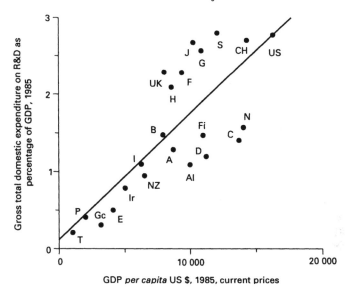

FIGURE 10.3   *GDP and total expenditure on R&D*

Gross domestic expenditure on research and development as percentage of GDP, 1981, comes from *OECD Science and Technology Indicators No. 3* (Paris: OECD, 1989). GDP *per capita* data come from *OECD Economic Surveys UK, 1987* (Paris: OECD, 1987). The equation is $y = 0.126 + 1.611e^{-4x}$, $r = 0.76$ ($P < 0.001$). See Figures 7.5 and 7.11 for a key to symbols.

a very dear season for cucumbers. I made him a small present, for my Lord had supplied me with money on purpose, because he knew their practice of begging from all who go to see them.'

We saw in Chapter 8 that, under *laissez faire*, some British and American scientists argued against government funding, and they only lobbied private sources of funding, but once the State had nationalised British and American science, those voices of disinterested responsibility were stilled, and all scientists everywhere now campaign for more government money. It is an inevitable but sad consequence of the nationalisation of any activity that it politicises, and therefore coarsens, the very people who originally embraced that activity from nobility or idealism. Scientists' pressure groups will no longer acknowledge a hierarchy of economic priorities; instead they will campaign ceaselessly, and that can be taken as a constant.

(ii) The politicians

Ministers of science will campaign as hard as the scientists to increase their budgets. It is often assumed that politicians try to cut their departmental budgets, but that is to misunderstand them. Ministers are judged by their peers by the size of their budgets, and all ministers will fight fiercely for money for their departments, as this translates into power and importance. So conventional is this behaviour that the odd ministerial ideologue who fails to conform occasions disapproval. Consider this complaint of Nigel Lawson's (British Chancellor of the Exchequer, 1983–9): 'the problem with John Moore [Secretary of State for Health and Social Security] was not the normal one of overbidding, but the fact that he had not asked for enough. This was to cause a number of problems.'[12] John Moore was soon sacked. The pressure from ministers of science, therefore, like that of the scientists, is a constant; it will not translate into economic priorities.

(iii) The Treasury

This pressure, too, is a constant – a constant no. All treasuries everywhere try desperately to rein in expenditure.

(iv) Industrialists

Unlike the scientists, ministers or national treasuries, the industrialists are pivotal; their support for science depends on their circumstances. In an advanced, technological country they will lobby aggressively for the government funding of science, because they will be looking for a return on their taxes in the form of basic science to exploit commercially, and they will be looking to recruit into their own laboratories researchers who have been properly trained in academic science. Industrialists in poor countries, however, do not lobby for academic science. Their technology is largely derivative, and they can generally copy or licence it. Their concerns are more basic, and they lobby for fundamentals such as infrastructure for transport or for good primary education so that they can recruit literate workers.

(v) The general public

This is the other pivotal lobby. The inhabitants of rich countries believe that they should support research, which is a sentiment that science lobbyists try to harness. Here is the Royal Society: 'A relatively rich country such as ours has an obligation to the whole of mankind for the continued advance of basic knowledge – of science and other disciplines – simply as a part of its contribution to civilisation.'[13] But the inhabitants of poor countries have more immediate concerns, and they will not agitate for the funding of pure science.

(vi) The lobbies: the conclusion

We can see, therefore, that the political system makes an economic judgement over the funding of academic science, with the industrialists making the pivotal decision on economic grounds, and the general public on cultural grounds. One hundred years ago, when the bulk of family incomes was spent on the barest necessities of food and clothing, a significant tax burden for science would have been democratically impossible – it took an *ancien régime* like that of France to force the poor to fund science. Only now that disposable incomes are so high (in some countries) can politicians tax for science – and that political decision is, in democratic practice, highly sensitive to the economic realities of the taxpayers.

None the less, different countries fund academic science remarkably similarly, which introduces a further influence – other countries. I once asked a British government minister for science to explain the rationale of a controversial decision: What was the economic basis of his thinking? He cheerfully assured me that he did not have one: 'We just look at what the French and Germans do, and then we try to copy them.' They, of course, do the same, and so the governments of the OECD pursue each other, like the Gadarene swine, up each other's fundamentals (to mix metaphors) in a collective chase of ignorance. It is the same fashionable unthinkingness that once gave us the European chase for colonies, the European thirst for war in August 1914, the European twentieth-century lust for totalitarianism, the European exchange rate mechanism and the European Union. The government funding of academic science will never, fortunately, yield disasters on that scale – but is that so fortunate? It is precisely because the consequences for academic science are so muted that the same unthinkingness is perpetuated generation after generation.

Sometimes, it is the international bodies that whip nations into consensus. In 1991, the OECD issued a public denunciation of the Government of Switzerland for spending the least of all OECD governments on civil R&D, private industry accounting for some 80 per cent of it. This so shamed the Swiss Government – that most gloriously independent of all – that it promptly capitulated and promised to do better in future, specifically committing itself to increasing its support for civil R& D by 16 per cent annually for at least three successive years.[14] The public denunciation by the OECD acknowledged that the overall Swiss expenditure on civil R&D was amongst the very highest in the world, but it escaped the OECD that the two phenomena might be linked, one causing the other, or that they might link with Switzerland's extraordinary economic success.

The Success of the OECD

We will explore the sad consequences of the nationalisation of academic science in Chapter 12; here let us acknowledge a mitigation: the tightly economic basis of the political decision over basic science funding encapsulates the secret of the success of the OECD countries: they have learnt to couple capitalism to *dirigisme*. Let us explain this point.

Laissez faire works. The historical (and contemporary) evidence is compelling: the freer the markets and the lower the taxes, the richer the country grows. But *laissez faire* fails to satisfy certain human needs. It fails the politician, who craves for power; it fails the socialist, who craves to impose equality on others; it fails the businessman, who craves for security; and it fails the anally fixated, who craves for order. It also fails the idle, the greedy and the sluttish, who crave for a political system that allows them to acquire others' wealth under the due process of law. This dreadful collection of inadequates, therefore, will coalesce on *dirigisme*, high taxes and a strong state. They will seek to organise society politically, not economically.

But politics are dangerous. *Laissez faire* is rational, and it embraces capitalism. In its turn, capitalism uses money as an objective signal. If a factory is redundant, it loses money; if corn is scarce, its price rises. But if politics supersede, and loss-making factories are subsidised, or prices are fixed, or people are taxed to empower politicians, then reason is lost and economies underperform or even collapse. This is the trap into which most countries fall. Remember Figures 7.6 and 7.10. Figure 7.6 illustrates the performance since 1950 of the 16 most successful economies today, but if the next richest 34 economies are added, to create Figure 7.10, we see that few other than the original 16 maximised their potential. Every single one of those underperforming countries has embraced *dirigisme*, a strong state, high taxes and powerful politicians.

Consider Argentina as a telling example. In 1900, when it was effectively run by the British as a *laissez-faire* economic colony, it was one of the ten richest countries in the world (the British built the railways, exported the beef, ran the banks – and introduced polo). But after the British were expelled, the Argentinians overode the free market. The major industries were nationalised, the social services (including science) were funded generously by the State, and all major economic decisions were subordinated to politics. The result? Standards of living that now approximate to those of the old USSR.

A crucial aspect of the mistake that Argentina, the old USSR and the 34 other countries in Figure 7.10 made, and continue to make, is specifically that of overgenerosity towards science and the other modernist icons beloved of progressive thinkers. Consider the Soviet Union. In *Science Policy in the Soviet Union*[15] and in *Science and the Soviet Social Order*[16] Stephen Fortesque and Loren R. Graham respectively showed that the USSR, before its collapse, possessed about a quarter of all the research scientists in the world (1.5 million researchers) and about a half of all the qualified engineers in the world. So huge were its science budgets, and so honoured were its scientists, that the highest paid official in the whole Soviet Union was the president of the Soviet Academy of Sciences. Researchers were so well paid, and were held in such esteem, and had access to such privileges, that even government ministers would urge their children to become scientists. But it was all wasted money. The Soviets would have done better to have simply copied the imperialist lackeys and running dogs of bourgeois oppression.

OECD politicians, who are attuned to the needs of industry, and who are democratically sensitive, will not pour money into pure science if industry or the mass of the populace do not clamour for it; but *dirigiste* politicians might, even in the absence of economic need. Having overriden the free market, they simply do not receive rational, money-based signals about the real needs of the market for science and scientists. Even if they did, *dirigiste* politicians would ignore them because they pursue political or ideological programmes, not ones based on empirical economics. They are obsessed with supply, and they ignore actual demand.

The safest protection against this common error is *laissez faire*, but failing that, Figures 7.11 and 7.13 do show that a contemporary OECD type of *dirigisme*, one that mirrors *laissez faire*, will work almost as well. That type of *dirigisme* requires that the politicians be obsessed with economic growth, that they be highly sensitive to the needs of industry, that they be highly sensitive to the taxpayers' burdens, and that they do not undermine the rationality of the residual free market. OECD politicians meet those conditions. To their credit, they will even face down considerable unpopularity from pressure groups who, not without reason, feel betrayed.

Nationalisation Betrays those it Nationalises

Industrialists resent nationalisation, but many areas of society court it. Britain, for example, has not only nationalised the universities and their science, it has also nationalised the schools, the hospitals and social security, to name but three chunks of society; each was nationalised with the active support of many of their workers (many scientists, teachers, nurses and social workers agitated for it). Implicit in their agitation was the expectation that the state would prove more generous than had the free market (also implicit was the expectation that nationalisation would empower the workers by relieving them of their accountability to private funders or clients). But, in their *Calculus of Consent*, the book that helped win Buchanan his Economics Nobel Prize in 1986, Buchanan and Tulloch showed that the OECD state is never more generous than the free market.[17] At best, the schools, hospitals and social security under nationalisation enjoy as much funding as they had under *laissez faire*, but they generally receive less (so, for example, health care in the USA, which still largely operates under a freeish market, enjoys a larger share of GDP than elsewhere – to the huge benefit of American doctors' and nurses' salaries and, in the main, patients' care). Nationalisation under the OECD, therefore, merely transfers power from workers, private funders and clients to the politicians; it does not increase funding – and this book shows that academic science and civil R&D are no exceptions to that rule.

OECD politicians are correct in not funding these icons more generously than would *laissez faire*. No one knows the 'correct' level of funding for any social good, but it is more likely that *laissez faire*, which empowers the vast mass of individuals to make their own financial dispensations on economic, cultural, philanthropic or social grounds, will approach the 'correct' level than would a handful of distant politicians.

Who would Fund Academic Science under *Laissez Faire*?

It is too easy to romanticise the state. Hegelians, for example, view it as the essence of society, constituting the forum where all interests, all knowledge, and all culture are represented, to act according to the wisest instincts of mankind to the general benefit of everybody. But one only has to encounter a few politicians to realise what, in practice, the state actually is. (Politeness forbids me from spelling it out. It is, of course, very shocking that Adam Smith wrote of 'that insidious and crafty animal, vulgarly called a statesman or politician').

The state differs from all other human collectives in only one characteristic – coercion. The state can make you do what it wants, and if you do not obey, it will send the police after you to put you in prison. This is why, of course, activists of all types become politicians, because they hope to force people to do things they would not choose to do voluntarily. But if people do actually wish to do something, they can do it, either individually or collectively, in the absence of the state. Thus we have seen that the two pivotal lobbies that demand the funding of academic science are the industrialists and the general public. But we can also see how university science would be funded (and was funded) under *laissez faire*, since both industrialists and the general public are capable of funding science without the mediation of politicians. Britain, for example, was *laissez faire* until 1914, yet its universities could boast, in the seventeenth century alone, names like Newton, Harvey or Boyle. Britain's nineteenth-century scientific flowering was wonderful, and university-based: we saw in Chapter 6 how the numbers of university-employed scientists in Britain doubled every $12\frac{1}{2}$ years between 1870 and 1900[18] (and continued expanding thereafter, though no formal academic study has yet chronicled exactly how much). When Britain was eventually overtaken, economically, it was by another *laissez-faire* country, the USA. *Laissez-faire*, we can therefore note, seems not to prevent countries from dominating the world both economically and scientifically. Who funds academic science in the absence of politicians? Let us explore the non-political options.

(i) How would industrialists support academic science under laissez *faire?*

Mansfield[19] and Griliches[20] have shown that industrialists fund a considerable amount of basic science within their own companies. This is not a trivial point since the overall R&D budgets of companies are so large that even a relatively small industrial contribution to in-house basic science becomes large. For example, see the list on p. 264 for Japan's top ten private R&D investors for 1994.[21]

Let us ask a separate question: would companies under *laissez faire* fund academic science, i.e. science performed within universities? Historically they did: remember, for example, the growth of university physical science that occurred in the USA after 1890 that we discussed in Chapter 7. When the USA overtook Britain economically, and therefore technologically, American engineers increasingly needed to be academically trained. By 1900, the majority of new engineers were college-educated – and the numbers of colleges had dramatically increased.

Japan's top ten private R&D investors for 1994

NTT	£2 000 000 000
NEC	£1 860 000 000
Toshiba	£1 830 000 000
Fujitsu	£1 800 000 000
Mitsubishi Electric	£1 100 000 000
Mitsubishi Heavy Industries	£780 000 000
Sharp	£730 000 000
Sanyo Electric	£540 000 000
Fuji Photo Film	£486 000 000
Mazda Motor	£480 000 000

This, in turn, fed academic science, because employers sought graduates of the best colleges, which were those with the best teachers. The best teachers did research, so colleges competed to foster research. Economically, the colleges were sustained by their students' fees, each student commuting his current fees against future earnings, like those American medical students today. Thus did the free market foster academic science during the nineteenth century.

Since industry needs to recruit more trained scientists than ever before, the same market mechanisms would still obtain today were governments to withdraw their funding. Consider the London School of Economics (LSE). The British universities have long agreed to charge a common fee of its students – one that the Government in its turn has agreed to pay on their behalf. But many universities now believe that they are underfunded and at least one, the LSE, has contemplated charging additional fees, which the students will have to meet privately. The LSE does not doubt that students would meet the fees, to enable the college to raise its standards of teaching and of research. Thus will the market support academic science, at least indirectly.

But under *laissez faire*, industrialists would also support the universities directly. There is an obvious danger to industrialists in supporting university science directly, namely their loss of monopoly control of the findings, but the factors that make the universities dangerous (their openness, their exchange of ideas and their catholicity) are the same factors that make them the best vehicles for the training of scientists. Some industrialists have explored other options. Many major Japanese companies, for example, run large research laboratories that are, effectively, private universities; they engage in pure science and they take training seriously. But the collective experience seems to show that nothing can beat a contemporary Western university for the training of scientists. In the absence of government funding, therefore, the industrialists would step in. Consider the case of Canada.

The Canadian government has long been overstretched, and it has repeatedly asked industrialists for help in funding academic science. In response, a number of joint government/industrial schemes have arisen. In 1981, for example, the Canadian Institute of Advanced Research was created. Half of its funding comes from the private sector (from companies that span the alphabet from Accurex Technologies Inc. to the Valleydene Corporation Ltd) while the other half has come from the Government of Canada. By 1990 the Institute had spent C$13 000 000 on academic studies, in fields ranging from population health, economic growth, artificial intelligence and evolutionary biology. In 1986, a similar scheme was created to foster the life sciences within Canada's universities. In this latter scheme, the Government only spent one dollar for every two spent by industry, yet by 1993 the scheme was so successful that companies, mainly those in pharmaceuticals, were spending C$20 million a year, and the Government a further C$10 million.[22] Thus will private industry foster academic science when government support fails.

None the less, there will always be areas of science – very pure research, for example – which industry will never fund generously (although it would fund some, both for training and for exploration). The funding of that science, under *laissez faire*, falls to the private individual.

(ii) How would private individuals support academic science under laissez-faire?

Science will attract generous patrons – if taxation spares them. Almost everyone in a wealthy country believes that pure science should be funded. Under *dirigisme* this sentiment is transmitted to the politicians (who, being human beings, will share the general values) but under *laissez faire* people, especially rich people, keep their money. The historical evidence shows that the empowerment of wealthy men by money breeds a sense of responsibility, which inspires them to endow science and universities. As a percentage of GDP (and corrected for the GDP *per capita* of the day) the empirical evidence shows that private donors are actually more generous to pure science than is the State (remember that doubling every 12.5 years between 1870 and 1900 of the numbers of science academics in Britain; that privately funded rate of increase was greater than any subsequently funded by the state). But the two sources of funding displace each other; it is the taxation of the rich that empowers the politician.

Historically, the major beneficiaries of patronage have been, successively, the Church and the poor, the arts and sciences, and the secular

charities. The historical record shows that the problem of private patronage for these beneficiaries has been overgenerosity, not underfunding.
Let us consider them in turn.

(a) The Church: mortmain and the problem of overgenerosity   Medieval
Europeans were fantastically generous to the Church, and by the beginning of the second millennium after Christ, the Church was threaten-
ing to own almost all of the property in Europe. Every European state,
therefore, passed laws of mortmain. Mortmain (from the French, *mort*
dead, *main* hand) referred to the dead hand of the Church because,
being a permanent body, the Church never sold or passed on its property. This aroused the jealousy of secular society which provoked the
mortmain laws that returned church property to the laity.

Such generosity was not restricted to Europe – many Islamic countries did not introduce mortmain legislation until the twentieth century, by when over 75 per cent of their property was owned by the
mosques.[23] We can see, therefore, that the fundamental problem over
gifts to the Church, in more than one culture, was overgenerosity. Such
gifts were not, of course, entirely disinterested. The donors hoped to
buy their way into heaven. But they also looked for some sort of im-
mortality here on earth, as their names lived on in their gifts, and they
also looked for recognition and approval from their peers. As we shall
see, such motives later actuated the benefactors of science.

(b) The arts   During the later medieval period, however, as society
gradually became more secular, the need to give did not atrophy, but
was re-channelled, *inter alia*, into the arts. The rich seem to feel a
positive pressure to distribute their wealth: a recent history of the private patronage of the arts was entitled *The Need to Give*[24] after
Nietzsche's observation that Man feels an overwhelming need to give.
As early as the fourteenth century, a poet like Petrarch could find
patronage from Francesco da Carrara, or even from the Church itself
as the clergy adopted the increasingly secular values of society. Such
patronage from rich men flowered into the Italian Renaissance of
Brunelleschi, Alberti, and Leonardo da Vinci. It continues to this day
as can be seen, for example, in the great contemporary American art
galleries which are largely the products of private donations.

(c) The poor   Societies that embrace the Judaeo-Christian ethic acknowledge a duty to the poor:

It is the duty of the wealthy man
To provide employment for the artisan. (Hilaire Belloc)

But although people have been giving money to the poor for millennia, the economics of philanthropy are still not widely appreciated. Let us elaborate them, using the history of England as an example, since they mirror the economics of the private support of academic science.

Much has been written, savagely, in denunciation of the English Poor Laws – and quite rightly. Writers such as Dickens may have been sentimental, but they were describing real distress caused by punitive and harsh laws. Yet it is not widely understood that the State only undertook philanthropy in the first place because it, the State, had destroyed the private sector. The evidence is that the private sector was, and recovered to be, kinder and more generous than the State.

The first English Poor Laws were enacted under Queen Elizabeth I, and that was no accident because it was her father, the frightful Henry VIII, who had destroyed the monasteries. Since it was the clergy, monks and friars who had distributed alms, it was only on their dispersal that the state needed to intervene – with the dreadful results that writers like Dickens chronicled.

Yet Stuart, Georgian and Victorian England witnessed the resurgence of private philanthropy. Private outrage over the meanness of government assistance to the poor manifested itself in much greater private than public largesse. Thus Fraser, for example, has shown that in 1861 the 640 largest London charities spent £2.5 million in the capital – more than was disbursed under the Poor Law – and that survey actually overlooked many of the parochial charities or the Lord Mayor of London's Mansion House funds, which only opened during recessions.[25] Yet who has heard of the work of Fraser, or of similar studies? Remarkably, commentators rarely 'see' private philanthropy, and they debate as if only the State had ever protected the poor. We are all Hegelians now. Nor do commentators see that private philanthropy was the kinder – one need only compare an almshouse (many still survive) with a workhouse.

The decline of private philanthropy coincides with the decline in the private support for academic science – and for the same reasons. The taxing of the rich to empower the State's support of the poor put paid to the private sector but, and let us repeat it, the empirical evidence is that, relative to the GDP *per capita* of the day, the private sector was kinder and more generous than the State has subsequently proved to be.

Dr Thomas Plume: a representative pivotal figure

As we approach modern times, we see a redirection of the philanthropic impulse. The poor remained an object of charity, but the church was increasingly supplanted by the arts and sciences. Consider Dr Thomas Plume (1630–1704). Dr Plume was a cleric, the Archdeacon of Rochester, but when St Peter's, a church in his home town of Maldon, Essex, fell into disrepair, he restored it – not as a church, but as a library and grammar school.* He also endowed £1000 to establish a college of weaving to provide sustainable employment for the poor of Maldon, and £200 for accommodation for them – and he further endowed £1902 to create the Plumain Professorship of Astronomy and Experimental Philosophy at Cambridge University. Thus did even clergymen during the seventeenth century transfer their charity to secular ends, endowing equally science, education and welfare for the poor.

The private endowment of science

The private endowment of science in England long predated Dr Plume – indeed it was co-terminal with English science itself. The first major scientific work by an Englishman was Gilbert's *De Magnete* (1600). Gilbert was a protégé of Mary Herbert, Countess of Pembroke (1555–1621). This is how Aubrey (1626–97), in his *Brief Lives*, described Wilton, Mary Herbert's house: 'In her time Wilton house was like a college, there were so many learned and ingeniose persons. She was the greatest patronesse of witt and learning of any lady in her time. She was a great chymist and spent yearly a great deale in that study. She kept for her elaborator in the house Adrian Gilbert (vulgarly called Dr Gilbert), halfe brother to Sir Walter Ralegh, who was a great chymist in those dayes. 'Twas he that made the curious wall about Rowlingtonparke, which is the parke that adjoyns to the house at Wilton. Mr Henry Sanford was the earle's secretary, a good scholar and poet, and who did penne part of the Arcadia dedicated to her (as appears by the preface). He haz a preface before it with the two letters of his name. 'Tis he that haz verses before Bond's Horace. She also gave an

---

*The Plumian Library still stands in St Peter's, Malden, exactly as Dr Plume left it in 1704. One of the oldest lending libraries in the world, it continues to house Dr Plume's collection of books, which anyone can read. I recently had the thrill of taking off the shelf and reading Dr Plume's own copy of Francis Bacon's denunciation of Essex. Unfortunately, the library does not own any of Bacon's scientific works, but if readers of this book wish to write to me (c/o Department of Clinical Biochemistry, Cambridge University, Addenbrooke's Hospital, Cambridge CB2 2QR, UK, we will set up a fund to buy such a book and donate it to the Plumian).

honourable yearly pension to Dr. (Thomas) Mouffett, who hath writt a booke De insectis. Also one . . . Boston, a good chymist, a Salisbury man borne, who did undoe himselfe by studying the philosopher's stone, and she would have kept him but he would have all the gold to him selfe and so dyed I thinke in a goale'.*

The scale of endowments and gifts for science (and the arts and the charities) accelerated throughout the eighteenth and nineteenth centuries (as was discussed in Chapter 6) and reached its apotheosis in the early twentieth century. Let us consider five of the more famous benefactors. Andrew Carnegie (1835–1919) made his money ruthlessly in steel, but having made it, he considered it a duty to spend it wisely, a duty he elaborated in his famous *Gospel of Wealth* (1889). The Carnegie Foundation was created with an endowment of $10 million in 1905, and it has funded much of the best academic science of America.† This transformation from ruthless businessman to wise benefactor (which Petrarch, who had seen it all before, described with 'aristocrats are not born, they are made') was also evidenced by John Rockefeller (1839–1937) of Standard Oil and various railways. He created the University of Chicago in 1891, the Rockefeller Foundation in 1913 (which funded, amongst many crucial discoveries, Avery's discovery of the function of DNA and Florey's discovery of penicillin) and, in all, he spent $530 million on various benefactions. In Britain, Henry Wellcome (1853–1936) the pharmacist, created the Wellcome Foundation in 1924, and this now spends £77 million a year on medical research (it is only second in size to the Medical Research Council at £200 million). Samuel Courtauld (1876–1947) the textile magnate, who elaborated his ethical beliefs in *Ideals and Industry* (1949) created a number of Courtauld Institutes, and although his Courtauld Institute

---

*Like many artistic persons, Mary Herbert embraced dubious morals: 'She was very salacious, and she had a Contrivance that in the Spring of the yeare, when the Stallions were to leape the Mares, they were to be brought before such a part of the house, where she had a *vidette* (a hole to peepe out at) to looke on them and please herselfe with their Sport; and then she would act the like sport herselfe with her stallions. One of her great Gallants was Crooke-back't Cecill, Earl of Salisbury' (Aubrey, *Brief Lives*).

Aubrey also had his doubts about Gilbert: 'very Sarcastick, and the greatest Buffoon in the Nation'. Aubrey was casual over detail, and his 'Adrian' Gilbert was 'William'. Moreover, it was Humphrey Gilbert who was Ralegh's step-brother. Gilbert was not 'vulgarly' called Dr Gilbert, he actually was an MD who was elected president of the College of Physicians in 1600. Mary Herbert later married Sir Matthew Lister, another member of the College, whom she may have met through Gilbert.

† Although Carnegie-funded archaeologists did terrible, if inadvertent, damage during the 1930s by dynamiting the Mayan monuments of Central America.

for Fine Arts is the best known, his Courtauld Institute for Biochemistry, in the University of London, continues to do good work (see the postscript on philanthropy and beauty). Lord Nuffield (1877–1963), the car manufacturer, created the Nuffield Foundation for Medical Research in 1943, but he also spent his money on other wise donations, such as the Nuffield Professorships of Clinical Medicine, Surgery, Obstetrics and Gynaecology, Anaesthetics, Pathology and Clinical Biochemistry at Oxford University, as well as the Nuffield Institute for Medical Research in Oxford.

And then . . . and then nothing, at least in Britain. The private endowment of science, which started during the Renaissance, picked up in the seventeenth century, accelerated during the eighteenth and nineteenth centuries, and crescendoed in the early twentieth with these gargantuan donations, suddenly stops. Why? The arrest coincides, of course, with the abandonment of *laissez faire* and the imposition of high taxes. After 1913 the British Government prevented men from accumulating wealth as they had, and it assumed the responsibilities for science and other cultural and good causes that rich men had earlier adopted for themselves. This obviously benefited the politicians, who now controlled a vast share of national GDP and who could luxuriate in their important distribution of huge sums in patronage to schools, hospitals, universities and laboratories. But it did not benefit the beneficiaries, as we shall see in the next chapters.

The situation is healthier in the USA, where taxes are lower, and rich men continue to endow the arts and sciences. The Howard Hughes Medical Institute spent $332 million on research in 1991, the W. M. Keck Foundation $95 million, the Lucille P. Markey Charitable Trust $43 million, and the John D. & Catherine T. MacArthur Foundation, the Pew Charitable Trusts, the David and Lucille Packard Foundation, the Alfred P. Sloan Foundation and the Robert J. Kleberg Jr & Helen C. Kleberg Foundation spent over $20 million each. These foundations all outspent the Rockefeller's $18 million and the other 700 or so US research foundations. Most of those testaments to philanthropy were themselves outspent by the big charities such as the American Heart Association ($105 million) or the American Cancer Society ($94 million).

There are still massive private donations to American science such as the $77.4 million gift in 1994 to Stanford University of William Hewlett and David Packard of computer fame or the $60 million gift to Harvard for neurobiology research of Isabelle and Leonard Golderson, but even in the US the private funding of science has been weakened by the combination of ever-rising taxes and the direct government funding of science. The displacement of private by public funding was well described in 1993 by Martha Peck, the executive director of the Burroughs Wellcome Fund who said: 'We've seen foundations turn away from research toward support for social and community issues. The perception has been that science is getting from other sources [i.e. Government] the kind of funding it needs.'[26] Martha Peck's comment illustrates that the government funding of science does not just damage because of the taxes it incurs (for these are relatively small for science itself, though in practice such taxes for science are usually imposed within a culture of high taxes generally, which do damage philanthropists) but because government funding apparently destroys the need for private support.

But a reversal is now occurring. Between 1980 and 1994 academic science budgets in the US grew by about 1 per cent a year in real terms, but the bulk of the increase came from the private sector, with the share of government funding falling from 67 per cent to 55 per cent.[27] The share of the industrial funding of academic science doubled. We can see, therefore, that the US during the 1980s, like Britain over the same period, demonstrated the same mutual (if largely overlooked) displacement of the private and public funding of academic science – to the universities' advantage.

We also see from the private foundations an originality that escapes government bodies. In February 1994, for example, the Rockefeller Foundation offered a prize of $1 million to anyone who developed a simple kit for the diagnosis of the sexually transmitted diseases, chlamydia and gonorrhoea. What a clever idea! Grants of millions are routinely given to scientists who promise to achieve something – how much better to reserve the rewards for those who do actually achieve.

Some countries such as Denmark, where government funding for science has historically been weak, have witnessed the rise of private science foundations such as the Carlsberg Foundation, and Germany can boast of the Volkswagen Foundation (DM 158.8 million in 1991). Remarkably, even the French are now beginning to subscribe as individuals to science, and Dr Cohen's Genethon raised FF80 million last year (to be fair to the French, their best scientific institute, where Luc Montagnier discovered the aids virus in 1983, is the Pasteur Institute which has always been largely funded privately, receiving for example the equivalent of £55 million in 1907 from the will of the financier Daniel Iffla-Osiris, and £30 million from the posthumous sale of the Duchess of Windsor's jewels in 1987). Even more remarkably, the fall in the top rate of tax in Britain, which is now allowing individuals to become rich again, is creating a new breed of philanthropists. George Soros, who made over £1 billion from John Major's folly over the ERM, has endowed his new International Science Foundation with over $100 million to spend on science in the erstwhile Soviet bloc. He has, of course, also endowed non-scientific charities with his well-gotten gains.

Perhaps Hong Kong best illustrates a contemporary example of private philanthropy towards science. Hong Kong's economy is one of the most vibrant on the planet (see Figure 7.10) and that has been achieved under one of the lowest tax regions on the globe. But is science neglected? Oh no. In 1994, the Royal Hong Kong Jockey Club announced a donation of over $220 million towards the building of the Hong Kong University of Science and Technology, an institution that is modelled on MIT and Caltech.[28] Hong Kong's 6 million people now need pure science and hi-tech research, and private philanthropy is providing it. Extrapolate from that $220 million for 6 million Hong Kongers to the sort of sums that the 56 million Britons or 220 million Americans would command, and you can envisage the scale of their private funding of science under a free market.

Yet one does not have to envisage anything in the case of Japan, since Japan demonstrates one stunning fact: whereas most US university science is funded by the government (85 per cent) as is the case with most OECD countries (see the figures in Table 12.1), in Japan over half of all university science is funded privately, yet that has not stopped Japan's economy from exploding, that has not inhibited the growth or quality of Japanese academic science (see Figures 7.11 and 7.12) nor has it distorted Japan's science towards applied rather than pure research. Indeed, Ben Martin and his colleagues have actually shown that Japan's science is the purest of all the major science countries' (see the discussion in Chapter 12).

The private funding for Japanese universities' science is derived from many sources including students' fees, industry and endowments. Japan, which is astonishingly *laissez faire* in all aspects of education (compulsory, free secondary education stops at the age of 15, yet 99 per cent of children stay on in schools at the age of 16, paying fees) and which has amongst the lowest tax rates of all OECD countries, thus confirms that pure science, economic growth and education will flourish under *laissez faire,* and will indeed outstrip the achievements of more *dirigiste* countries like the US, the UK, France or Germany.

Table 12.1, moreover, also highlights the high private funding of

academic science in Canada and the UK, confirming that when governments refuse to fund academic science generously, the private sector
will indeed compensate – copiously.

Remarkably, however, in the face of vast public funding, the private
sector today continues to perform major pure science. *Current Contents*, 11 July 1994, vol. 37, p. 4, recently reviewed the institutions
that produce the largest number of cited papers in biology, and of the
top 7, two were private companies (Genentech and Chiron, which shows
how industrial funding must support fundamental research), one was a
charity (the Howard Hughes Foundation), three were private institutions (but which now receive government grants; the Salk Institute,
the Cold Spring Harbor Laboratory and the Whitehead Institute) and
only one was a wholly government-founded, government-funded laboratory, the Institute de Chimie Biologique in Strasbourg. How much
better our science would be today if its private funding had not been
displaced!

POSTSCRIPT ON PHILANTHROPY AND BEAUTY

Courtauld's endowment of institutes for both the arts and the sciences
illustrates a general point; private benefactors of science and of higher
education are motivated by a love of culture. They tend, therefore, to
create beautiful endowments. Consider Thomas Holloway, the pharmaceutical chemist who, in 1883, created Holloway College, London
University, for the education of women. For his College, Holloway
erected a lovely, and spectacular building in Egham, Surrey, and he
left to it 79 equally lovely, equally spectacular paintings, including
Turner's *Van Tromp Going About to Please His Masters*, Constable's
Sketch for a View of the Stour, near Dedham, Gainsborough's *Peasants Going to Market, Early Morning*, Millais's *The Princes in the
Tower*, Frith's *The Railway Station* and Landseer's *God Proposes, Man
Disposes*. Bizarrely, in 1993 the College started to sell off his paintings. Professor Norman Cowan, the Principal, said that the College
needed money 'for education rather than art' (quoted in *The Times*, 24
February 1993) as if the two were opposed! Great art, of course, is a
source of education.

Professor Cowan's philistinism, tragically, is typical of university
academics today. Almost all the beautiful buildings of universities were
the gifts of non-academics. The buildings that universities build with

their own resources are almost invariably ugly. Drab utilitarianism is the hallmark of the modern academic, as a visit to his house or a glance at his clothes will generally reveal. (It was ever thus: in his satire on scientists in *Gulliver's Travels*, published in 1726, Jonathan Swift wrote: 'The first man I saw was of a meagre aspect, with sooty hands and face, his hair and beard long, ragged and singed in several places. His clothes, shirt and skin were all of the same colour'). Academics seem, sometimes, to actually hate beauty, resenting it as upper class (academic salaries being what they are in Britain, most university employees are now of relatively humble origin; the Lord David Cecils are largely extinct). During the 1960s and 1970s, London University tore down some lovely Georgian terrace houses to build some staggeringly brutal concrete lumps (many of which, I am glad to say, have weathered badly and require expensive maintenance) but perhaps nothing London ever built was quite as ugly as the horrors Oxford foisted on itself, monstrosities like the University offices off Clarendon Street, or its hideous Law Library.

Drab utilitarianism is also the hallmark of the modern politician, and the buildings that governments build for academics are amongst the dreariest on the planet. Professional academics may deserve no better, but their students often possess decent feelings, and one of the tragic consequences of the supplanting of private endowment by public grants for higher education is that generations of bright, eager young people have to submit to years of exposure to ugliness. The nationalisation of the universities seems to have aggravated the academics' dislike of beauty. Because it has not needed them, Holloway College has been irritated by its responsibilities to its paintings; its money comes from the State according to quotas and formulae of students' numbers that do not recognise beauty. But Dulwich College in contrast, an English so-called Public School (i.e. private school) which houses an equally lovely art collection, positively revels in its gallery (built by Soane). Dulwich, in the private sector, needs to market itself, and so it uses its art. Holloway, a nationalised institution, needs to market itself about as much as did a collective farm in Siberia in 1952.

The academics' hatred of beauty was never better illustrated than by Oxford's rejection of Harold Acton's bequest of his lovely Florentine villa. It went to New York University instead, which will use it as a centre for Renaissance studies. Oxford claimed that its refusal was prompted by an inability to raise the necessary funds for its repair and maintenance, as if half of Europe would not have subscribed to it! In reality, of course, Oxford academics prefer to see themselves as gritty students of 'relevant' subjects, preferably those redolent of 1940s' austerity and social engineering, and they will never forgive Acton for having been more distinguished as an undergraduate than they will ever achieve as dons. Not for nothing did Belloc dismiss dons as 'remote and ineffectual'.

11 The So-called Decline of British and American Science

During the 1980s, Britain's scientists exploded with anger. They believed that a mean government was destroying their science and their universities. Their professional journals published articles with titles such as: 'Britain over the Hill' (*Nature*); 'Sorry, Science has been Cancelled' (*New Scientist*); and 'Bye-bye Britain' (*British Medical Journal*). Meanwhile, distinguished scientists made frightening public statements. Sir George Porter, the President of the Royal Society, claimed that 'the morale of the scientific community has fallen to its lowest point this century'; Sir David Phillips FRS, the Chairman of the Advisory Board of the Research Councils, maintained that 'a British scientist can hardly travel anywhere without being subjected to pity by others'; while Professor Robin Weiss, the Director of Virology at the Institute of Cancer Research, explained that 'we were able to respond to the AIDS challenge only because of the excellent groundwork laid in the UK. But the erosion of our science has been so bad that the next virus along will beat us'.

British scientists regularly explode with anger. During the 1960s, Quinton Hogg, the Tory Minister with responsibility for science, was nearly refused an honorary degree at Cambridge, and in 1984 Mrs Thatcher really was refused an honorary degree at Oxford, a public rejection which Oxford had only made once before (over Zulfikan Ali Bhutto of Pakistan, a known murderer). The Oxford vote against Mrs Thatcher was orchestrated by Professor Denis Noble FRS, who went on to found the pressure group Save British Science (SBS). Professor Noble summarised its concerns when he wrote in The *Independent* of 13 January 1987: 'that when I ask my colleagues in almost every discipline . . . the talk is less about how to save the situation, but whether it is possible to do so'.

Save British Science (SBS) has since published a large number of documents claiming that the numbers of scientific papers published from Britain have declined, as have the numbers of citations they attract, as has the Government's financial support. SBS have also claimed that there has been a 'brain-drain'. Although SBS has had vast impact on public opinion, it has not performed any research of its own; instead it has largely relied for its data on the work of two academics from the Science Policy Research Unit, Sussex University. These two, Mr Ben Martin and Mr John Irvine, published throughout the 1980s a series of papers in *Nature* and *New Scientist* whose titles and subtitles are self-explanatory: 'The Writing on the Wall for British Science' (subtitled 'Britain's basic science is rapidly declining in quality and quantity'),[1] 'Charting the Decline in British Science';[2] 'Is Britain Spending Enough on Science?' (subtitled 'Britain is falling behind the major industrialized countries in its investment in academic and related research, which may explain the nation's declining contribution to world science')[3]; and 'The Continuing Decline of British Science'.[4] *Nature*, the most prestigious of science journals, gave these pieces generous space, and supported them vigorously in its editorials (such as the leader entitled 'Bringing Research Back to Life' which was subtitled 'Having all but killed it off, the British government is now brooding on the question of how best to restore a once-successful research enterprise to good health').[5] Indeed, so vigorously did *Nature* champion the declinists, it even allowed the Royal Society's Science and Engineering Policy Studies Unit, directed by Dr Peter Collins, to publish a paper[6] which largely duplicated the earlier one of Martin and Irvine's 'Charting the Decline in British Science'. In that latter paper, Dr Smith and his colleagues concluded that 'UK performance has deteriorated since the early 1970s . . . in absolute terms'.

These three academics, Martin, Irvine and Smith, not only supplied SBS with its ammunition, they also helped write reports for influential bodies such as the Advisory Board of the Research Councils (ABRC) which coordinates the Government's funding of academic science (see, for example, ABRC Science Policy Studies No. 1, *Evaluation of National Performance in Basic Research*, 1986, or ABRC Science Policy Studies No. 2, *An International Comparison of Government Funding of Academic and Academically Related Research,* 1986). These apparently authoritative reports confirmed that British science was indeed in decline.

Yet it is all nonsense. British science grew massively during the 1980s and continues to grow into the 1990s. Martin and Irvine only did two pieces of research – bibliometric and financial – and then they misinterpreted their own data.

TABLE 11.1   *Percentage changes in world publication share of major industrial countries, 1973–82*

Canada	-8%
France	-9%
West Germany	+2%
Japan	+40%
UK	-10%
USA	-3%
USSR	-15%
Rest of world	+9%

SOURCE: This table comes from Irvine *et al.* (*Nature* (1985) vol. **316**, pp. 587–90). Irvine *et al.* obtained their data from Computer Horizons Inc., who, in turn, obtained it from the Institute of Scientific Information, the publisher of *Current Contents*.

The Bibliometric Data

Martin and Irvine counted all the scientific papers that were published between 1973 and 1982. Actually, they did not do it themselves, they bought the information (on a research grant provided by the Economic and Social Research Council, an organisation supported by the British taxpayer) from Computer Horizons Inc., an American company which, in turn, obtained its data from the Institute of Scientific Information, the publisher of *Current Contents*, a weekly listing of the scientific papers and journals published worldwide. Table 11.1 summarises the findings of Martin and Irvine. Since it was this one table, more than any other, that fuelled the whole debate in Britain over its decline in science during the 1980s, we will consider it in detail.

The table is blunt: between 1973 and 1982, Britain's share of scientific publications fell by 10 per cent. Decline! But is there not a lot wrong with Table 11.1? First, is it not intuitively improbable? Can it really be that not only British, but also Canadian, French and American science are in decline? A Soviet decline is not, perhaps, surprising, but it seems unlikely that, amongst the major science nations, only Germany and Japan expand.

Second, the table gives no absolute figures. This is important. If, for example, the Queen of England were to lose 10 per cent of her wealth, and one of her footmen were to increase his by 40 per cent, she might not be too concerned for her position. She would still be much richer than him. Absolute or quantitative figures are important if we are to assess nations' standing against each other. Table 11.2

TABLE 11.2   *International comparison of scientific papers published for 1982*

Country	Total number of papers published	Population in millions	Numbers of papers published per million
Canada	11 700	24.4	483
USA	106 900	229.8	465
UK	23 900	56.0	432
Germany	17 900	61.7	295
France	14 600	54.0	279
Japan	21 200	117.7	174

SOURCES: The data for the number of papers came from the ABRC Science Policy Studies no. 1, *The Evaluation of National Performance in Basic Research* (London, 1986). Population data for 1981 came from the *OECD Economic Surveys 1983–1984* (Paris: OECD, 1984). The number of papers published per million are the author's calculations. The papers, although for 1982, came from the 1973 fixed journal database.

provides that assessment, and it shows that, *per capita*, the three great science nations in 1982 were Canada, the USA and the UK. The British published significantly more papers *per capita* than the Germans or the French, and many more than the Japanese. To have claimed, as Martin and Irvine did in 1985, that British science had the 'writing on the wall' when the British were publishing a third more papers *per capita* than the Germans or the French, can only be described as misleading.

There is yet another problem with Table 11.1. Although Irvine *et al.* studied Britain's papers in journals published between 1973 and 1982, throughout that period they only studied the papers in journals that existed in 1973. Table 11.1 does not catalogue any papers that were published in journals founded after 1973. But a large number of journals were founded between 1973 and 1982. Indeed, in 1973 ISI only covered 2364 journals, but by 1982 this had risen to 3246, nearly half again. This expansion reflected the expansion in science which has proceeded ineluctably for over two centuries. In his *Little Science, Big Science*[7] Derek da Sola Price showed that the numbers of journals, papers and researchers have doubled, worldwide, every 15 years since 1750 (it was Derek da Sola Price who observed that the statement '90% of all the scientists who have ever lived are alive today' has been true for every date since 1750).

Between 1973 and 1982, therefore, the numbers of journals increased by around 50 per cent; if we assume that the new journals are as large as the established ones (not necessarily a valid assumption, but the best we can readily do) and if we assume that British papers are representatively distributed (and the evidence is that English language papers are actually over-represented in the new journals[8]) then it can be seen that Britain published some 30–40 per cent more papers in 1982 than in 1973. A 30–40 per cent increase over a decade can only be described as healthy growth. It was, therefore, truly extraordinary of Martin and Irvine to describe their findings as 'a catastrophe that we face as a scientific and educational nation'.*

It could be argued that the new journals are not as good as the old ones, but this is improbable. It is precisely the new, expanding, important fields that spawn new journals. *Cell*, for example, the most prestigious of life-science journals, was founded in 1974. The journals of long dead subjects such as anatomy, however, plough on for century after century, publishing increasingly unimportant papers.

Neither Martin nor Irvine understood that they were looking at a relative British decline, which is why they claimed 'Britain's basic science is rapidly declining in quality and quantity.' Dr Smith and his colleagues at the Royal Society also failed to understand, hence their 'UK performance has deteriorated since the early 1970s . . . in absolute terms.' Professor Noble also failed to understand, hence his 'when I ask my colleagues in almost every discipline . . . the talk is less about how to save the situation, but whether it is possible to do so'. Moreover, the quality of British science remained remarkable. Every two years, in *Current Contents*, Eugene Garfield determines the national origins of the 300 most cited papers annually. As Table 11.3 shows, during the years that British scientists were claiming that their science was in decline, it consistently came second only to that of the USA. Some writing. Some wall.

Numbers of Academic Staff

It became fashionable during the 1980s for all British academics, not

---

*David Pendlebury claims that the doubling of world journals every 15 years started to slow down in 1970, and that during the period 1973–82 journals were only increasing in numbers at 0.7 per cent a year (D. Pendlebury, 'Science's Go-go Growth: Has it Started to Slow?', *The Scientist*, vol. 3, pp. 14–16, 1989). Unfortunately, no-one really knows how many journals are now published internationally (ISI only covers a small fraction of the world's journals), though perhaps there are 40–55 000. It is certain, however, that: (i) journals are expanding in size and in the numbers of papers they publish; (ii) the numbers of scientists are increasing; and (iii) science budgets continue to increase. The argument presented here is, therefore, essentially correct.

TABLE 11.3   *National origins of the 300 most cited papers during 1983 or 1984 (for 1985 or 1986 see Notes, below)*

Country	100 life science papers (1984)	100 chemical science papers (1983)	100 physical science papers (1984)	Total citations
USA	79	66	74	219
UK	13	8	9	30
Germany	1	11	14	26
France	5	5	8	18
Switzerland	5	6	5[a]	11
Canada	6	4	1	11
Japan	4	1	3	8
Australia	4	1	0	5
Israel	2	2	1	5
Holland	0	2	3	5
Italy	0	0	5	5
USSR	0	1	3	4
Spain	0	0	4	4
Belgium	3	0	0	3

NOTES: The total number of papers exceeds 300, because some come from more than one country and so are counted two or more times. These data have since been updated in *Current Contents* of 28 November, 12 December 1988 and 6 February 1989. This shows that Britain remains second only to the USA as the nation of origin of the most cited life-science papers for 1986. For 1985, chemical-science papers Britain remained third, after the USA and Germany. But for physical sciences in 1986, Britain actually climbed up into second place.
[a] Papers from CERN, an international body based in Switzerland have not been included. Currently, 1992/3, Britain contributes just under £100 million to CERN, the European Laboratory for Particle Physics, the European Space Agency, and the Institute Laue Langerin in Grenoble. If papers from these organisations are considered, then Britain's world share of papers and citations rises even higher).

SOURCE: The data came from *Current Contents* of 24 November 1986 (physical sciences), 8 December 1986 (life sciences) and 22 December 1986 (chemical science).

just scientists, to complain of decline. Professor Malcolm Bradbury, the Professor of American Studies at the University of East Anglia, wrote during 1988 in *Unsent Letters*[9]: 'Our present political masters regard British universities in much the same way as Henry VIII regarded the monasteries. My own university, for example, far from looking for new people, is paying distinguished colleagues considerable sums to depart.'

TABLE 11.4   *Numbers of full-time British academic staff*

Academic year	Total full-time staff financed	Full-time staff wholly university financed	Full-time staff not wholly university financed
1938–9	5 000	–	–
1954–5	11 000	–	–
1962–3	15 682	–	–
1967–8	26 839	–	–
1976–7	40 246	32 738	7 508
1981–2	43 924	33 735	10 189
1982–3	43 080	31 642	11 441
1983–4	43 149	31 096	12 053
1984–5	44 210	31 043	13 167
1985–6	45 743	31 412	14 331
1986–7	47 038	31 432	15 606

NOTES: These data come from the annual *University Statistics*, published by the University Grants Committee; they do not distinguish between numbers of scientists, social scientists and those in the humanities, but at least half of those in column 4 are in the sciences.

But the numbers of British academics have consistently risen throughout the nineteenth and twentieth centuries, and they continued to rise during the 1980s (see Table 11.4).

Between 1976–7 and 1986–7 the numbers of full-time staff increased from 40 2246 to 47 038. Of particular importance to scientific research, this expansion did not come in the category 'full-time staff wholly university financed' (these are the tenured professors and lecturers whose time is divided between teaching, administration and research). The expansion came in the category 'full-time staff not wholly university financed'. This category covers the professorial research fellows, senior research fellows, post-doctoral fellows, research associates, research assistants and research technicians, all of whom are employed solely to do research. Although they work in the universities, their salaries derive from a variety of sources including the research councils such as the Medical Research Council (MRC) or Science and Engineering Research Council (SERC), the medical charities (such as the Wellcome Trust or the Imperial Cancer Research Fund), or from industry.

Thus, at the very time that British researchers were claiming that their numbers were shrinking, the numbers of full-time researchers in British universities had doubled between 1976–7 and 1986–7. The numbers of PhD students rose almost as fast.[10] That rate of expansion has been maintained since, and even the Association of University Teachers, the trade union, has now admitted that between 1987 and 1993 the total numbers of academic staff in British universities and polytechnics increased by over 10 000, from 98 400 to 109 800.[11] But the intriguing question is: why did the academics not see the 1980s expansion at the time? The UGC data, after all, is published annually, and Martin and Irvine only had to look out of their office. Their own unit, SPRU, is the largest science policy research unit in the world. It was only founded 25 years ago, but it has grown fantastically. It now boasts four professors, 20 readers or senior fellows, 24 lecturers or fellows and 11 research officers.[12] Professor Noble's own Department of Physiology in Oxford is also huge (in 1989 it boasted of over 110 scientists, to say nothing of technical and support staff) and it continues to expand, yet these distinguished academics seem incapable of counting their colleagues down the corridor.

Financial Data

One reason the academics failed to see the expansion around them was that they were obsessed with government funding. Ever since the First World War, the expansion of government funding for the universities and their science had outstripped the expansion in GDP *per capita*, but this disproportionate expansion had to stop, which it did around 1970 (remember that in 1971 Mrs Shirley Williams MP, latterly the Secretary of State at the Department of Education and Science, announced that 'for the scientists the party is over'). But 1981 actually saw a cut, of between 5 and 8 per cent, in the Government's grant to the University Grants Committee (UGC) and to academic science.

The universities never forgave Mrs Thatcher (Prime Minister 1979– 90) for those cuts. Yet was that bitterness deserved? Mrs Thatcher was as much a victim as the instigator of circumstances, and government policy was much more bi-partisan than the universities acknowledged. It is hard to uncover many discontinuities in policy between the Labour Government's last two years (1977–9) and the Tories' first few after 1979. It was Denis Healey, the Labour Chancellor of the Exchequer, who first introduced a monetarist budget (in 1977). In 1978, the Labour Government halved the NHS building budget, and its resistance to the trades unions' campaign for greater public spending precipitated the so-called 'Winter of Discontent'.

Retrenchment was forced on both Labour and Tory Governments by sheer economic circumstances. Throughout the 1970s, Britain's economic performance was dreadful, and by the late 1970s, all the comfortable nostrums had been exhausted. Harold Wilson's 1960s White Heat of the Technological Revolution had not wrought relative economic growth, the British economy was so weak that an IMF loan had to be negotiated on humiliating terms and simple Keynesian economics had failed. It was James Callaghan, the Labour Prime Minister, who told the Labour Conference of 1978 that 'you cannot spend your way out of recession'

The 8 per cent cut in the UGC budget was easily justified. In 1981, Britain, a rapidly impoverishing country, relatively, enjoyed higher staff: student ratios in its universities than any country in the world but Holland (which was also trying to cut them). To a government desperately needing to cut expenditure, it seemed reasonable to save money by trying to bring British university staff : student ratios closer to West German or US levels. But British academics responded with outrage, and it was not uncommon in senior common rooms during the early 1980s to hear Mrs Thatcher being compared with Hitler – to her disadvantage. The *Oxford Magazine*, the dons' house journal, used to compare her publicly with Hitler.

The situation was not eased by Martin and Irvine's simplistic endorsement of the linear model, which led them to claim that the government science cuts would hinder economic recovery: 'We are convinced that the development of new technologies depends on the results of basic research. . . . Yet, such research has been hard hit by [British] government cuts over the last few years.'[13] One of the benefits to Martin and Irvine in not calculating the numbers of papers published *per capita* by different countries is that they were spared the problem of explaining how Japan began to overtake the UK economically while only producing one third as many scientific papers (Table 11.2).

The academics were so outraged by the government cuts of 1981 that, collectively, they appear to have lost any objectivity in their political campaign. In an odd way, they failed to see any funding for science that did not come from Government. Thus Martin and Irvine entitled one their papers 'Is Britain Spending Enough on Science?' and subtitled it 'Britain is falling behind the major industrialized countries in its investment in academic and related research . . .'[14] but they had not actually looked at *Britain's* expenditure, they had only looked at that of the British *Government*, and the paper's proper title and subtitle should have read 'Is the British Government spending enough on science?' and subtitled 'The British Government is falling behind. . . .'

This is not a pedantic point because Martin and Irvine, by not look-

TABLE 11.5   *The research funding of the major British medical charities,*
1987–8

British medical charities	£ million
Imperial Cancer Research Fund	38.8
Wellcome Trust	27.8
Cancer Research Campaign	24.6
British Heart Foundation	11.8
Arthritis and Rheumatism Council	7.0
Ludwig Foundation	5.0
National Foundation for Crippling Disease	3.0
Muscular Dystrophy Foundation	2.0
Multiple Sclerosis Society	1.8
Tenovus	1.4
Cystic Fibrosis Research Trust	1.26
British Diabetic Association	1.2

SOURCE: Data collated by the Association of Medical Charities UK.

TABLE 11.6   *Research expenditure by the Cystic Fibrosis Research*
Trust UK

Year	£
1965	20 000
1970	50 000
1975	120 000
1980	450 000
1987	1 260 000

SOURCE: Data collated by the Cystic Fibrosis Research Trust UK. The sums
are those actually spent; no correction has been made for inflation.

ing at science funding as a whole, missed the source of the huge expansion in university science in Britain during the 1980s – the private sector. In a beautiful demonstration of the Third Law of Science Funding proposed in Chapter 10, the stagnation in government funding prompted a resurgence of private funding. The two major private sources were the medical charities and industry. The medical charities doubled their support in real terms to university science during the 1980s. In 1987–8 their support amounted to £128 million, which nearly equalled the Government's Medical Research Council budget of £146 million for that academic year, and during 1989–90, the charities actually overtook the MRC in the money they spent on academic science.

Table 11.5 demonstrates the research funding of the major British medical charities for 1987–8 and Table 11.6 demonstrates the steady growth in the research expenditure of a typical small-to-mid size medical charity over the last 25 years.

Industry, too, doubled its support for British university science during the 1980s.[15,16] As the Committee of Vice-Chancellors and Principals itself admitted in its *State of the Universities* (1991): 'Increases in funding from industry and medical charities have more than made up for the lack of government funding over the past decade.'[17] Between 1984 and 1988, the research income per tenured academic had risen from £12 300 to £16 900 per year (1984 prices, adjusted for inflation), a rise of 35 per cent.[18] Money from industry rose by 150 per cent (i.e. more than twofold) during the 1980s (adjusted for inflation) and money from the charities by 130 per cent. Research council money only rose by 25 per cent. It will be seen, therefore, that the British universities' and the scientists' post-war expansions continued unabated during the 1980s, but because the expansion was fuelled privately and not publicly, Martin and Irvine overlooked it.*

The source of the funding affected the nature of the expansion. The charities, industry and other non-UGC sources of university research usually fund discrete projects or fellowships for up to five years, but they do not routinely endow permanent university positions. The numbers of professorships and lectureships, therefore, fell in response to the 8 per cent cut in the UCG budget, but the numbers of research fellows doubled. Science *per se* did not gratuitously suffer, but scientists' aspirations for secure permanent jobs did (see below).

International Comparisons

Martin and Irvine's papers are so odd that it is extraordinary that they made such impact. For example, in their 'Is Britain Spending Enough on Science?' they conclude with: 'What conclusions, then, can be drawn from the results? From the point of view of the United Kingdom, it is clear that overall government funding of academic and related research is falling increasingly behind that of our nearest European competition.' Yet their data actually show the UK was doing better than West Germany or the USA (Table 11.7). Indeed, nowhere in the conclusion do Martin and Irvine actually address the most interesting fact that emerges from any international comparison of science funding, namely

---

*Recent CVCP data acknowledges £36 000 research funding per academic, a tripling since 1984, despite a fall in government support (*Nature*, **375** (1995) p. 96).

TABLE 11.7   *Percentage change in government funding for academic and academically related research, 1975–82, as percentage of GDP*

France	25
Japan	11
Holland	3
UK	-3
USA	-5
West Germany	-15

SOURCE: Irvine and Martin, *Nature*, vol. **323**, pp. 591–4, 1986.

TABLE 11.8   *Government expenditure on academic science in 1982*

Country	Percentage of GDP
West Germany	0.49
France	0.44
UK	0.38
USA	0.31
Japan	0.25

SOURCE: *International Comparison of Government Funding of Academic and Academically Related Research*, ABRC Science Policy Studies, 2, 1986.

that the British Government outspends that of both the USA and Japan (Table 11.8) yet the USA dominates world science and Japan is the fastest-growing scientific nation.

Moreover, it was simplistic of Martin and Irvine to have claimed in their subtitle that 'Britain is falling behind the major industrialized countries in its investment in academic and related research, which may explain the nation's declining contribution to world science' when all they showed was that the British government's investment in academic and related research fell by 3 per cent of GDP between 1975 and 1982, while Britain's share of papers fell by 10 per cent over the same period (1973–82). A 3 per cent change in a budget that dates from 1975 could not have acted retrospectively to reduce papers from 1973. It takes several years for budget changes to translate into the numbers of scientific papers.

The Brain-drain

During the 1980s, the academic community was so desperate for political ammunition that it abandoned proper standards of scholarship. Nothing demonstrated this abandonment better than the scare over the brain-drain. Academic after academic bemoaned the terrible brain-drain. The philosophers Professor Bernard Williams, Provost of King's College, Cambridge, and Dr Steven Lukes, Fellow of Balliol College, Oxford, in particular, generated great publicity over their own departures, although just why they were leaving was left unclear. Each enjoyed a tenured, important, well-paid job, and each was well-buffered from any putative government cuts by the nature of their research (philosophers need little more than pencils and paper). Bernard Williams told the *Observer* (18 January 1987) that he was leaving because he found Mrs Thatcher's Government too right wing, and Cambridge too middle class – though, one may ask, why did he leave for America, the land of unregenerate capitalism, Ronald Reagan, and university fees? Why, moreover, did he return in 1989, with no fuss and with minimal publicity, to Oxford (that well-known proletarian university) still groaning under the Thatcherite terror?

A brain-drain was not, in itself, an impossibility. There had been a bad one during Harold Wilson's Labour Government of 1964–70 (in 1968 alone 491 British scientists emigrated permanently to the USA), but during the 1980s there was no objective evidence of one so, in 1986, the three leading representative academic bodies, the Royal Society, the Economic and Social Research Council and the British Academy investigated the loss in science, social science and the humanities respectively. But, curiously, the drain proved difficult to measure. Individuals do of course emigrate to America, but an embarrassing number of foreigners travel in the other direction. Finally, both the Economic and Social Research Council and the British Academy concluded that they could find no drain of brains in either social science or the humanities, but as their findings could not be used to embarrass the Government, they chose to suppress rather than publish them. This is the sort of scholarship of which Lysenko would have approved, but why should the taxpayer who funds both the Economic and Social Research Council and the British Academy?

Fortunately, the Royal Society, which understands the meaning of truth, did publish its findings,[19] but they were a great disappointment to the *bien pensants*. Between 1975 and 1985, 931 university scientists emigrated, while 685 immigrated. The net emigration of 246 scientists over a decade, or 24 a year, when there are over 128 000 scientists in Britain supported by a further 155 000 technical staff (a total of 283 000 scientific staff)[20] can hardly be described as a serious brain-drain (but see the postscript on the brain-drain and employment practices).

Conclusion

What is the truth? What is the real situation with British science? The answer seems to be thus: as Britain, throughout the 1980s, continued to enrich itself, so British science expanded, both in quantity and in quality. But government support for academic science did not increase markedly in real terms. The private sector compensated for that, however, by increasing its support for academic science. That, in turn, introduced its own problems. Because private support for academic science is generally short term, private money has financed a huge expansion in the number of university scientists holding three-or five-year contracts. Good though that might be for science, it is bad for individual scientists because they are deprived of a secure career structure. It is easy for young scientists to stay within the universities on a succession of short-term research contracts, but hard for them to obtain permanent university posts as lecturers. Because their long-term expectations are not met, these young people may become disillusioned and bitter; moreover, if they stay too long within the universities, they may not easily find jobs in industry, because companies generally like to recruit relatively recent PhD graduates.

This crisis must be met by leadership from the universities. They need to act in at least two different ways. First, all short-term research employees should receive regular careers advice and guidance, so that expectations can be made realistic. Ultimately, more universities might follow the example of the University of Newcastle upon Tyne, and forbid certain research fellows from staying on (although I believe that would be unacceptably *dirigiste*). Second, and more positively, the universities should refuse grants from external sources unless sufficient overheads are also provided to enable permanent career posts to be endowed. As early as 1970, the Committee of Vice Chancellors and Principals warned that, with the ending of the Robbins expansion, a block of fellows' careers was inevitable, and it is shameful that, over two decades later, the universities have still not implemented a proper policy.

The shift in the funding of academic science from the Research Councils to the medical charities or industry need not occasion alarm; the linear model has been thoroughly discredited, and major fundamental advances are as likely in applied as in pure work. (Thus, for example, Mike Waterfield of the Imperial Cancer Research Fund discovered, during the early 1980s, that many cancer-inducing oncogenes were derived from growth factors or their receptors – a discovery which has illuminated areas in fundamental biology as disparate as embryology and comparative zoology.) Nor need the expansion in the numbers of research fellows occasion alarm, economically. Historically, the universities were created primarily to train recruits for industry, and this continues to be their main function, economically. Remember the study of the economic value of university science, that of Langrish and his colleagues in 1972 which concluded that: 'although scientific discoveries occasionally lead to new technology, this is rare ... the most important benefit ... of basic research is ... the output of highly qualified men and women educated in science and its methods'.[21]

Although there has been an absolute expansion in British science, the evidence shows a relative decline. That evidence should be treated with circumspection. In a recent study, Martin and Irvine showed that the relative British decline had halted, while Germany and France, those erstwhile paragons, had declined by 5 and 4 per cent respectively[22] (this study is not well known, because Martin and Irvine tend to publish bad news in *Nature* or *New Scientist*, but they bury good news in obscure journals; this paper appeared in *Science and Public Policy*). Much of these so-called declines in countries' science amounts to little more than short-term fluctuations or 'noise' in the numbers of papers published. But, long-term, a relative British (and German and French) scientific decline is not only inevitable, it is actually desirable.

One hundred years ago, there were only three major science nations, Britain, Germany and France, and each accounted for nearly a third of the world's science. Since then, the USA, Canada, Japan and other industrial giants have also become scientific giants. Britain, Germany and France now do much, much, much more science than they did 100 years ago (absolute growth) but their share of the world's total science is now around 8 per cent each (relative decline). In a totally just world, when Gabon or Somalia will be as rich *per capita* as Germany or France, a country like Britain, with 1 per cent of the world's population, will only do 1 per cent of the world's science. But that day may be up to two centuries away.

Poor countries grow faster, economically, than rich ones; so, therefore, does their science. This is illustrated in Figure 11.1 which charts the relative increase in the numbers of scientific papers published *per capita* between 1979 and 1983 for the OECD countries. Figure 11.1 shows that, between 1979 and 1983, the rich OECD countries witnessed only small increases in the numbers of papers they published *per capita*, whereas the poor OECD countries witnessed up to 50 per cent increases in just those four years. Rich countries' science, therefore, grows slower than that of poor countries, so they enjoy absolute growth but relative decline.*

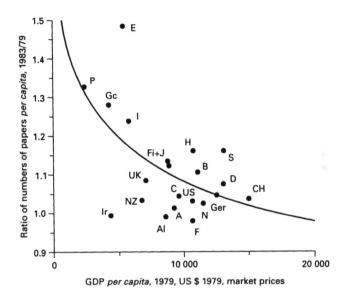

FIGURE 11.1   *Wealth and rate of scientific growth*

GDPs *per capita* come from *OECD Economic Survey, UK 1980/81* (Paris: OECD, 1983). The average increase in numbers of papers published *per capita* come from the author's calculation on the figures in T. Braun, W. Glanzel, and A. Schubert, 'One More Version of the Facts and Figures on Publication Output and Relative Citations Impact of 107 Countries 1978–1980', *Scientometrics*, vol. **11**, pp. 9–15, 1987, and T. Braun, W. Glanzel, and A. Schubert, 'The Newest Version of the Facts and Figures on Publications Output and Relative Citations Output of 100 Countries 1981–1985', *Scientometrics*, vol. **13**, pp. 181–8, 1988. See Figures 7.5 and 7.11 for a key to the symbols.

---

*I have published many of the arguments presented above as a paper in *Scientometrics*, vol. 20, pp. 369–94, 1991, to which Ben Martin has replied in *Scientometrics*, vol. 29, pp. 27–56, 1994. To prove my scrupulous fairness, I refer my readers to Ben Martin's reply.

UNITED STATES OF AMERICA

Scientists in the USA now complain of decline. In 1990, in a paper entitled 'Declining American Representation in Leading Clinical Research Journals', Stossel and Stossel bemoaned that the major medical journals in the world [*New England Journal of Medicine* (USA), *Journal of Clinical Investigation* (USA) and *Lancet* (UK)] carried a smaller and smaller proportion of American papers (between 1980 and 1988, the share of American papers in these journals fell from around 70 per cent to 50 per cent).[23] Stossel and Stossel complained that the United States 'may be becoming a consumer rather than a producer of medical research [and] scholars have begun to warn of serious consequences if subsidies for the national biomedical enterprise do not improve'.[24] Such scholars include J. B. Wyngaarden who wrote a paper entitled 'The Clinical Investigator as an Endangered Species'[25] and B. Healy, the author of 'Innovators for the 21st Century: Will We Face a Crisis in Biomedical Research Manpower?'[26]

This is all nonsense. The National Institutes of Health, the major US Government funder in this area, increased its budget in real terms by about 45 per cent during the 1980s, and the medical charities and pharmaceutical companies made the same increases.[27] The numbers of clinical investigators rose proportionately, as did the numbers of papers they submitted to journals.

Let us consider the numbers of papers submitted to journals: unfortunately, we do not know the numbers of original research papers submitted between 1980 and 1988 to the *New England Journal of Medicine*, *Journal of Clinical Investigation*, or *Lancet*, but if we look at *Blood*, the leading haematology journal, the picture becomes clear. Between 1980 and 1988, the percentage of US papers published in *Blood* fell from 80 per cent to 65 per cent, but during that time the numbers of US papers submitted rose by 50 per cent; however, the numbers of non-US papers submitted rose by 110 per cent, i.e. US-derived papers submitted to the journal increased considerably between 1980 and 1988, but non-US derived papers increased at twice the rate (they mainly came from Japan and Europe). The share of American papers that were published fell, therefore, not because their quality or quantity fell, but because so many more papers of non-US origin were offered to the journal. Since *Blood* expanded considerably, anyway, between 1980 and 1988, the total numbers of American papers that were published actually rose, even though their share fell – a classic case of absolute growth and relative decline (and very much to be welcomed: in a fair world, US papers would only account for 5 per cent of *Blood*'s contributions).

All other contemporary American claims of decline in basic science should be dismissed. David Kleppner's warning that 'American science is on the verge of a long, downward slide'[28] is just not true; nor is the claim in the *New York Times* of 29 September 1992, by Harold Varmus and Marc Kirscher, that American 'basic research faces the ax'. The response in 1991 to President Bush's budget proposals by the Federation of American Societies for Experimental Biology that 'The level of support called for in the Administration budget proposal under-utilizes the national resources in the biomedical research community at a time of opportunity to reap the benefits of past investment' should be examined in the light of the figures: the 1991 NIH budget was no less than $8.8 billion, having grown throughout the 1980s at around 6–7 per cent annually, and the Federal Government's support for all academic science in 1991 stood at over $13 billion. In his 1990 State of the Union Message, President Bush had exclaimed 'The money is there. It's there for research and development, R&D, a record high' and 1991 was even better.

Yet the academics are unhappy. On 7 January 1991, Leon Lederman, an American Nobel Laureate in physics, and President-Elect of the American Association for the Advancement of Science (135 000 members) published the results of a survey of the morale of 250 academic scientists. He found 'a depth of despair and discouragement that I have not experienced in my 40 years in science'.[29] In his speech to the National Academy of Sciences in Washington, Lederman acknowledged 'the paradox of continuing increases in federal research and growing dissatisfaction in academic laboratories' but argued that 'my concern is not for the unhappiness of my colleagues in science, much as I love them. My concern is for the future of science in the United States and for the profound cultural and economic benefits that science brings.' Lederman's solution was simple: the Federal Government should immediately double its budget for academic science from $13 billion to $26 billion a year, and then continue to increase it at 8–10 per cent a year indefinitely.

Some American scientists disagreed with Lederman; why not triple or quadruple the budget and thereafter increase it by 16, 20 or 35 per cent a year? Those were the scientists who were angered by the response of Dr Press, the President of the National Academy: 'No nation can write a blank cheque for science. . . . American science is the best supported in the world. . . . We may have to make choices'.

Those scientists were even more angered by Rep. George E. Brown Jr (Californian Democrat and Chairman of the House Science, Space and Technology Committee) who told the annual meeting of the AAAS on 15 February 1991:

> One could easily document a similar level of despair among 250 Medicare recipients, 250 disabled veterans, 250 soldiers in Saudi Arabia, or even among 250 Members of Congress. If we are going to justify the privileged treatment of research and development by the Federal Government – and we will have to justify it if we hope to sustain it – then we must present a case that is based not on the frustration and discomfort of individual worthy scientists. Rather, we must present a case rooted in the welfare of our nation.
>
> One can easily demonstrate that strong federal support for basic scientific research is neither a sufficient, nor a necessary, condition for vigorous economic growth and societal vitality. This is clearly demonstrated in the case of Japan and Germany, whose phenomenal economic growth occurred in the absence of significant government or private funding for basic research.
>
> As a politician, I must tell you that unlimited Federal funding for basic research is no longer viewed by the US Congress as a birthright of the scientific community, and I must warn you that the generous support you enjoy today was part of the fall-out of the creation of nuclear weapons, not because of the great contributions of science to a more humane society.[29]

Some American scientists dislike Rep Brown even more than they dislike Congressman John Dingell.

Do American scientists, therefore, have a case? Or are they just a bunch of greedy whiners? Actually, they have real grievances. First, it is increasingly difficult for individuals to raise federal money. Although the federal allocation for research increases all the time, the numbers of researchers increases even faster, so there is not enough government money to meet all the worthwhile claims on it. Only one-third of grant applications that have been judged, scientifically, to be excellent, can be actually funded – just like in the UK. This is dispiriting, not only because the weeks spent writing a grant are wasted, but also because it feels so horrible. Rejection is horrid.

We noted in Chapter 8 how William Flower, the Director of the Natural History Museum in London, explained in 1880 that he refused to join the Royal Society's Grant Committee because 'this method of subsidising science, accompanied as it is with the (as it appears to me) humiliating necessity of personal application in each case, must do

TABLE 11.9   *International comparisons of research and development for 1985*

Country	Percentage of GDP spent on all R&D
USA	2.8
Germany	2.7
Japan	2.6
UK	2.3
France	2.3
Italy	1.1

NOTE: These figures include both military and civil R&D, funded either by Government or the private sector.

SOURCE: The figures come from the *OECD Science and Technology Indicators* as quoted in the *Annual Review of Government Funded R&D* (HMSO, 1988).

much to lower the dignity of recipients'. One hundred years later, Flower's assessment has been confirmed. It is humiliating – it does lower dignity – to fail to get a grant. Since grants are judged, anonymously, by one's peers, it is very easy to become paranoid and depressed – whose are the smiling faces that, privately, marked down one's grant as 'excellent' but not excellent enough? Is such a judgment fair: if the work is excellent what more can one do? If the work is excellent, surely it *should* be funded?

The federal grant bodies aggravate the situation by deliberately ranking grants as 'excellent, but unfunded' to put political pressure on Government. That is a cruel exploitation of researchers' aspirations, since the classification of grant applications is largely a value judgement but, underneath the politics, there is a real problem: science is so Malthusian. Scientists breed at a much greater rate than does money. An active researcher can produce a trained PhD every year (it means running a team of three or five PhD students, and many teams are bigger than that). In consequence, the numbers of trained researchers in a country could double annually. Not every PhD graduate wants to do independent academic research, and not every active researcher produces a trained PhD every single year, but the demography is dreadful. Academic science, therefore, faces a Malthusian crisis. More and more scientists chase relatively less (if absolutely more) money. What to do?

Governments will do nothing. They have funded enough science to satisfy the teaching requirements of the universities, the potential needs of defence, and the recruitment needs of industry, and they will leave

TABLE 11.10   *International comparisons of defence research budgets*

Country	Percentage of GDP spent on defence research for 1985	Percentage of total R&D budget spent on defence for 1983
USA	0.88	28
UK	0.60	29
France	0.45	21
Germany	0.13	4.3
Italy	0.06	–
Japan	0.01	0.05[a]

NOTE:[a] Japan's figures are for 1981. The percentage of total R&D budget spent on defence for 1983 are the author's calculations.

TABLE 11.11   *Total civil R&D as a percentage of GDP (i.e. industrial and academic research combined)*

Japan	2.5 (1983)
West Germany	2.5 (1985)
USA	1.9 (1985)
France	1.8 (1984)
UK	1.7 (1985)

SOURCE: *Annual Review of Government Funded R&D* (HMSO, 1987).

it to the scientists to sort out their own *angst*. The free market, of course, would also have met the national needs – at any rate the civil ones – but it would also have looked after its researchers better. Anyone in Britain who observes the sense of responsibility to its scientists felt by the Wellcome Trust or the Cystic Fibrosis Trust, and compares that with the arbitrary, bureaucratic, impersonal decisions of the Research Councils, will regret the displacement of private funds by the State. The same is true in the USA: compare the Carnegie Foundation's treatment of its scientists with the NIH's or the NSF's.

The private foundations are kinder employers than the State for at least two reasons. First, they feel their prime responsibility to be to science and to scientists, while State organisations use scientists to serve the State. The Director of the NIH, for example, is appointed politically, and he or she is replaced by each incoming President. This is not good for science, as was witnessed by the repeated criticisms of Bernadine Healy, President Bush's appointment, that she was more concerned to please the White House than to serve science. But the economics also favours the private foundations; their expansion tends to be regular and steady (consider Table 11.6). Steady growth facilitates the forward planning of careers, but government funding for science fluctuates with shifts in the political mood (see Figure 8.6) and the succession of cuts and expansions in State support aggravates the effects of the underlying Malthusian problem in science (although even private foundations are not completely immune from economic cycles – and rightly not, one should trim back on research during recessions).

Rather than address these fundamental issues, however, science lobbyists on both sides of the Atlantic have focussed on a softer target – their countries' expenditure on defence R&D. The percentage of GDP spent on R&D by the major industrialised, scientific nations look similar (Table 11.9), but the USA, UK and France spend so much on military R&D (Table 11.10) that their civil R&D falls below that of the two surging industrial nations of Germany and Japan (Table 11.11).

Science lobbyists, therefore, argue for a transfer of funds from defence to academic research. Here is a typical argument from Kleppner:

The federal government lavishes some $70 billion . . . on R&D, but the NSF budget was only $2.1 billion. . . . Approximately two-thirds of federal R&D expenditure is for defense. The long history of aborted or failed high-technology defense systems cannot but raise doubts about whether those funds are being invested sensibly: the list includes the B-1 bomber, ninety-six of 100 of which are grounded; the B-2 bomber, which is about to be scuttled; and the Navy 1–12 plane, which has been abandoned. More than $25 billion has been spent on the Strategic Defense Initiative, whose goals have constantly shifted as one technology after another has had to be scrapped. Charged-particle beams, chemical lasers and X-ray lasers are among the more costly SDI failures. Another sink for federal funds is NASA. . . . Industrial interest in the space shuttle has been nil. . . . NASA has been vigorously pushing its proposed space station [which] has no scientific purpose aside from keeping people alive in it, and it is no more likely than the space shuttle to succeed technically. Nevertheless, $1.9 billion will be spent on it this year, almost as much as the total NSF budget. It seems incredible that the government can spend billions on such flawed projects while allowing the world's greatest scientific institutions to decline for lack of relatively modest funds.'[30]

There is one minor, and one major, problem with Kleppner's argument. First, the NSF at $1.9 billion is not the only source of academic funds. The NIH at $8.8 billion is another, and other government agencies, university endowments, industry and the medical charities are also important. But the major flaw with Kleppner's argument is his suggestion that funds be transferred from defence R&D to civil (preferably academic) R&D. This is a *non sequitur*. The two budgets are not interchangeable. Defence research is customer-led, not curiosity-led. The Federal Government does not sit down one morning to decide how much it will spend on R&D, and to then apportion 28 per cent of that to defence. But the Government does sit down one morning to apportion the Defense Department a budget, from which the Defense Department commissions its research. If the Defense Department were to spend less on R&D, that would not necessarily boost civil R&D. That would be a quite different political decision. Kleppner could have as relevantly complained of waste in the education or welfare budgets.

Kleppner's concerns over defence R&D were, however, justified, but only in economic terms. Even if America's defence or space projects were more successful, they would still be intrinsically profligate; studies show that the economic benefits of defence R&D are only about 10 per cent of civil R&D.[31] But countries do not invest in military R&D for economic reasons; they do so for strategic and historical reasons, and America's defence R&D needs to be judged strategically and historically as well as economically.

But Kleppner is wrong to argue for more academic R&D in America. There is no evidence that an increased government investment in academic science in the USA would boost economic growth, it would merely add to the tax burden and boost even more the numbers of unhappy, unemployable PhD graduates. Unless Kleppner can produce evidence of demand – complaints from industry of a shortage of trained scientists or of new science to exploit (there is no such evidence) – then his pleas for more supply must be dismissed as self-serving.

Nor should the Federal Government boost civil R&D. The National Science Board, in a recent report, argued for more federal industrial R&D.[32] The Board complained that industrial funding for R&D had hardly risen in real terms since 1986, whereas Japan now spends 16 per cent more as a percentage of GDP on civil R&D. So what? If the American economy did falter during the late 1980s, being overtaken by Japan's, then American industry made the correct decision to economise on R&D and to concentrate on adopting Japanese practices such as their management techniques.

It may be, however, that the American economy did not falter, but that it only appeared to. America is increasingly becoming a service economy but, as the economist M. R. Darby has noted, it is hard to measure service. Japan, however, is not moving so fast into services; its manufacture is maintaining much of its relative strength. Since manufacture is easier to measure than service, it may only appear that Japan is overtaking the USA economically.[33] If so, American industry made the correct decision to economise on R&D, since civil R&D largely supports manufacture, and manufacture does appear to have suffered. Either way, the National Science Board's case for higher taxes to fund a government-expansion in civil R&D is economically inappropriate.*

Research pressure groups always go for easy targets. They can always find at least one other country's government that spends more on civil R&D or on academic science (either absolutely or as a percentage of GDP, whichever is the easiest figure to exploit), but they never try to correlate each country's government's expenditure with that country's real need as expressed by its GDP *per capita*, nor do they ever assess any country's actual demand for research. The pressure groups exploit the public's ignorant faith in the linear model and the consequent assumption that economic growth can be pushed by increasing the supply of science. It cannot, and it is disgraceful to play on the public's fears, ignorance and goodwill.

As Adam Smith noted, arguments by pressure groups for public funding should always be viewed with suspicion, but scientific pressure groups often behave with a dramatic lack of regard for dispassionate analysis.

---

*This is a book on science funding, not federal policy, but it is obvious that many of the US's problems do not derive from inadequate economic growth, or from inadequate science funding, but from inappropriate decisions made in Washington, DC. America's federal budget deficit, or its profligate expenditure on defence and space, are well-known problems, but consider its bizarre immigration policies. In 1989, the last year for which we have figures, the USA accepted over a million immigrants, most of whom came from the Third World (405 000 from Mexico; 57 000 from Salvador; 57 000 from the Philippines; 37 000 from Vietnam; 34 000 from Korea; 32 000 from mainland China; 31 000 from India; 26 000 from the Dominican Republic; 24 000 from Jamaica, etc.). Many more arrived illegally. America is a land of immigrants, and immigrants generally work hard, so this policy may appear to make good economic sense, but does it? These immigrants are poor, and they are poor because they are relatively unskilled. Are they not inappropriate, therefore, for a lead service-based economy? Even if the immigrants' work ethic and educational ambitions revolutionised their economic importance, they would still be destabilising. These immigrants are not Anglo-Saxon, and a homogenous culture is a great facilitator of wealth creation (witness Japan, Germany, France or Scandinavia). Immigrants, moreover, compete successfully for the low-paid jobs, and so rob the existing underclass of any hope, and the employers from any incentive of helping the underclass.

Save British Science, for example, often complains that, between 1981 and 1989, Britain's share of GDP devoted to civil R&D fell, the only country in Europe to have so suffered.[34] But SBS never note that, between 1981 and 1989, Britain's economy grew faster than any other in Europe: indeed, between 1981 and 1989 (the trough and peak respectively of the economic cycle) Britain's economy grew at over 3 per cent a year – the fastest rate of British growth for over 50 years. Britain's share of GDP devoted to civil R&D fell, therefore, because GDP grew so explosively, not because civil R&D budgets contracted (they rose).

These figures were discussed very publicly in the run-up to the British General Election of 1992, when Professor Denis Noble and a further thousand scientists made their claims in a famous letter to *The Times*, and I pointed out the fallacy then (see *The Times*, letters for 23, 25 and 27 March 1992) and it is dispiriting for those who believe that scientists are rational that their activists continue to make these discredited claims (see Professor Noble's speech to the British Association for the Advancement of Science, Southampton, August 1992,[35] or the speech, reported in *Hansard*, of Anne Campbell, Labour MP for Cambridge, to the House of Commons, 11 June 1992).

On 18 February 1993, *Nature* claimed that the numbers of young scientists in training had shrunk[36] even though I had twice, in *Nature* itself, exposed that fallacy.[37] British science activists, in their disregard for truth, illustrate Adam Smith's dictum that: 'People of the same trade seldom meet together, even for merriment and diversion, but the conversation ends in a conspiracy against the public' (*Wealth of Nations*).

Americans can be as bad. In 1992, the Investigation Subcommittee of the House (US Congress) Science Committee reprimanded the National Science Foundation for publishing a report that was 'pseudoscience' and 'nonsense'.[38] In 1987, the NSF had claimed that a study by one of its policy analysts, Peter House, predicted that by the year 2005 the US would be short of 675 000 scientists. But the House Subcommittee found that the so-called shortfall amounted to no more than unwarranted extrapolations of inadequate statistics. Damningly, NSF statisticians had already exposed the study's faults privately, but internal NSF memoranda had tried to cover up these faults, referring to the need to 'protect the foundation from damage' and worrying about 'losing this discussion'. The Chairman of the US Congress Subcommittee, Rep. H. Wolfe (Democrat, Michigan) found that NSF officials had peddled misleading statistics to obtain federal funds. He commented 'nobody expects NSF to play that game. Everyone around here assumes that NSF's numbers are good science.'[39]

One of the NSF's most effective critics at the hearings of the Sub-committee was John Andelin of the Congressional Office of Technology Assessment (OTA). Asked about the NSF's study, he commented: 'if you want press attention, yes, it's useful. If you want intelligent discourse, the answer is no'. The OTA is emerging as a useful counterweight to the science propagandists. It is often claimed, for example, that investment in science has to outstrip inflation because science becomes progressively more expensive, but in 1989 the OTA commissioned a study. Reporting in 1991, it concluded 'the cost of producing a published paper or performing a given scientific measurement has decreased: with less than double the investment per year since 1965, more than double the number of papers are published today in academia, and more than double the number of PhD scientists are employed in the academic sector'.[40] The OTA suggests that although equipment can get more expensive, its increasing power produces more data, which is cost-effective and ensures that the rate at which papers are published is still determined by scientists' time and effort, not technology.

One should not be too surprised at the scientists' perfidy, because it can be genuinely difficult for them to view their own activities in perspective. Science is desperately important to them, and they can fail to understand how politicians and economists must prioritise between competing, equally crucial claims. In the *Road to Serfdom*, Hayek noted how difficult it was for non-economists to accept prioritisation:

> It is the frustration of his ambitions in his own field which makes the specialist revolt against the existing order. We all find it difficult to bear to see things left undone which everybody must admit are both desirable and possible. That these things cannot all be done at the same time, that any one of them can be achieved only at the sacrifice of others, can be seen only by taking into account factors which fall outside any specialism, which can be appreciated only by a painful intellectual effort – the more painful as it forces us to see against a wider background the objects to which most of our labours are directed, and to balance them against others which lie outside our immediate interest and for which, for that reason, we care less.[41]

As long as science is funded publicly, science lobbyists will fight a propaganda battle with the custodians of public funds. But, however much money the scientists extract, they will remain the ultimate losers. Science is nothing if it is not truth, and truth is hard to reconcile with politics.

POSTSCRIPT ON THE BRAIN-DRAIN AND EMPLOYMENT
PRACTICES

Though the number of scientists who emigrated during the 1980s was
tiny, their quality was high. A significant proportion of emigrants were
FRSs. Unfortunately, the incentive for top scientists to forsake Britain
for the USA flowed directly from the employment policies of British
universities, not from their overall budgets. Thus, in America, good
scientists get paid more, several times more, than average ones, whereas
British universities have been tied to a national pay scale that, within
the rank of professor, has made little recognition of distinction. Moreover,
in Britain, half of the Government's research budget has been distributed equally, across all universities, across all academics, to foster the
pretence that all academics are equally gifted and equally deserving of
support (the so-called system of dual funding). American universities
labour under no such illusions, and they direct resources away from
poor labs into good labs. This focusing of resources on excellent scientists has been facilitated by the reluctance of American universities
– especially the good ones – to bestow tenure (a job for life) while
British universities have given all lecturers, let alone senior lecturers,
readers or professors, tenure. Thus, in Britain, taxpayers' money has
been squandered on the idle, stupid and incompetent, while American
universities have strained to reward the able, industrious and gifted.

These abuses in Britain were corrected, at least in part, by the Education Reform Act of 1988, an Act the universities resisted with vit-
riol. British universities seem to believe that the taxpayer is a cow to
be milked without compassion. British universities also seem to believe that good scientists should be punished, not rewarded, for their
wicked inegalitarianism.

12 Dr Pangloss was Right

We have constructed this book as a debate between Francis Bacon and Adam Smith. Who won? The world's answer, of course, is Francis Bacon: The governments of all the industrialised countries now support their universities and their science much as Bacon prescribed, and if one asks about funding in a contemporary laboratory, university or ministry of science, the responses come straight out of the *Advancement of Learning*. To paraphrase, modern science policy is little more than a series of footnotes to Bacon.

But the world can be wrong. Indeed, it often is, particularly when it flouts Adam Smith. Consider some of the collective errors which much of the industrialised world has made over the last one hundred years. The thirst for colonies, which crescendoed during the latter part of the nineteenth century and early part of the twentieth – a thirst which we now recognise as having been self-defeating for the colonisers, let alone the colonised – was specifically abjured by Adam Smith in his most celebrated quotation: 'To found a great empire for the sole purpose of raising up a people of customers, may at first sight appear a project fit only for a nation of shopkeepers. It is, however, a project altogether unfit for a nation of shopkeepers.'[1] Yet, despite Smith's warning, a forward imperial policy was for decades, if not centuries, advocated by the chancelleries of Britain, France, Germany, Holland, Belgium, Spain, Portugal, Italy, Russia, Japan and even the USA. Similarly, Smith would have been appalled by the popular lust for totalitarianism that once swept much of Europe, as did the lust for socialism, the lust for war (particularly in 1914) and Europe's current and absurd preoccupations with protectionism, the ERM and subsidies. Let us, therefore, recognise that the world is frequently and collectively in error, and let us examine, free of prejudice, Bacon's arguments for the funding of science. Let us review Adam Smith's objections, and let us then assess the evidence that has since accrued.

In the *Advancement of Learning*, Bacon made three specific proposals: (i) that academic science bred wealth; (ii) that governments were required to fund academic science; and (iii) that academic science was morally ennobling. Let us assess these propositions in turn.

303

(i) Academic Science Breeds Wealth

Adam Smith conceded that academic science did breed wealth, but he ran it a poor third to the two major types of commercial research:

> A great part of the machines made use of in those manufactures in which labour is most subdivided, were originally (i) the invention of common workmen. . . . All the improvements in machinery, however, have by no means been the inventions of those who had occasion to use the machines; (ii) many improvements have been made by the ingenuity of the makers of the machines, when to make them became the business of a peculiar trade; (iii) and some by those who are called philosophers.'[2] [my numerals]

Subsequent academic studies have vindicated Smith, not Bacon. The work of Mansfield, Gellman *et al.* and Langrish *et al.* (see Chapter 9) has shown that at least 90 per cent of wealth-creating research emerges from applied, not pure science. Bacon's linear model is largely irrelevant.

(ii) Governments are Required to Fund Academic Science

Smith certainly did not subscribe to this. He believed that a few activities – very few – do require government money, but that academic science is not one of them. Indeed, the obvious superiority in his day of the Scottish over the English universities strengthened his belief that both academic teaching and academic research thrived under the free market.

Subsequent academic studies have, again, vindicated Smith, not Bacon. The work of Mansfield, Griliches and Rosenberg (see Chapter 9) has shown that the free market drives companies into funding pure science, both in-house and within universities. Moreover, the historical evidence of the endowment of British and American universities and of their science by rich patrons under *laissez faire* disproves Bacon's suggestion that governments need to do it.

(iii) Science is Morally Ennobling

Bacon elevated science above all other activities: 'This glory of discovery that is the true ornament of mankind' (*Novum Organum*). He also believed that science improved the mind: 'Science also must be known by works. . . . The improvement of man's mind and the improvement of his lot are one and the same thing' (*Novum Organum*). He also believed that study conferred wisdom: 'Learning. . . . will make learned people wise in the use and administration of learning' (*Advancement of Learning*). Bacon believed that science was also a moral good; his *New Atlantis* (1627) describes how an ideal society would consist solely of scientists, whose pursuit of science would create a morally good society: 'a picture of our salvation in heaven' and morally good people: 'a land of angels'.

Yet again, experience has disproved Bacon's vision. There now exist many large collections of scientists, both in universities and in research institutions. Sadly, these organisations are not peopled by 'angels' nor do they 'picture our salvation in heaven'. Rather, they picture our fate in another place. Scientists may not necessarily be much worse than other people, though their intense, self-centred and narrow competitiveness can be unattractive, but if not much worse, they are certainly not much better. Angels they are not. Personal experience aside, the eagerness with which German scientists served Bismarck, the Kaiser and Hitler, the eagerness with which Soviet scientists served Stalin, and the eagerness with which Japanese scientists served Tojo, should put paid to any fantasies Bacon may have elaborated about scientists adumbrating our salvation in heaven. (Perhaps Montaigne had Bacon and the scientists in mind when he wrote: 'In trying to make themselves angels, men transform themselves into beasts.')

Adam Smith, of course, never confused an abstract, intellectual activity such as science with a moral or ethical activity. In his *Theory of Moral Sentiments* (1759) Smith proposed that morality and ethics flowed from sympathy for other people, not from understanding Euclid's *Geometry*. Indeed, Smith specifically denied that reason has anything to do with goodness: 'Though man be naturally endowed with a desire of the welfare and preservation of society, yet the Author of Nature has not entrusted it to his reason to find out that a certain application of punishments is the proper means of attaining this end; but has endowed him with an immediate and instinctive approbation of that very application which is most proper to attain it.'[3]

Yet this silliest of all Bacon's ideas has proved his most durable. To understand why, let us break it down into two categories: (a) science as a cultural good; and (b) science as a moral good.

(a) Is science a cultural good?

Yes, of course it is – and it did not take Francis Bacon to point that out. Men who had died before he was born, practical scientists like

Copernicus (1473–1543) and Columbus (1451–1506) or thinkers like his namesake Roger (1214–92) had long disproved the medieval church's belief in revelation as the source of knowledge about the world. Observation, induction, deduction and experimentation – in short, science – make better tools. Contemporaries of Bacon were widely embracing the scientific method. Consider Philip Sidney (1554–86) who in his *Defence of Poesie* (published in 1591) wrote of 'the balance of experience' and of 'proof, the overruler of opinions'.[4]

In as much, therefore, as science has improved Man's mastery over the world, it has represented a cultural advance, since an improved intellectual technique must, by definition, be a cultural benefit. But that is no argument for governments having to fund it. Science is not a delicate little flower, whose fragility demands the protection of the state. Science emerged out of medieval Europe despite the opposition of the authorities, as was witnessed by the sufferings of researchers such as Roger Bacon, Vesalius and Gallileo. Obviously the scientific method would never have emerged had every single powerful person in medieval Europe set out to destroy it – science always had friends as well as enemies – but its history reveals that the triumph of the scientific method did not require the active support of the state.

Today, the scientific method is so ingrained in Western culture that the attitude of the state is irrelevant. Indeed, where governments are unscientific, like the regimes of Eastern Europe who worshipped Marx, the people bring them down. The scientific method is stronger than the state, and it is employed universally in the West. The housewife whose door lintel falls on her head sues her builder; she does not propitiate evil spirits. As Thomas Huxley remarked in his *Man's Place in Nature* (1863), 'Science is nothing but trained and organized common sense', but he could only say that because common sense has become so scientific.

(b) Is science a moral good?

Intellectual techniques are just that, techniques, and techniques are intrinsically amoral. On occasion, Bacon admitted as much: 'Light is in itself pure and innocent; it may be wrongly used, but cannot in its nature defile' (*Novum Organum*). More generally, however, Bacon argued the opposite: 'the rule of religion, that a man should show his faith by his works, holds good in natural philosophy too. Science also must be known by works. . . . The improvement of man's mind and the improvement of his lot are one and the same thing' (*Novum Organum*).

Bacon believed that because science would raise standards of living, it must intrinsically be morally good; but that argument is false, as his contemporary, the poet Philip Sidney (1554–86) had already noted: 'the mathematician might draw forth a straight line with a crooked heart'.[5] The consequences of science may be good, but that does not make the activities of scientists intrinsically good, just as the bad consequences of science like poison gas, Hiroshima or pollution, do not make the activities of scientists intrinsically bad.

Scientists like to portray their early predecessors as martyrs in the cause of truth – victims of a wicked Church. The reality was more complex. The Church itself actually funded much science to bolster its Aristotelian and Platonic doctrines, much as the Sumerians had funded astronomy. Many of the early Western scientists were curiously reactionary. It appears, for example, that Copernicus, a canon of Frauenburg Cathedral, wanted to believe in heliocentricity to match his Platonic Ideal of the universe. William Harvey, too, may have been attracted to the circulation of the blood for similarly Platonic reasons. And the early scientists were certainly no better, morally, than the Church. The ethical balance of that most famous of all confrontations, between Galileo and Urban VIII, is about equal. Urban should not have condemned Galileo, but his sentence was light by the standards of the day, and Urban, a serious reformer of the Church, was incomparably the nicer man.

Bacon's conflation of science, a cultural good, with science as a moral good has, however, struck deep roots within Western culture. Why? Because it serves the interests of those who propagate culture. Let us explore the history of this self interest.

(c) Science, art and culture as social escalators

The moment intellectuals emerged as a distinct group, they evolved self-promoting philosophies. Thus Plato, who founded the Academy in 388 BC, elaborated the concept of the Philosopher-King, which maintained that (surprise, surprise) kings should be philosophers, and vice versa. This was a silly idea, since the qualities required of kings and of philosophers are so different. Unsurprisingly, Dion of Syracuse, the very first Philosopher-King (he personally studied under Plato) proved so dreadful a ruler that his enraged subjects murdered him in 353 BC. The idea of the Philosopher-King still attracts academics (politicians are not so enamoured of it) and I can remember, as an undergraduate at London University during the 1970s, listening to my teachers dilate over it.

Artists in antiquity also made grandiose claims for themselves and for their art; it was the Athenian, Denis, who said that: 'The dull mind rises to truth through that which is material' and this remains the basis of the belief in the value of art to this day. But by medieval times, the artists had lost their special status. They were just artisans executing their patrons' commisions. Nor were artists admired as individuals; rather, their work was interchangeable and often anonymous. Artists, in short, were indistinguishable from the vast mass of humanity in the medieval period – a period which belittled the individual. But the increasing wealth of Italy spawned increasing amounts of patronage, and as patrons competed for the best artists, so an old idea was rediscovered – some artists really are better than others. Thus, within the world of art and of culture, was the concept of individuality gradually reclaimed.

As artists started to flex their social and economic muscles in the movement we now call the Renaissance, so increasingly did they reestablish their claims to personal recognition. The poets were the first to break the social barriers when the Roman Senate, in 1341, crowned as Poet Laureate the impoverished son of a Florentine notary, Francesco Petrarch (whose *De Viris Illustribus* we were rude about in Chapter 3). Other artists sought to emulate his social success. Michelangelo claimed to be an intellectual, not an artisan: '*si pinge col cervello, non colla mano*', and other painters aspired to the status of *pictor doctus*, 'learned painter'. To promote their claims to gentility, the post-medieval secular painters specialised in scenes derived from ancient history, from literature or from other sublime sources. Often, as in the Dutch genre paintings, artists set themselves up as moral guides. And artists painted themselves into pictures of aristocrats, to emphasise their social equality. Some painters went further: Raphael, painted himself alongside Apollo, Aristotle, Plato and Euclid in the Pope's Stanza della Segnatura in Rome, and in his *Trattato della Pittura* (1651) Leonardo da Vinci claimed that artists were inspired by God (so passionately did Dürer believe this that he painted himself on the Cross as a latterday Jesus Christ – the painting is now in the Prado – while Matisse was one of those artists who actually believed he was God[6]).

As we have seen, Francis Bacon made similar social claims for scientists, condemning the 'opinion, or inveterate conceit, which is both vainglorious and prejudicial, namely, that the dignity of the human mind is lowered by long and frequent intercourse with experiments. . . . since such matters generally require labour in investigation' (*Novum Organum*). The scientists loved Bacon for his social championing, and indeed for his social position which conferred lustre on them. In a socially aware age, lineage could be as important as achievement, hence the absurd description of Robert Boyle as 'the father of chemistry and the son of the Earl of Cork'.

In war, the victors write the history, and the battles of ideas we call the Renaissance, Reformation, Counter Reformation and the Age of Reason culminated in the triumph, in mainland Europe, for a time at least, of the intellectuals of the Enlightenment. To raise their social standing, their incomes and their claims to power, those thinkers embraced the artists' and scientists' claims that their activities were intrinsically virtuous, to disseminate the self-serving propaganda that intellectual or cultural activities are intrinsically virtuous. The German neohumanists coined a term for this, *Wissenschaft*.

That artists are intrinsically virtuous, or that they have been so rendered by their exposure to art, is comic to anyone who knows them. A more egomaniacal, promiscuous, and downright silly group of people does not exist on the planet, except perhaps for actors, whose intimate exposure to literature seems to have made them even worse. Obviously, individual artists, actors or writers can be tolerable, but as a group they are frightful. The artists' pretentions to virtue were punctured as early as 1550 by Giorgio Vasari, whose *Vite de piu Eccellenti Pittori, Scultori e Architettori (Lives of the Artists)* revealed some shocking goings on between artists and their models; but the scientists did not get their comeuppance until 1750 when Rousseau published his *Discours sur les Arts et les Sciences*.

Scientists had been criticized earlier, of course. Philip Sidney had mocked those who: 'persuading themselves to be demigods if they knew the causes of things, become natural and supernatural philosphers'.[7] Sidney went on to note how: 'the astronomer, looking to the stars, might fall in a ditch, and the inquiring philosopher might be blind in himself, and the mathematician might draw forth a straight line with a crooked heart'.[8] Even earlier, in his *Praise of Folly* (1509) Erasmus had mocked those philosophers who 'speak with confidence about the creation of innumerable worlds, measuring sun, moon and stars, and never hesitating for a moment, as though they had been admitted into the secrets of creation with whom and with whose conjectures nature is mightily amused'; but such scoffing by Sidney or Erasmus was essentially benign. Rousseau's assault was more serious. Since Rousseau has made such an impact on Western thought, we will dedicate a separate section to him.

(d) Jean-Jacques Rousseau (1712–1778)

Rousseau, the son of a Swiss watchmaker, left Geneva as a young man to hover around French intellectuals and aristocrats. A failed gigolo, musicologist and artisan, his career did not flourish until 1749 when Diderot suggested that he enter the essay competition of the Dijon Academy of Letters on the theme 'Whether the rebirth of the sciences and the arts has contributed to the improvement of morals'. Diderot hoped that Rousseau would defend the intrinsic morality of science and art – every other entrant maintained that view – but, as Rousseau later described in his autobiographical *Confessions*: 'The moment I read this [the notice of the competition in the *Mercure de France*] I beheld another universe and became another man.... My feelings rose with the most inconceivable rapidity to the level of my ideas. All my little passions were stifled by an enthusiasm for truth, liberty, and virtue.'[9] In a letter to M. de Malesherbes, the censor, Rousseau confided that he declared: 'Virtue, truth! I will cry increasingly, truth, virtue!' and he described how his tears soaked his waistcoat.[10]

As might be expected of something written by an author who is crying increasingly truth! virtue!, the *Discours* explains how Man was a Noble Savage before civilisation came along to enslave and corrupt him. For Rousseau, nature was good but science, the midwife of civilisation, was bad. Rousseau noted that most science was financed either for profit or from nationalism or through ego. He claimed that, in practice, scientists were cold-hearted, impersonal and competitive. And he claimed that their discoveries simply fed an exploitative, capitalist, oppressive state. Rousseau concluded, therefore, that science destroyed morals.

The *Discours* enjoyed a fabulous success. It won the prize of the Dijon Academy, it sold in thousands, and it made Rousseau famous throughout Europe overnight. Its unexpected success spoke of a widespread popular unease over the moral vacuum that underpinned Bacon's vision of society, a vision that the Enlightenment had adopted for its own. Let us, therefore, view Bacon's impact on concepts of salvation before returning to Rousseau.

(e) Bacon and salvation

Francis Bacon was, more than anything, the Father of Materialism. In the *Advancement of Learning* he wrote of: 'matter, which is the contemplation of the creatures of God'. During his life, Bacon had witnessed the failure of either religion or politics or philosophy to create happiness, so he offered economic growth instead – an economic growth that was to be achieved through technology. Consider this quotation from *Novum Organum* (1620): 'Inventions come without force or disturbance to bless the life of mankind, while civil changes rarely proceed without uproar and violence.' Bacon was truly prescient, because this has become the central belief of the industrialised countries. No longer do nations or their governments concern themselves collectively with theology, ideology or even political principles very much; they have become empirical machines for delivering economic growth.

Although Bacon never attacked Christianity directly, he effectively supplanted it with the worship of technology. In his *New Atlantis*, his description of a Utopian, ideal society, he invented a culture that venerated science, and statues of scientists, in a procession of rites that both mimicked and substituted for those of the Church. Bacon believed that he was introducing the final chapter in Western Man's quest for the meaning of life.

Not every culture has been galvanised by the quest for the meaning of life. Neither the ancient Greeks nor the feudal Japanese, for example, worried excessively over sublime questions of ultimate purposes, the majority of their citizens having generally concerned themselves with the practical problems of daily life. Anthropologists have described these as 'shame cultures' whose members worry predominantly about their public reputations. But the Judaeo-Christian societies have been racked by fundamental questions over ultimate purposes. Anthropologists have described these as 'guilt cultures' whose members worry predominantly over their private adherence to moral codes.

The Jews, the first great monotheists, were obsessed by questions of right and wrong, and they believed that God would reward the virtuous, and punish the wicked – on this earth, during their lifetimes. By the time of Christ, however, it had become only too clear that such a belief was unsustainable; the best jobs in biblical Judea, like the best jobs everywhere else since, were invariably grabbed by the rapacious and greedy, to leave the meek, the mild and the good with the crumbs from the rich man's table. By Christ's time, under the influence of prophets like Isaiah and Ezekiel, the Pharisees and the Essenes, two of Judaism's three major sects, had embraced the concept of the afterlife, where God could judge the dead safe from the empirical gaze of the quick; only the Sadducees still did not believe in the afterlife.

But who would be saved? Who would actually go to heaven? Everyone knows they have sinned; what if, on the dreadful day of judgement, one sin negated ten virtues? Jesus's contribution was the popularisation of the concept of salvation by faith: everyone sins, but if you only believe, you will be saved.

By Bacon's day, however, the Church and its attendant beliefs were being increasingly discredited. Its dogmatic attachment to Aristotelian cosmology discredited it intellectually, and its cruelties, schisms and greed discredited it ethically. As Bacon said: 'the rule of religion' was 'that a man should show his faith by his works' and one and half millenia of Christian works had failed to create a decent university, let alone a heaven on earth.

To hell with it all, suggested Bacon. The only surety we have is that wealth makes one happier than the alternative, so let us all be rich. Before the days of technology, the love of money may indeed have been the root of all evil because it could only be accrued by extorting it from the poor, but technology really is a free lunch, which enables man to create new wealth, to allow everyone to be rich.

(f) Rousseau and the Romantic reaction

Such frank materialism offended Rousseau, who deplored the damage that the Enlightenment was doing to religion. In *Emile* (1762) he attacked intellectuals for robbing ordinary people of the consolation of faith: 'They destroy and trample underfoot all that men revere, steal from the suffering the consolation they derive from religion and take away the only force that restrains the passions of the rich and powerful.' But Rousseau was just as discouraged by the failures of Christianity (and he was persecuted for his criticisms by both the Catholics and the Protestants, being forced to flee both France and Geneva on occasion).

Disappointed by both the Enlightenment and the Church, Rousseau elaborated the doctrine that spawned the Romantic Movement. Since this swept Europe – we are all, today, the children of the Romantic Movement as much as of the Enlightenment – and since the Movement changed forever our views on science, we must consider it here.

Rousseau believed that, in the natural state: 'Man is born free; and everywhere he is in chains' (the opening sentence of his *Social Contract*, 1762). The enslavement of Man occurred through capitalism, which originated in private property: 'The first man who, having enclosed a piece of ground, bethought himself of saying "This is mine", and found people simple enough to believe him, was the real founder of civil society. From how many crimes, wars, and murders, and from how many horrors and misfortunes might not any one have saved mankind, by pulling up the stakes, or filling up the ditch!' (*Discourse on the Origins of Inequality*, 1755).[11]

Rousseau was not the first to claim salvation through poverty. The Buddha, Jesus Christ and St Francis had all preached and lived its virtues, but Rousseau provided a historical model of how wealth had destroyed the native virtues of man. According to Rousseau, the ownership of property encouraged its owners to improve it, which caused them to foster science and technology: 'it was iron and corn which first civilized men and ruined humanity'.[12] With civilisation came kings, who further fostered science to strengthen their tyranny. 'Necessity raised up thrones . . . the sciences have made them strong' (*Discourse on the Arts and Sciences*). Tyrants also use the arts and the sciences as opiates, since the arts and the sciences 'fling garlands of flowers over the iron chains that weigh them [the people] down. They stifle in men's breasts that sense of original liberty, for which they seem to have been born; cause them to love their own slavery, and so make of them what is called a civilized people.'[13] Rousseau wrote that: 'Astronomy was born of superstition, ambition, hatred, falsehood and flattery; geometry of avarice; physics of idle curiosity; all science of human pride'[14] (note how this mimics, but subtly alters, Aristotle's comment reproduced in Chapter 2. Rousseau was well read).

According to Rousseau, man in the native state is actuated by two major motives, *amour de soi* (the healthy selfish instincts such as hunger, thirst or sex) which is balanced by *pitie* (goodwill to others). But under the distortions induced by reason (which is emotionally sterile), by capitalism (which inflames greed) and by private property (which isolates and destroys sympathy) a man's personality degrades, and he becomes psychologically alienated, both from himself and from other people. His selfishness grows into a devouring monster of vanity and of egotism, *amour-propre*.

Rousseau and his followers prescribed two solutions to this problem. The first solution, the private one, was that individuals should seek self-fulfillment through a return to nature. This was not an entirely original suggestion. Theocritus and Virgil had both written pastoral eulogies, and although a love of nature was lost on medieval man who lived too close to mud to appreciate it, Petrarch resurrected the idyll when he climbed a mountain in 1340 to admire the view. Later Italians like Sannazaro sustained an interest in nature (his *Arcadia* appeared in 1485) though, interestingly, it was the English who championed it through books such as Sidney's own *Arcadia* of 1598, through the work of the eighteenth-century English nature poets like William

Collins and James Thompson and of the English landscape painters like Gainsborough and Wilson, and through the iconoclasm of the English gardeners who broke the European tradition of formal gardens with their quasi-natural ones.*

But it was Rousseau's adoption of Nature that was to transform the Western concept of culture, because Rousseau suggested that any art form that stretched the imagination or that sparked an aesthetic response helped heal the alienated mind by encouraging it to explore and thus know its natural self. Moreover, artists had to divorce themselves from the corruptions of society, to produce pure, natural art. Thus was born the poet as social rebel, the painter in his garrett, and the artist as bohemian.† Thence, too, emerged the concept of art as a moral, social, psychological and cultural good in its own right – one that would always be opposed to capitalism and its attendant science and reason.

Artists love making grandiose claims for art, but Rousseau gave them the contemporary twist that still endures. Let us see how he changed the claims that poets, for example, have made. In his *Defence of Poesie* (1591) Sidney claimed that through poetry there occurred a: 'purifying of wit – this enriching of memory, enabling of judgement, and enlarging of conceit – [whose] final end is to lead and draw us to as high a perfection as our degenerate souls, made worse by their clayey lodgings, can be capable of'. For Sidney, therefore, culture improved men by elevating their 'degenerate souls' from their 'clayey lodgings', i.e. Man's enemy is his own innate, flawed personality, his original sin.

But after Rousseau, the poets put the blame on society, capitalism and science. So, in his own *Defence of Poetry* (1821) Shelley made similar claims to Sidney, writing that poetry: 'awakes and enlarges the mind itself by rendering it the receptacle of a thousand unapprehended combinations of thought . . . the great instrument of moral good is the imagination, and poetry administers to the effect by acting on the cause'. But for Shelley, the enemy does not lie within, but with capitalism: 'Poetry and the principle of Self, of which money is the visible incarnation, are the God and the Mammon of the world.'

---

* Kenneth Clark believed that England embraced naturalism because it was the first irreligious country. Montesquieu had noted in 1730 how: 'There is no religion in England. If anyone mentions religion people begin to laugh' [quoted in K. Clark *Civilization* (London, 1969, p. 188)].

† Rousseau prided himself on being 'dressed in my usual careless on style'. In his *Confessions*, he wrote of 1752: 'Considering my unkempt state an act of courage, I entered the theatre where the King, the Queen, the Royal Family, and the whole Court were shortly due to arrive.' Naturally, he 'trembled like a child . . . about the curiosity of which I was the object'.

Capitalism meant science. Once, the poets had praised science. In *Paradise Lost*, Milton applauded Galileo as 'the Tuscan artist with the optic glass' and Milton's contemporary Henry Vaughan (1622–95) could write:

> I saw Eternity the other night,
> Like a great ring of pure and endless light,
> All calm, as it was bright,
> And round beneath it, Time in hours, days, years,
> Driv'n by the spheres
> Like a vast shadow mov'd' in which the world
> And all her train were hurl'd.
>
> (*Silex Scintillans, The World*)

A little later, a painter like Joseph Wright of Derby could celebrate physics in paintings such as *The Air-pump* (1768, in the National Gallery in London). But, after Rousseau, the artists turned on science. In his *Lectures and Notes of 1818*, Coleridge wrote that 'Poetry is opposed to science', and William Blake (1757–1827) who knew his sources (*Mock on, Mock on, Voltaire, Rousseau*) believed that 'Art is the Tree of Life. . . . Science is the Tree of Death.' Wordsworth shared his generation's distaste of research:

> Sweet is the love which nature brings;
> Our meddling intellect
> Mishapes the beauteous forms of things:–
> We murder to dissect.
>
> (*The Tables Turned*)

This reverence of nature and of the Noble Savage was, of course, nonsense. As the Marquis de Sade (who knew) said in 1792: 'Nature averse to crime! I tell you that nature lives and breathes by it, hungers at all her pores for bloodshed, yearns with all her heart for the furtherance of cruelty.' Voltaire, the leader of the Enlightenment, wrote in a famous letter to Rousseau that 'No one has ever used so much intelligence to persuade us to be stupid. After reading your book [*Discourse of the Origin of Inequality*] one feels one ought to walk on all fours. Unfortunately, during the last sixty years I have lost the habit.' Dr Johnson, the great empiricist, on being exposed to 'a gentleman who expatiated to him on the happiness of savage life' replied' 'Do not allow yourself to be imposed on by such gross absurdity. It is sad stuff. If a bull could speak he might as well proclaim: "Here I am with this cow and this grass; what being can enjoy greater felicity!".' But the clear thinkers were swamped by Rousseau's disciples, who were ultimately to range from Goethe (barely one of whose pages does not refer to Nature) to Ruskin, whose *Modern Painters* maintained that Nature illustrates Moral Law.

Science has never fully recovered from the success of Rousseau and of the Romantic Movement. No popular writer has ever rescued scientists from Rousseau's description of them as deficient in human warmth, spirituality or *pitie*.* Indeed, the Romantics turned on science with a vengeance. It was Shelley's wife, Mary, who wrote the first of the many novels that have since portrayed scientists badly, *Frankenstein* (1818). The Romantics made it fashionable to despise scientists as over-rational, megalomanic, emotional cripples, in hock to materialism, vainglory and greed. That view persists; a 1962 survey of American college students' attitudes to scientists revealed that they are perceived as intelligent, but socially withdrawn and possessed of a 'cold intellectualism'.[15] That survey may have reflected prejudice, but it may also have reflected reality, since prejudice becomes self-fulfilling. If young people believe that the humanities, not science, represent the fullest expression of the human spirit, then the most ambitious and successful of young people will be drawn to the humanities. The socially clumsy will, in contrast, be drawn to science as a refuge.

Rousseau and the Romantics never extinguished science, of course, since at least four powerful forces continued to operate in its favour: (i) the Enlightenment; (ii) *laissez faire*; (iii) nationalism; and (iv) socialism. Let us elaborate.

(i) The Enlightenment   On mainland Europe, where the Enlightenment was most cherished, reason continued to be venerated, and its thinkers tried to harmonise Rousseau's visions with those of Bacon or of Descartes. Thus the cosmopolitan Shelley wrote in 1811 of how: 'I vehemently long for the time when man may dare live in accordance with Nature and Reason, in consequence with virtue.'[16] Rather than adopt Rousseau's dismissal of reason, therefore, continental Europe tried to graft Rousseau's doctrines of Nature (and the State) onto the Enlightenment view of Reason (and the State), and continental governments continued to support science.

---

* Nor did any scientist apparently return the poets' scorn, and one has to search in the pre-Rousseau literature to find the robust Thomas Spratt FRS who wrote in his *History of the Royal Society* (1667) that: 'poetry is the parent of superstition'.

(ii) Laissez faire The Anglophonic societies also continued to support science, not because they subscribed to Bacon, or to the State, or because they disagreed with Rousseau – because, in truth, the anglophones were never very interested in any of them – but simply because of the free market. The anglophones subscribed to *laissez faire*, which bred capitalism, which bred the private funding of technology, which bred private funding of science.

English gentlemen rarely had much time for the Romantic Movement and as late as 1832 an English gentleman like Lord Morpeth, Yorkshire MP for the West Riding, could write poetry in praise of capitalism, and of science, which would have seemed 50 years out of date to his clever continental contemporaries:

> Still, commerce, thine unfettered track pursue,
> Court torrid zephyrs, brave the icy gale
> Rivet creation's severed links anew,
> With they light rudder and they roving sail.
>
> Crowned with the myrtle, vine and olive leaf,
> Before they peaceful keel chase gory strife,
> Waft to each want that visits man, relief,
> The lamp of knowledge, and the cross of life.[17]

(iii) Nationalism The rise of nationalism worked powerfully in favour of science. We have already seen how enthusiastic nationalists like Bismarck or Avery tried to develop science as a weapon in superpower rivalry, and how reluctant nationalists like Kilgore or Asquith were forced to adopt the same policies in defence. Nationalism still dominates science policy, and the science activists of all countries use statistics about other countries to force their own governments to increase domestic expenditure. The greatest contemporary exponent of science as a national tool is Corelli Barnett, whose *Audit of War* was so named because he believes that the ultimate test or audit of any human activity is the degree to which it contributes to national victory in war – a miserable philosophy.

Nationalism as we understand it today is a surprisingly modern development, as is witnessed by the title of one of its recent histories, Eric Hobsbaum's *Nations and Nationalism since 1780*.[18] It apparently arose in response, in part at least, to the loss of certainty occasioned both by the collapse of the authority of the church and also by the collapse of feudalism. It represents, therefore, a psychological response to the cultural vacuum of capitalism and of the Enlightenment, and so it competed with the other psychological responses to that vacuum, Romanticism and, especially, socialism and communism (see below). The early communists studied nationalism, the better to fight it. Thus the best definition of nationalism is Joseph Stalin's: 'A nation is a historically evolved, stable community of language, territory, economic life and psychological make-up manifested in a community of culture.'[19]

Nationalism still dominates. Robert Reich, who is now President Clinton's Labour Secretary, tells how when he taught economics at Harvard he used to ask his students this question: 'What would you prefer: that both the US and Japan grow economically at 1% a year; or that the US to grow at 2% a year, but Japan at 3%?' Overwhelmingly, his students preferred the first option, i.e. people prefer to be absolutely poor but relatively rich, rather than absolutely rich but relatively poor. That nationalistic obsession with relative wealth, both economic and scientific, ensures that all economic and science activists, the world over, are continually dissatisfied and paranoid. Nationalism turns one country's success into everybody else's failure. It is a recipe for chronic misery. Yet, in both scientific and economic terms, it is a nonsense, since any country's advance can be immediately copied or imported by every one else, to everybody's benefit. Technology and capitalism are the most benign forms of competition; everybody gains. There are no losers. (Consider, for example, the fax machine. No matter who invented it, no matter his nationality, and no matter his profits, these are as nothing compared to the benefits and ease it has brought to innumerable millions of people worldwide.)

Nationalism has caused little but misery since 1780, and it has rightly been condemned by the thoughtful like Dr Johnson: 'Patriotism is the last refuge of the scoundrel.' It is wry-making that the science activists, who uniformly portray themselves as enlightened and socially responsible, degenerate so speedily into Colonel Blimps at the hint of foreign success in research. Such chauvinism can only be justified by the threat of war, but few OECD countries now plan to invade each other (though one can never be sure about the French, who are still smarting over Waterloo). The world remains a dangerous place but the potential warmongers are no longer the richer countries.

(This is, actually, an important point. The world does remain a dangerous place, but the potential aggressors – men such as Saddam Hussein or Vladimir Zhirinovsky – come from relatively poor countries. This means that, for the first time in human history, the lead countries no longer need to distort their economies to develop military technology

– they can easily sustain, at relatively little cost, a superiority in military technology over their poorer potential aggressors. That is the true peace dividend of the fall of the Soviet bloc. There remains, therefore, no reason for the governments of rich countries to fund academic science for military reasons, since the civil funding of academic science can hereafter be trusted to meet national needs. Since military reasons have hitherto been the major stimulant of the government funding of academic science, we can expect a vast diminution in this over the next few decades. One only needs to look at the US Congress's abandonment of the SSC project to glimpse the future.)

(iv) Socialism The great restorer of science's image after Rousseau's demolition was Karl Marx. He worshipped technology: 'The bourgeoisie, during its rule of scarce one hundred years, has created more massive and more colossal productive forces than have all preceding generations together. Subjection of nature's forces to man, machinery, application of chemistry to industry and agriculture, steam navigation, railways, electric telegraphs, clearing of whole continents for cultivation, canalisation of rivers, whole populations conjured out of the ground – what earlier century had even a presentiment that such productive forces slumbered in the lap of social labour?' (*Manifesto of the Communist Party*, 1848).[20]

Through all the fuss over Marx and Marxism, his essential achievement has been largely overlooked. It was trivial, just one more progression in Western man's futile quest for meaning in his life. Neither of his two preceeding secular prophets, Bacon and Rousseau, had fulfilled that need. Bacon's technological materialism was attractive, but his philosophy denied the sublime, while Rousseau's romance was lovely, but it precluded the comforts of technology. So Marx simply shuffled their ideas about to produce a philosophy that combined Bacon's technology and *dirigisme* with Rousseau's egalitarian ethics and *dirigisme*. That Marx's philosophy was inherently flawed and internally contradictory need not concern us here; what matters is that it captured the intellectual mind – a sickly thing that has rarely encompassed Adam Smith's cool, correct minimalism – for a century.

Like Bacon, Marx characterised science as social progress. He also extrapolated the religious scepticism of Bacon and Rousseau to embrace atheism, to adumbrate a new psychology of salvation – salvation through society. He legitimised politics as the contemporary cult. 'Politics is everything', Rousseau had said; Marx agreed. Rousseau (and later Lenin) believed that politicians should be all-powerful, and ordinary citizens powerless, because only politicians could embody the perfect General Will.

After Marx, the poets – those social and intellectual weather vanes – become significantly less rude about science. Indeed, Soviet poets were known to write odes to chemical formulae.

THE NATIONALISATION OF SCIENCE

Under the influence of nationalism, socialism and the Enlightenment, science has been nationalised almost everywhere. What have been the consequences?

First, let us be pedantic and re-define our terms. Civil R&D has not been universally nationalised, but where it has, the consequence has been clear: underinvestment. Since civil R&D underlies long-term economic growth, we can conclude that the nationalisation of civil R&D has impoverished, relatively, those countries that have performed it.

For academic science, however, the situation is less clear; because it has been largely nationalised everywhere, we cannot correlate different countries' different outcomes with their different policies (Japan is the partial exception, see below). Yet, even with academic science, we can witness the workings of its fundamental economic laws; when governments cut back on it, as did the British government during the 1980s, the private sector grows to compensate (see Chapter 11). That contemporary evidence, coupled with industry's need for university-trained scientists (which therefore fosters university science), coupled with the economic forces that drive industry to fund basic science, and coupled with the historical evidence of the private endowment of academic science, allows us to conclude that, at the very least, the public and private funding of academic science displace each other. For lack of data, we cannot prove that the government funding of science actually overdisplaces private funding as it does with civil R&D, but it is likely. Since we can only safely conclude that the private and public funding of academic science displace each other, we must ask: so what? Does it really matter where the money comes from?

It does. First, it is apparent that the nationalisation of academic science has not actually increased countries' rates of economic growth (see Figure 8.6 for the American example). We see here, therefore, the first damaging consequence of the government funding of academic science – public disillusionment with science and with scientists. The science activists of all countries promised their governments that their funding of the linear model would yield wealth. It has not, and the subsequent disillusion with scientists and with their attachment to truth has been palpable (see the comments of President Johnson in Chapter 8 and of Rep. H. Wolfe and John Andelin in Chapter 11). Thus has the public standing of science been damaged by the politicisation of its funding.

But the government funding of academic science has actually been self-defeating, economically, because of governments' confusion between academic and basic research. As we have seen, the market demands a considerable quantity of basic science, and governments have attempted to supply this – but within the universities. Yet the universities are a long way from industry, and the distance has damaged the industrial exploitation of academic science. To try to correct this, Western governments have applied considerable pressure on the universities over the last two decades, to make their science more industrially relevant.

Britain, for example, witnessed during the early 1970s the Rothschild Report, which recommended the transfer of government funds from the independent research councils to particular government departments; these, in turn, would fund specific pieces of university research, in areas that the politicians and bureaucrats believed to be important, on the 'customer–contractor' principle. In 1993, William Waldegrave, the British science minister, re-organised the Advisory Council for Science and Technology (ACOST) and the Advisory Board for the Research Councils (ABRC) to create a new body, the Council for Science and Technology, to supervise Government science funding. That body is chaired by the minister, and is dominated by industrialists, not scientists, to push British science into applied work. Meanwhile, the University Grants Committee, which was dominated by academics, was superseded by the Higher Education Funding Council (via the University Funding Council) which is dominated by industrialists. These changes were introduced by the 1993 Government White Paper on science which stated that the prime function of university science was the generation of wealth.

But are such moves appropriate for university science? Do they not simply degrade it? A clue is provided by Japan, the one industrialised country whose academic science has been relatively neglected by the State. In 1987, for example, the Japanese government spent the equivalent of US $3735 million on academic science, while the British government, serving only half the population, and with a lower GDP *per capita*, spent the equivalent of $2914 milion.[21] But in a lovely demonstration of the Third Law of Research Funding, Japan's universities compensate for the lack of government funds for their research with a vast influx of private funds. Thus (see Table 12.1) Japan enjoys a unique (greater than 50 per cent) private funding. (See also on Table 12.1 on p. 353 how well the UK and Canada do in the face of public squeezing. The Third Law lives.)

And Japan actually publishes more basic, and less applied science, than Britain. Ben Martin and his colleagues, on classifying science journals as 'basic' or 'applied', showed that, in 1986 as a typical year, the ratio of basic to applied publications in Japan was 1.58, to Britain's mere 0.68.[22] Thus the government funding of academic science in Britain, which is portrayed as vital for pure science, may actually have distorted academic research into an over-emphasis on applied work; the relatively *laissez-faire* science of Japan is the purer.

Not all governments have so distorted their academic science; France enjoys a basic:applied ratio of 1.56, but Canada, Germany, Holland and the USA approach Britain with basic:applied ratios of 0.99, 0.97, 0.92 and 0.84 respectively. This is not to argue that basic science is 'better' than applied – the USSR's ratio was a staggering 2.34, and much good that did it – but Japan does demonstrate how the free market protects the purity of academic science better than do, in practice, the governments of the USA, Canada, Britain and Holland. Moreover, even the French government is changing. François Fillon, the French minister of science, gave *Le Figaro* of 2 February 1994, a chilling interview in which he said that he wanted scientists to apply their research more commercially in future. That same week, in an unpleasant echo from the other side of the Atlantic, Senator Barbara A. Mikulski (Democrat, Maryland) told a meeting of the Office of Science and Technology Policy that government funded research must create 'new technologies which must lead to jobs, particularly in manufacturing'.[23]

Such economic illiteracy, coupled to *dirigisme*, reminds one of the *ancien regime* of France. It will not work. Consider Britain. A fascinating UNESCO report of 14 February 1994[24] confirmed that the top ten Western European scientific countries, as judged by their number of papers published, were (in order) the UK, Germany, France, Italy, Holland, Sweden, Switzerland, Israel, Spain and Belgium (so much for Britain's so-called scientific decline). But, despite all the government pressure of the last few decades, this science failed in Britain to translate into patents. Europe's greatest patent filer is Switzerland at 40.44 patents per 100 000 population, followed by Germany and Sweden at 28.14 and 22.91 respectively. America, incidentally, came in at 19.12, but the UK, together with Norway, Denmark, Holland, Belgium, Greece and Spain could only manage between 7.7 and 4.4 patents per 100 000 population (Austria, Finland and France scored above 15). Since it is patents, not papers, that translate into economic growth, these figures represent a dreadful failure for the British government's policies.

The world's top patent filer is, of course, Japan at 43.44, and it is no coincidence that Japan and Switzerland are the two top filers since theirs are the two most *laissez-faire* civil R&D regimes. For all their nasty pressurising of academics, governments are not commercial, and their pressure only produces a simulacrum of commercialisation. Industry, of course, is commercial, and it understands exactly how to turn research into patents. *Laissez faire* works, *dirigisme* does not, but *laissez faire* robs the Fillons and the Mikulskis of their strutting on the stage.

We showed in Chapter 7 that patents obeyed the First Law of Research Funding, namely that the number of patents filed nationally *per capita* reflects national income *per capita*. Japan and Switzerland are both, of course, richer than Britain, but not *that* much richer. Their disproportionate numbers of patents, therefore, illuminates a further aspect of the Third Law of Research Funding (the disproportionate displacement of private funding for civil R&D by State funding) namely the disproportionate displacement of the commercialisation of civil R&D by State funding. This lesson, of course, will never be learnt by research activists on either side of the Atlantic, who will always urge more government funding for civil R&D, regardless of the national damage it produces.

Because the Japanese government has neglected basic science, Japanese companies have compensated by building their own basic science laboratories, to effectively create their own in-house universities. These have fostered the usual scientific values of open discourse, freedom of research and publication for the most empirical of reasons – they cannot flourish without them.* Not only has this been economically beneficial, since it integrates basic science with industry, but it also illustrates the Second Law of Science Funding (mutual displacement).

Had basic science not been nationalised in the West, Western departments of university science would be smaller than they currently are, and the industrial laboratories would be bigger, as in Japan today. The total quantity or quality of basic science would not have suffered,

---

*The value of openness in promoting excellence was first articulated by Vasari in his *Lives of the Artists* (1550). Trying to explain why it was in Florence, and not elsewhere, that men perfected art, he wrote: 'The spirit of criticism: the air of Florence making minds naturally free, and not content with mediocrity.' The same is true in science: openness fosters criticism, and thus excellence. Secrecy inhibits criticism, and so fosters mediocrity. Not all companies, or even scientists, understand this.

but university science would have been more free because, under the free market, it would be largely funded by endowments and by the income from fees. It would, therefore, be autonomous, to ensure a uniquely academic science base – one where science is academically driven – rather than the current unpleasant hybrid where governments force their concept of commercial research on the universities. Academic freedom does not require the huge universities we now see in the West; it requires, however, that they be genuinely free.

Because of government tyranny, however, academic freedom in Japan is not found within its universities, but in industry. Japan's universities are bizarre. Research is funded on the *koza* system, by which chairmen of departments regulate, fund and control the research of the junior academic staff.[25] This is intensely frustrating for the so-called juniors who would, in Europe or the USA, be independent. Because promotion to *koza* status is so hard, suicide is common in Japanese universities; and of the surviving academics, 85 per cent are dissatisfied.[26]

But Japan's universities have no freedom at all. Let one typical example illustrate Japan's oppressive government regulation. In January 1994, Toshio Yamaura of Japan's Ministry of Education, Science and Culture was arrested for having accepted a bribe of ¥500 000 (£3500) from the Sugiyama Jogakuen University in Nagoyo.[27] The university wanted to change the name of its 'Faculty for Home Economics' to the 'School for Life Studies'. Even though the university is private, like all other Japanese universities it is so controlled by the State that it could not change a faculty's name without the permission of the Ministry, a procedure that takes between two and five years (Japan's bureaucrats being notoriously inefficient). To speed the process, therefore, the university bribed Yamaura to rush it through in a mere six months. But that very speed betrayed the corruption, and the conspirators were exposed and Yamaura arrested. (Not for nothing do Japanese children go to school dressed in a uniform based on Bismarck's. Japan has modelled its education systems on those of nineteenth-century Prussia.)

Academic science is not free when governments fund it. The Haldane Principle is now dead in Britain, and all but the smallest research-council grants in Britain have to be directly cleared by the ministers for science. That clearance is sometimes refused. During the late 1980s, for example, Mrs Thatcher personally vetoed a £1 million investigation into sexual habits because she believed that the proposed benefits to the AIDS prevention programme were outweighed by the invasion of privacy. Both the scientific community (the project had been properly peer-reviewed) and the potential subjects disagreed (their cooperation was voluntary) and academic freedom was eventually safeguarded by the (private) Wellcome Trust, which stepped in to fund the project instead.

It was the scientific community, not the government, that was right. In his memoirs, Kenneth Baker, one of Mrs Thatcher's ministers, wrote that the government refused to fund the project because it would only show 'that Britain had become a more promiscuous society – which we knew – and more experimental in the realm of bisexual relationships – which we also knew'.[28] But we did not know that. The report actually revealed that Britain was an astonishingly restrained and monogamous country.[29] Half of all men are still virgins at 18, only one in twenty girls has sex before 16, and fewer than one in ten married men, and fewer than one in twenty married women, had been unfaithful to their spouse in the previous five years. Mrs Thatcher and her colleagues, in judging Britain by the louche standards of Conservative ministers, had completely failed to understand the morality of the country they were trying to govern – and even worse, they were trying to prohibit the scientific investigation of the question. Pope Urban VIII would have been proud of No. 10. Thank God for the Wellcome Trust and for the private funding of research. (Interestingly, American society outside Hollywood, the universities and Capitol Hill is also astonishingly monogamous.)[30]

In a more systematic undermining of academic autonomy, the 1993 British Science White Paper proposed that all PhD students take an MSc during their course. Few activities are more central to the universities than the granting of degrees, but the Government now proposes to overrule individual universities' different practices by *diktat*. I was astonished by the lack of opposition to the measure.

In America, government interference in science has become institutionalised. In 1992, for example, Congress overruled 32 projects that the NIH or NSF had decided to fund after proper peer review, on the grounds that projects with titles such as 'Holism in psychobiology in the twentieth century' were silly. But in 1993 Congress surpassed itself with the devastating cancellation of the Superconducting Super Collider (SSC) project. This project, which had already consumed $2 billion of its $11 billion budget, employed 2000 scientists and engineers who, on 21 October 1993, were given just 90 days' severance pay – insulting treatment. The termination costs will bring the total expense of this non-project to $3 billion.

The cancellation of the SSC gives the lie to the argument that private funders would never fund CERN-type projects, because the SSC was cut as part of the peace dividend; i.e. CERN, SSC *et al.* were funded primarily for military reasons, and by definition any of their intellectual or economic benefits would have been achieved at a grotesque cost. None the less, the economic laws of science do confirm that private philanthropists and foundations would fund CERN-type projects if the scientific pay-off were reasonable.

No one denies, of course, that if governments do fund projects, they must retain the right to cancel them, but the examples given above show: (i) that the scientific community has lost the freedom to run itself, a freedom which foundations like the Carnegie had once protected, and (ii) government is a monumentally bad manager of science; at the same time as $3 billion and 2000 careers were being wasted on the blighted SSC, the Navy was building a third, unnecessary, totally redundant, atomic clock to satisfy a pork-barrelling politician eager to bring jobs into his state.

At least the American government no longer actively destroys academic autonomy as it did during the late 1940s, 1950s and 1960s. We call that episode McCarthyism, and it cost many academics their jobs or their passports. (The Republicans get much of the blame for McCarthyism – McCarthy himself was the junior Republican Senator for Wisconsin – but the policy was pretty bipartisan. It was the Democrat President Truman who established the Federal loyalty oath in 1947 – the oath that the universities imposed so destructively – yet McCarthy did not make his famous speech alleging that the State Department was controlled by Moscow until 9 February 1950. Moreover, the House Committee on Un-American Activities was created in 1938, at the height of the New Deal, and it was not disbanded until 1968, after Nixon's election as President).

No one believes that peer review is the perfect way of distributing research money, though no one has yet devised a better one. But if peer review is to be overruled by someone, it would be hard to find a body less worthy than Congress, whose members are currently colluding with science and university lobbyists to the tune of no less than $60 million a year in fees and perks to by-pass peer review to pork-barrel hundreds of millions of dollars worth of research projects into their own constituencies.[31] One does not have to be H. L. Mencken to be astounded by American politicians. Moreover, the definition of 'silly' depends on one's intellectual background. As Keynes remarked: 'Practical men, who believe themselves to be quite exempt from any intellectual influences, are usually the slaves of some defunct economist.' Congress would undoubtedly have taken Pope Urban's side against Galileo and it would have found Darwin silly; who can forget the thundering congressional denunciations of evolution during the Scopes trial?

(The Italian government is even worse than the American Congress. Italian science is amazing. Money is not distributed by peer review but by small committees of politically appointed cronies.[32] Academic appointments and promotions are utterly corrupt, which has prompted a terrible despair within the universities,[33] and the Italian government lurches around making violent policy changes. In 1994, for example, Italy simply withdrew from the European Molecular Biology Laboratory (EMBL). In the words of *Nature*, this has thrown 'EMBL's plans into chaos'.[34] The purported reason for this withdrawal is that EMBL is not serving Italy's interests, but the real reason, of course, is that EMBL will not bribe Italian politicians.)

Had governments not funded academic science, young scientists would have enjoyed more fulfilling careers, at least in Britain, since the universities are such bad employers. Much of the malaise of young academic scientists derives from their frightful conditions of employment. They have to hop from short-term contract to short-term contract, with no security beyond two or three years, and no severance pay should their contracts fail to be renewed. If young scientists have to move between cities, universities do not pay their removal expenses. These conditions would be unthinkable in industry.

Bizarrely, private industry increasingly funds research within the universities specifically to exploit the sweat-shop contracts they would never impose on their own employees. Private industry, unlike the universities, values its workforce. Once upon a time, the government would have intervened to protect university scientists, as nineteenth century British governments did to force the universities to admit women, Jews and dissenters (and to teach science and to allow academics to marry), but one of the tragedies of nationalisation is that it removes Government from its role as society's referee – one that intervenes on behalf of the weak – by transforming it into a provider. As a provider of university science, government has an interest in exploiting its workers, and they have little recourse. A monopoly employer, in Britain at least, drives salaries down abysmally. Before 1914, a Cambridge University professor earned the current equivalent of around £80 000 p.a. Now he earns about £37 000. Since 1979, a professor's salary has fallen from three quarters of that of a senior (grade 2) civil servant, to half. Since 1979, moreover, the average salary of academics has only risen, in real terms, by 9 per cent, whereas that of school teachers has risen by

52 per cent and that of all non-manual workers by 54 per cent.[35]*

Further, the very fact of nationalisation damages science, because it implies (incorrectly) that society would not otherwise fund it. This makes scientists insecure. Instead of understanding that they are an integral part of the capitalist process – a part that capitalism needs and would fund – nationalisation makes scientists feel vulnerable, unappreciated and over-dependent on the political process. This alienates them from society, and fosters an isolating ideology. It fosters, in fact, the ideology of Rousseau.

Remarkably, we have to return yet again to Rousseau, since his theory of personality still underpins contemporary views of the State. So far in this book, we have largely concerned ourselves with the economics of science, but as we conclude our survey we have to review the psychological or non-logical pressures that drive men to entrust important aspects of society to an institution so inimical to individuals as the State, which forces us to focus on Rousseau and the Romantic movement.

Rousseau's Theory of Personality

Rousseau believed that man's natural selfishness would always cause him to behave badly. He believed that individuals would neglect their social obligations. But collectively, Rousseau claimed, they would know where their social duties would lie. Thus, for example, no individual would give money to the poor, or to pure science, because it could not benefit him personally, but in a parliament, individuals would vote collectively to tax themselves to support paupers or researchers. Parliament, unlike the individual, is perfect.

Here is the psychological basis of Rousseau's *Social Contract* (1762). The contract is made between the individual, who signs away his rights, and the State which, by embodying the General Will, is perfect in its morality. According to Rousseau: 'The people making laws for itself cannot be unjust', 'The General Will is always righteous', 'The Gen-

---

*The Government's abuse of its status as a monopoly employer was well illustrated by its PhD stipends which until the end of the 1980s were only £2975 a year, despite frequent complaints that they were too small. It was not political action that raised them, it was the rise of private, pluralistic funding for science. The Wellcome Trust and the Cystic Fibrosis Research Trust broke the long standing convention that medical charities should match research council stipends, and they started to pay £4000 a year. Since then the various medical charities and industrial sponsors have leapfrogged each other to retain the best students, and their PhD stipends can now be £12 000 a year or even more. Interestingly, on the physical side of science, where there are fewer competing bodies, PhD stipends remain low.

eral Will always favours the public interest'. And because the General
Will is always good, and the individual in society always bad, the
individual will 'assign himself, with all his rights, to the state'.

Rousseau had to laud the State as a psychological counterpoint to
his paranoia. His writings are permeated by a loathing of other people.
Just one, typical, chapter of his autobiographical *Confessions*, contains
such phrases as: 'those low-born men, deaf to the gentle voice of Nature,
in whose breast no real feeling of justice and humanity ever arise . . .
they would have been led to hate, and perhaps to betray, their parents . . . deliberately to dishonour the friends they have deceived . . .
utter baseness and infamy . . . senseless opinions of the vulgar herd . . .
the so-called great . . . so-called wise . . . so-called friends . . . bitter
enemy . . . most bitter enemy . . . dreadful libels . . . hundreds of enemies . . . jealous arrogance . . . connivance . . . cruel thought . . . tor-
tured me . . . persecuted . . . hatching in his inmost heart the plot . . .
an ugly little man . . .'[36]

Rousseau claimed that his distrust of other people was born of cruel
experiences, yet the reality was the exact opposite. From the time he
left home at the age of 16 for the protection of Madame de Warens,
he lived off a succession of rich, kind benefactors. As he admitted, he
did not need to earn a living: 'I renounced for ever all plans for fortune and advancement' by living off the extraordinarily generous charity of people like Mme d'Epinay, Frederick the Great, David Hume,
the Marquis de Mirabeau, the Prince de Conti and the Duc de
Montmorency-Luxembourg to name but a few of the patrons who lent
him chateaux, *cottages orné*, or who simply gave him money. Garrick,
for example, put on a special benefit performance for him in Drury
Lane during his visit to England in 1766–7. Rousseau, in short, lived
a gilded existence, funded by arisrocrats who sponsored his philosophy.

But Rousseau was paranoid and, loathing other people, he invented
the concept of the perfect State. But this is silly; what is the State but
a collection of individuals? If the individual in society is so irredeem-
ably corrupted by that society ('Man's breath is fatal to his fellow
man': *Emile*) how did Rousseau himself escape, to describe himself
thus: 'My warm-heartedness, my acute sensibility, the ease with which
I formed friendships, the hold they exercised over me, and the cruel
wrench when they had to be broken; my innate goodwill towards my
fellow men; my burning love for the great, the true, the beautiful, and
the just; my horror of evil in every form, my inability to hate, to hurt,
or even to wish to; that softening, that sharp and sweet emotion I feel
at the sight of all that is virtuous, generous, and lovable.'[37]

If society is constituted of corrupt individuals, how do they suddenly become so perfect through the General Will? Consider, again, help for the poor or for pure science. Is it not likely that a corrupt individual, who refuses personally to assist these disinterested causes, would also vote against their public funding, to keep his taxes down? If the General Will is indeed the general will, it will only be virtuous when the mass of people is virtuous, and vice versa. If not, then by definition it cannot be the general will. Thus a virtuous people will wish to help the poor, and a bad people will not, and it is obvious that a democratic parliament could only reflect that generality. It is also obvious, therefore, that if a good people chooses, as a practical matter, not to entrust charity to voluntary agencies, but to empower the State, then statutory agencies would only largely do what the voluntary agencies would have done anyway. This, it will be remembered, was established empirically by Buchanan and Tulloch who, in their *Calculus of Consent*, showed that the nationalisation of a range of social services from health, education and welfare, had not actually increased the share of GDP spent on them.[38]

The decision to fund science through the State rather than voluntarily is, moreover, morally flawed, since the State forces all people to support an activity to which some individuals are opposed. There are people such as religious activists, or believers in animal rights, or Rousseau himself, who find the activities of scientists offensive, and yet who are compelled by the State to support it through their taxes. That is an immoral imposition on individual will by the collective. Furthermore, *dirigisme* destroys individual morality, which thrives when it is exercised under *laissez faire*, but which atrophies with disuse when the State takes over the responsibility for moral decisions (see the postscript on Rousseau and morality).

But the nationalisation of science, like the nationalisation of compassion generally, has appeared to legitimise Rousseau's horrible vision of humanity. People now assume that governments provide for science and welfare because individuals cannot be trusted to do so. People rarely understand that governments have merely displaced private funding. People's personal experience, therefore, apparently confirms Rousseau's theory of personality: other people are corrupt, and only government is moral.

This is the dreadful Romantic legacy. By creating society in Rousseau's image, we have created a self-fulfilling prophecy whereby daily reality appears to assert the horribleness of other people. Thus it has become almost impossible to persuade scientists (or academics, artists or intellectuals generally) that under *laissez faire* people would choose to support pure science voluntarily. Every practising scientist to whom I have suggested that has stared back at me as if I were dangerous, mad and possibly rabid; i.e. every practising scientist I have encountered has believed, Rousseau-like, that he, almost uniquely, possesses a morally sublime view of society's need for pure science, but that the mass of people are blind to their responsibilities, and have to be dragooned by the State into funding research.

This is an unhappy philosophy, which can only make its protagonist miserable. It is also an absurd philosophy, one that was explicitly contradicted by the hero of this book. It was Adam Smith who, in his *Theory of Moral Sentiments* (1759), showed that people only acquired morality and ethics from the judgements of others: 'Were it possible that a human creature could grow up to manhood in some solitary place, without any communication with his own species, he could no more think of his own character, of the propriety or demerit of his own sentiments and conduct, of the beauty or deformity of his own mind, than of the beauty or deformity of his own face.... But bring him into society, and he is immediately provided with the mirror which he wanted before.... This is the looking-glass by which we can, in some measure, with the eyes of other people, scrutinize the propriety of our conduct.'[39] For Rousseau, therefore, other people are hell, but for Smith, society is redemptive – and of course he is right; how else does one learn one's morality except from other people?

Yet intellectuals are very Rousseauesque. Like him, they feel superior to other people. Like him, they wallow in self-pity. And like him, they are egotistical, selfish and mean. Obviously, there are individual exceptions to those generalisations, but no organisation fosters jealousy, bitterness and envy like a modern university. Industrial research laboratories are, in my experience, much more harmonious; the free market fosters group values.

Why are intellectuals so Rousseauesque? Many, of course, are just born that way. Not everyone inherits social skills, and awkward people will naturally be drawn to the marginalised, protected world of the laboratory and the library. Intellectuals, therefore, are not necessarily more intelligent than other men, but many of them are socially inadequate. That was precisely the path of Jean-Jacques Rousseau. His mother died following his birth, so he never knew maternal love. His only brother was sent to a reformatory for his criminal behaviour, and his father was a tyrant. Rousseau's childhood was lonely, loveless and dreadful, and he emerged a social misfit. Throughout his adult life, he

failed to keep a single friend. His was, in fact, a sad and lonely existence, and he became paranoid towards the end. Desperate for love,
attention and fame, only through his writings could he achieve the
social distinction that his boorishness prevented him from attaining
through more conventional channels.

But the intellectual's paranoid personality can be made as well as
born. A charming young man who becomes a professional scientist
through a love of discovery will soon discover that he will not succeed without working very long hours. I do not know any successful
biochemist who has not systematically worked at least 60 hours a week
for years at a time. Nor do I know any successful biochemist who has
not immersed himself in the culture of science, and who has not, in
consequence, drawn most of his friends from other scientists. Science
is a vocation, not a job or even a career. It assumes one's whole life
outside the family.

Professional science, therefore, is an isolating existence. That isolation, unfortunately, breeds distrust, for it is a sad aspect of human
nature that ignorance does breed distrust. Common clinical experience
confirms that – witness, for example, deafness. Deaf people become
paranoid. They see people having conversations, but failing to understand what is being said, they assume they are being criticised. So it is
with intellectuals and scientists. Failing to understand ordinary people,
they grow to distrust them. Only the State, an artificial construct which
the intellectual mind can understand, seems to the intellectual to be
trustworthy, since it appears – unlike all those frightful people out
there – to be transparent, logical and accountable (and because academics are so horrible they find it even harder to understand, or believe in, the philanthropic impulse).

Nationalisation has aggravated scientists' isolation. Under *laissez faire*,
scientists have to straddle more than one world. They have to pene-
trate and understand the worlds of industry or of medicine or of charitable trustees or even of the parents who pay their pupils' fees. But
under nationalisation, scientists need only relate to other scientists. Their
money comes from government grants which are judged and awarded
by other scientists; government grants support the students; and scientists need only leave their laboratories to attend meetings (government
assisted) with other scientists. State-funded reseachers inhabit a govern-
ment-created asylum, peopled solely by other scientists, where the only
values that matter are those of scientists. No wonder researchers become narrow and intense and, often, moral cowards, terrified to of-
fend. If you only inhabit one tiny world, and you are desperate for the

recognition of that world, you are careful to be politically correct.

This artificial, self-referential, government-funded asylum has bred a horrible competitiveness that has helped to destroy science as a civilised activity. That scientists are competitive often shocks the general public, which cherishes the illusion of researchers as calm, selfless beings, cooperating in the noble search for truth. The reality is harshly different, as was illustrated by the vignette James Watson told in his *Double Helix*. In the early 1950s, Watson and his colleague Francis Crick were trying to discover the structure of DNA, the central molecule of life (they were not doing their own experiments, merely modelling the X ray studies of two London scientists, Maurice Wilkins and Rosalind Franklin to which they had privy access). One day, Watson and Crick heard that Linus Pauling, the great Californian chemist, doing his own experiments (!), had just published his own model of DNA. They rushed to read his paper and, oh joy! he had got it wrong. Instead of a double helix, he had proposed a triple helix, which their private knowledge of Franklin's data enabled Watson and Crick to know was impossible. This made Watson and Crick so happy that, instead of sending him Franklin's results which would immediately have enabled him to correct his error (a course of action that the general public might have supposed was proper) they roared off to the Eagle pub to celebrate.[40]

And who is to say Watson and Crick were wrong to celebrate and to compete? With one bound, on publishing their own double helix, they were transformed from unsuccessful, struggling, unpopular members of Cambridge's scientific community into world stars. They won the Nobel Prize, they picked plum jobs, and they had fun forever more. Nor did they particularly harm anybody, since Pauling was to win two Nobel Prizes himself, and Wilkins shared that of Watson and Crick's (Franklin, sadly, had earlier died).

Watson and Crick had understood the cruel truth that no individual in science matters. They could hardly be blamed if the world, misunderstanding the true nature of science, chose to elevate them to 'superhuman rank'. That world includes the world of science, which is obsessed with celebrity. Researchers so crave for recognition that they create institutions which have no other role but to bestow it. The Royal Society in London is one, and its only real function (the others are camouflage) is to elect, every year, forty or so scientists into its fellowship, thus enabling them to place the magic letters FRS after their names.

The competition for FRS is desperate. One scientist I know, repeatedly disappointed in election and fuming with envy and with resentment, left Britain for two or three weeks holiday over Easter (the fellow-ships are announced at Easter) every year for nearly 20 years, until he finally made it. His first comment to me, on hearing the news, was to thank God that he could finally spend an Easter at home. Yet, so desperate is the craving of scientists for recognition, they even rank different FRSs.

The only time I visited the Royal Society, the first object I encountered in the hall was a plaque explaining the different grades of FRS. Inferior FRSs have to make do with just that, the Fellowship, but the better ones win medals, though only the very good ones win the top medals such as the Copley or the Royal or whatever. These medals are themselves carefully ranked. But some rankings are secret. One FRS told me that only after years of fellowship, following his own election to a medal, did he discover the Royal Society Dining Club, whose membership is limited to the President and ex-Presidents, the Vice Presidents and ex-Vice Presidents and the holders of the grander medals. I wonder if Dr X FRS is a member of the dining club. The letter he sent to the *Independent* on 9 October 1992 exposes the scientists' bizarre obsession with the Royal Society:

From Dr X, FRS
Sir: Thank you for publishing my letter (6 October), criticising the Government's plans for the closure of London hospitals. I am not normally a stickler for formality when it comes to forms of address. However, when the form of address is relevant to the impact of the letter, I am a stickler. The omission of the FRS after my name seriously blunts the impact of my letter. Please restore it to its rightful place.
Yours faithfully,
X   London, NW3

Since, under nationalisation, scientists work for recognition, it is important that institutions arise to meet that need, or they might stop working. It is also right there should be graduations in recognition to keep researchers researching, since some honours induce complacency. So many FRSs relax after their election that many people believe that the initials FRS represent 'Further Research Suspended'. But there is a price to be paid for the system of recognition. Many scientists' lives appear to be ruined by the misery they experience by their failure to attain certain honours. Obituaries of distinguished, but not sufficiently distinguished, scientists often include statements like 'he did not receive the recognition that many felt he deserved'. These statements speak of real unhappiness.

To avoid disappointment, some scientists lobby and scheme and plot for honours in ways that, did they but stop and think, degrade the honour. Thus friends will be quietly approached with the suggestion that they invite candidates for the FRS to deliver well-publicised lectures or research seminars at venues known to be attended by influential electors. The friends of one candidate recently organised a public dinner in his honour, to publish the details in the following day's *Times*, to nudge the Royal Society. No wonder some people believe that 'FRS' represents 'Friends of the Right Sort'. Undoubtedly, there are many non FRSs as good as those who have made it.

Not everyone suffers terrible *angst*, however, as some make it absurdly easily into the Royal Society. It is, after all, a club, and under the 'general' list anyone can get elected if they are perceived to be a good chap (which is how Mrs Thatcher, many senior professors and various captains of industry have made it).

The intense, narrow competitiveness of science is horrible. I remember a particular episode from 1980. A good friend had just completed his PhD at Oxford (they call it a DPhil there) and he had won a post-doctoral position with a famous biochemist in East Scotland. Since British universities, then as now, do not pay the costs of moving, his friends were helping, and we drove a convoy of three trucks containing his furniture. We left Oxford early on a Saturday morning, to arrive in Scotland at 8.00 p.m. It had been a long day. We were met by the famous scientist (FRS before he was 40 years old) who took us, exhausted though we were, straight to his laboratory. That was itself a bizarre place, with entire walls covered in newspaper cuttings, lecture notices and photographs of the famous scientist, who then gave my friend the rules of the laboratory:

(1) The working day starts at 9.00 a.m. Anyone who arrives after 9.30 a.m. will automatically lose a half day's holiday;

(2) Members of the laboratory may only take 2, or at most 3 weeks' holiday a year [despite the university rules which actually allowed for 5 or 6], and these may only be taken between July and September. In the event of the work being at a critical state, the laboratory member will be expected to forgo the holiday.

(3) Any time lost for reasons beyond the laboratory member's control (such as illness) will count against holiday.

(4) Coffee, lunch or tea breaks may not last for more than 10 minutes.

(5) Newspapers or non-scientific reading matter may not be read in the laboratory.
(6) Your laboratory task is to keep the balances clean.
(7) Your journal for the journal club is the *Journal of Biological Chemistry*.

These rules went on and on, to finish with:

I will note any infringement of these rules on your references.

Naturally these rules 'worked' in the sense that the laboratory produced lots of research papers, but in a fundamental sense they are a denial of the human, let alone the intellectual spirit (they made my friend's wife cry). They have no place in a university, yet more and more university researchers across the globe are writing and circulating their own. Obviously many of the best scientists need to be obsessional, but their unfortunate tendencies to authoritarianism can be fostered by the State. This has a long history. Tycho Brahe, for example, exploited his licence under King Frederik II to beat and even imprison his technicians if their astronomical measurements proved inaccurate.

Since science is no longer qualitatively different from other Western human activities, it can only be distinguished by its style: are scientists more honest than other people? Are they kinder and more gentle? Or are they competitive swine who run their labs like Prussian sergeant majors? If so, they dishonour science.

Such frenetic competitiveness cannot be blamed solely on the nationalisation of science. It derives from scientists' values, which are a mixture of the Judaeo-Christian desire to make a permanent, individual contribution to society, coupled to a Baconian/Marxist materialism. Thus scientists are obsessed with publishing first; what else do they have? Ironically, researchers aspire to make contributions to history – contributions that posterity will value – yet scientists themselves are impatient of history. Their Baconian obsession with 'progress' causes them to dismiss the history of their own subject as largely irrelevant, and an interest in the history (or economics) of science or of science policy is generally believed to be the mark of a scientific failure.

But this narrow, technocratic and philistine aspect of the modern scientist is one that nationalisation has helped foster. By isolating researchers from economic, industrial and social realities, governments have created an ignorant research hothouse, where ignorant philosophies can spread unchecked by wider ideas. Nor has the death of religion helped. Until the nineteenth century, most scientists were devout, believing they were doing God's work in revealing nature. Newton, for example, studied the Book of Revelation as assiduously as he studied gravity. Occasionally, this would lead to farce as in the case of Philip Gosse FRS who, shocked by Darwin, published *Omphalos* in 1858 to propose that fossils had been created by God to test our faith (the episode is brilliantly retold in Edmond Gosse's *Father and Son*, 1907). But in the main, a belief in God's work informed scientists' morality. Now, however, scientists have no ethical framework other than the strictly professional. Thus a breach of the professional code such as cheating, is, rightly, damned, but general acts of horribleness go unjudged, because personal ethics or morality are not seen as relevant in science.

Such moral blindness was well illustrated by the relationship between Otto Warburg and Hans Krebs. Warburg was a great scientist, the Father of Metabolism and, naturally a Nobel laureate (1931) but he was a truly horrible man. So nasty was he that Adolf Hitler, recognising a fellow spirit, funded him generously to sustain him in Berlin throughout his Chancellorship, even though Warburg was Jewish. (In 1945, with typical malice, Warburg sacked all his colleagues to cloak the evidence of his collaboration with Hitler. Thus the very colleagues who loyally stuck by him throughout the war were cruelly ejected onto the streets at the precise time that millions of Germans were literally starving, jobless and homeless.)

Somewhat earlier, if less life-threatening, Warburg had also been horrible to Krebs, his PhD student, because the moment Krebs started to develop independent ideas, Warburg expelled him from the lab (many scientists do that to bright young students of whom they are jealous; it is a mean trick, and often succeeds in derailing their careers). Krebs recovered his research, ultimately to win his own Nobel prize, and then what did he do? He wrote a hagiography of Warburg.[41] Krebs, the very model of a modern scientist, refused to judge Warburg in any terms but as a researcher – wider ethical judgements are irrelevant.

Being pathologically competitive, many scientists just love the State funding of research because it provides yet another forum for recognition. Under the State (in Western countries, except for Italy) funding is competitive. Scientists submit their grant applications to a central committee of scientists which balances one against another, to produce a supposed hierarchy of scientific excellence. The 'top' grant applications pass, the 'bottom' ones fail to get funded. Successful scientists value this because it provides an illusion of national success; funding from central bodies allows them to feel that they have proved themselves against all other scientists. But this is nonsense. One cannot truly judge a proposal in molecular biology, say, against particle physics. The process of judging or peer review is highly subjective, and certainly faulty. A plurality of funding bodies would make better decisions, even if it did deprive scientists of their lust to be regularly ranked. But scientists seem drawn to science partly because it is just an extension of school, where good pupils are rewarded with high marks, and poor ones reviled.

Non-national or specialised funding bodies, which frustrate the scientists' need for universal ranking, are now considered second rate, and it is becoming routine to rank scientists higher if they obtain money from the Research Councils or from certain national, general charities like the Wellcome Trust, than if they obtain their money from committed bodies such as industry or disease-specific charities. The Royal Society, for example, now requires that potential fellows list their grants as well as their papers on their applications, and some university faculties will only promote to senior positions the holders of certain types of grant.

This is noxious. A scientist should only be judged (if he has to be judged) on his output. Was Einstein only a second-rater because he failed, while describing special relativity, to be funded by a national, competitive, peer-reviewed body? If so, then we will have to degrade our opinions of Aristotle, Galen, Roger Bacon, Newton, Harvey, Darwin, Maxwell *et al.* Their ghosts might argue in their defence that they had no access to peer-reviewed, centrally-funded bodies in their day, but the modern scientist can have no truck with such feeble excuses: scientists are to be judged by their success in national competitions for grants, and those who fail to achieve such grants must be second rate. Any suggestion that peer-review is faulty – that it is violently subjective and sometimes even corrupt when groups of scientists conspire to support each other – is clearly the suggestion of a subversive maniac who naively believes in that out-dated nineteenth-century concept of judging scientists by what they do, not by whom they know.* Moreover, is Dr Mengele now to be revered as a great scientist, worthy of promotion to the highest academic position, be-

---

* About the only valuable service those industrious bibliometricians Irvine and Martin ever performed for science was their demonstration that the two British radioastronomy laboratories in Cambridge and Manchester used to routinely judge each other's grant applications very favourably whereas in reality one was much better than the other. Which one? The Cambridge one, of course. There was no corruption involved, just an understanding between the two laboratories that radioastronomers should hang together against all those other frightful physicists trying to grab their rightful money.

cause he enjoyed peer-reviewed grants from the Rockefeller and Loeb Foundations for his research into eugenics?

It may even be that peer review which, by definition, is a collective and pseudo-predictive process, militates against the very greatest science, which will always be individual and daring. None of Britain's recent Nobel Laureates worked in the universities (where research money is awarded by competitive peer review), they worked in industry (Black, 1988; Hounsfield, 1979), or in private research foundations (Vane, 1982; Mitchell, 1978) or in the MRC's Laboratory for Molecular Biology, where funding has historically been constitutive rather than competitive (Klug, 1982; Sanger, 1980). Moreover, neither of the two recent British discoveries which have made the greatest impact worldwide, *in vitro* fertilisation (IVF) and the gap in the ozone layer, was made by peer-reviewed researchers (indeed, the MRC repeatedly refused to fund IVF research). The current obsession with peer-reviewed grants is merely a further reflection of scientists' obsession with competition and ranking, and an absurd misuse of an already faulty mechanism of awarding research money.

But, under the influence of government, British universities are becoming less and less pleasant. Consider, for example, the contemporary mania for 'research assessment exercises'. During the 1980s, I and others pointed out that much of the money disbursed by Government on university science was misspent.[42, 43] In particular, about half of all research money, under the system of 'dual funding', was being distributed near-equally across all universities regardless of their respective scientific merits. I recommended the abolition of dual funding, and the transfer of the UGC's block research grant to the Research Councils, so that their research grants could then carry 100 per cent overheads (i.e. be doubled) to direct the money for the infrastructural support of science to the universities actually doing the science.

The fuss this caused! I knew at once it was the correct solution to the problem because the scientific establishment condemned it so vehemently, with Wellcome Trustees, Knights of the Realm, Fellows of the Royal Society and Regius Professors writing long passionate defences of dual funding. The professors, of course, hated the possible abolishment of dual funding because under my scheme the 100 per cent overheads would have gone directly to the researchers (who are often mere lecturers) who could then buy in to those university services (such as libraries) and those departments or professors or deans whose merits they appreciated. I would, in short, have helped democratise university research by empowering the researchers, unlike dual funding which empowers vice-chancellors, deans and professors. (There does have to be, of course, some central coordination and even planning of university research, but there is currently far too much, particularly as basic science is quintessentially individualistic. Science is built up out of the creativity of researchers – it does not come down from central committees.)

[Though I sometimes wonder why I bother to worry about academic democracy. Scientists are so odd and so hierarchical they positively love being subordinate. Consider Oxford and Cambridge. These two universities are unique amongst British universities in that the academics *are* the university. They are sovereign. There is no supervising senate or other body of non-academics to direct the dons. In Regent House [the academics' parliament in Cambridge] all are equal, and a lecturer's vote is as good as the vice-chancellor's. A few years ago, the professors clubbed together to propose a dramatic transfer of Regent House's powers to various committees of professors on the grounds of 'administrative efficiency' (the classic anti-democratic argument of tyrants since time immemorial) and to 'respond more quickly to government initiatives' (the bloody Government again). Naturally, one voted against this preposterous and shameless motion, but it won easily, the great bulk of scientists rushing to divest themselves of any remaining trace of individual autonomy or, God forbid, responsibility for themselves.]

The government, however, shared some of the professors' fears. Unlike them it wanted to abolish dual funding, but like them it did not want to empower the researchers. So it established regular research assessment exercises under which teams of senior professors roam the land to inundate innocent scientists with complex, otiose forms. These forms are then collated to assess research, to score departments on a scale of 0–5, just like those happy days at school for which so many academics so obviously hanker. Since, of course, research cannot be assessed so simplistically, the research assessors have engaged in exercises which would have thrilled the Schoolsmen: is a paper in *Cell* worth one and three-quarters or one and seven-eighths more than a paper in the *Journal of Cell Science*? Is a grant from industry (ugh, tainted) worth only two-fifths or three-fifths of one from a proper centralised, bureaucratic, nationalised body like the Medical Research Council?

The answers to these questions vary widely as governments shift in response to electoral whims or fashions. After the 1993 White Paper, for example, which decided that the major function of university science was to promote economic growth, the government suddenly discovered that industrial money was as good as that of the research councils, and the deans and professors who had been previously urging their subordinates to shift out of industrial research promptly engaged in an abrupt *volte face*. Industrial money had become good. PhD students, too, have suddenly become good, carrying oodles of Government infrastructural money, and we have all been told to take on as many as possible, irrespective of the actual needs of our research programmes.

None of this would really matter, except that the academics themselves actually seem to believe in these absurd exercises. Consider their enthusiasm for personal assessment. Under the terms of the 1988 Education Reform Act, academics have to be regularly assessed by their superiors (lecturers by professors, professors by deans, deans by vice-chancellors and vice chancellors by God). These assessments are otiose as they serve no formal function (promotions and other career decisions are taken elsewhere). But these assessments are also odious, as they do nothing but formalise bullying *de haut en bas* – which is why professors have taken to them like ducks to water. An academic – even a Nobel Laureate – can always be criticised for not publishing more papers in better journals or for not raising more money in grants or for not teaching well enough or for not doing more administration; and too many seniors use assessments to exercise the arrogance and sadism that in earlier, more civilised times, they were trained to curb under the dictates of good manners. Thus we have lost the pleasant collegiality of old.

Who has shown that this inexorable, incessant critical pressure on academics is the best milieu for fostering great science? The government has introduced it, and the universities have adopted it wholeheartedly, but no one has justified it (important people like politicians or professors never apparently have to justify any of their policies to the poor bloody researchers). Ironically, one cannot blame government *per se*, for some governments have been very enlightened. Here, for example, is how the great German organic chemist Emil Fischer (1852–1919) described the conditions of work at the Kaiser Wilhelm Gesellschaft in a letter he wrote in 1911 to recruit Richard Willstätter: 'You will be completely independent. No one will ever trouble you. No one will ever interfere. You may walk in the woods for a few years or, if you like, may ponder over something beautiful.'[44]

This was the policy under which giants like Carl Correns (1864–1933, the plant geneticist who rediscovered Mendel's Law), Richard Goldschmidt (1878–1958, another leading geneticist), Fritz Haber (1868–1934 who won the Nobel Prize in 1918 for the synthesis of ammonia),

Otto Hahn (1879–1968, 1944 Nobel Laureate for discovering the fission of heavy nuclei), Michael Polanyi (1891–1976, the polymath), as well as Otto Warburg, Albert Einstein, Lise Meitner and many others flourished.

But, on examination, one discovers that that successful German Government policy was only an imitation of the successful conditions of work at the private British laboratories of the Royal Institution and of the then privately endowed universities. And it was the British government that helped destroy those institutions as civilised centres of research. When the universities were small, elite bodies, their faculty was highly motivated, highly hard working and highly gifted; and when students personally paid fees, the free market ensured a high quality performance from the teachers.

But the Government has now so expanded the universities in Britain (436 000 students in 1993, a third of school-leavers) that thousands of academics have been hired who are, in truth, mediocre. Thus has the culture of the secondary school been introduced to degrade once-lovely organisations to ensure that mediocre staff, who lack self-motivation, perform acceptably.

We have devoted much space to Rousseau, but his thoughts have spread, like corrupt tentacles, into the heart of Western culture. We can trace a descent of disciples that leads to the Nazis and to the Communists. Rousseau's distrust of science inspired such proto-fascist tracts as Spengler's *Decline of the West*, published in 1918 and 1922, which glorified a German romantic nationalism that inspired Hitler. Rousseau's worship of the State fed the philosophies of a stream of influential thinkers including Hegel ('The state walks with God') and Marx, it fed the philosophies of tyrants like Robespierre ('Rousseau is the teacher of mankind'), Hitler of course ('The Age of Reason is Dead'), Lenin, Mao and Mussolini ('Everything within the State, nothing outside the State, nothing against the State') and it fed the fantasies of poets like Southey ('There can be no health, no soundness in the state till government shall regard the moral improvement of the people as its first great duty')[45] and Arnold ('We [English] are in danger of drifting into anarchy. We have not the notion, so familiar on the Continent and to antiquity, of the State – the nation in its collective and corporate character, and trusted with stringent powers for the general advantage, and controlling individual wills in the name of an interest wider than that of individuals').[46]

Matthew Arnold's view of the State

Let us stay with Matthew Arnold for a few paragraphs, as he was highly influential. A major poet, an Oxford professor, an Inspector of Schools, the son of Thomas Arnold and a friend of Tolstoy, he wrote a series of widely read books and reports including *Schools and Universities on the Continent* (1868) and *Culture and Anarchy* (1869). Arnold's thesis was simple if unoriginal: the Germans are coming. He believed that Bismarck's *Polytekniks* and *Hochschulen* would thrust Germany into scientific, technological and industrial dominance. He did not believe that Britain's free market could provide these because, you see, schools and universities have to be planned on a national scale, and the free market is quite incapable of planning anything on a national scale, oh no.

This is, of course, rubbish. Scale is no argument for public funding. Huge organisations, if based on private need, do not require the State to create or support them. General Motors or Unilever or Shell seem to struggle on without government control. Indeed, the multinational corporation is now bigger than many of the states in which it trades. Adam Smith's 'invisible hand' shows how the collective but free decisions of millions of individuals can create huge institutions if they are appropriate.

But Arnold did leave one enduring legacy; he justified *dirigisme* under democracy. Nineteenth-century Britons believed in *laissez faire* because they feared dictatorship. But Arnold claimed that, under democracy, the State had become tamed, and that the British could thus safely enjoy all the lovely advantages of central State planning and central State control without losing their freedom. Indeed, he claimed they actually gained in freedom because, under democratic *dirigisme*, all the major functions of society such as education, health and science became democratically accountable.

The reality, of course, is the exact opposite. Once activities are controlled by the State, the individual is steamrollered, however democratic the State might try to be. The individual only has power under the free market; it is only when the potential customer approaches a school or a university or a shop with money in his pocket that the institution arches to satisfy his needs. Under the free market the individual is a customer, and the customer is King. One could call it a market democracy. Under *dirigisme* the individual is a supplicant and he has no power. To achieve anything, he has to lobby, to act politically or act collectively.

THE CONCLUSION

If this book has a message it is this: relax. Economic, technical and scientific growths are free lunches. Under *laissez faire* they just emerge, like grass after the rain, through the efforts of individual entrepreneurs and philanthropists. Once the State has initiated the rule of law and sensible commercial legislation, the goodies will flow – and *laissez faire* is morally superior to *dirigisme* as it maximises the freedoms and responsibilities of the individual.

What are the most important determinants of technological development? Are they the so-called supply factors of new science and new discoveries as thinkers like Musson and Robinson,[47] Parker,[48] Rostow[49] or Rosenberg[50] believe, or are they the so-called demand factors of the search for greater profits as Marx, Schmookler[51] and Zevin[52] believe? It does not matter, as under *laissez faire* both needs are met. Is technical development largely a matter of giant leaps forward as Schumpeter[53] or Kuznets[54] believed, or does it emerge from the cumulative, anonymous accretion of thousands of small improvements as Usher,[55] Gilfillan,[56] Landes[57] Fishlow,[58] Hollander,[59] Hall[60] and Enos[61] believe? It does not matter, *laissez faire* meets both needs.

Is technology dependent on science as Bernal[62] believed, or is the opposite true as Gillispie[63] and Henderson[64] claim? Is technical innovation largely directed at saving labour or at other costs? How important is the rate of diffusion of new technology, and what determines that? Was Fogel[65] right to deny any overwhelming importance to any particular invention? (he maintained that, had railways never been invented, the economic history of North America would hardly have changed). These and other questions are fascinating, but for the science policymaker or for the politician they are irrelevant. *Laissez faire* meets all needs.

This is not a fashionable position, but it is an old and respectable one. Leibniz, the great physicist, popularised philosophical optimism in his *Theodicée* (1710), but in *Candide* (1758) Voltaire set up Dr Pangloss as a naive optimist ('this is the best of all possible worlds') only to destroy him. Thereafter, the Enlightenment adopted pessimism – a pessimism that could be partly alleviated by *dirigisme* – for its own, if only because it justified the State-funding of intellectuals' salaries; and we are all children of the Enlightenment. But Dr Pangloss was right, this is the best of all possible worlds – or rather, it would be if only politicians left it alone. Under *laissez faire,* the combined, individual judgements and ethical impulses of the entire populace can be liberated, to provide the optimal mix of supply, demand and kindness.

Through that mix, technological development and scientific growth will emerge, at no general cost, to fructify the external and internal lives of every one of us.

I believe that Francis Bacon, who categorised some intellectual errors as Idols of the Tribe or Idols of the Market Place, would agree with me were he to return to judge today. His linear model and his *dirigisme* have failed and he could recognise that. The Market Place does not worship false Idols, it makes empirically correct judgements. It is the government funding of science that is an Idol of the Tribe.

POSTSCRIPT ON ROUSSEAU AND MORALITY

Rousseau's doctrines are immoral. By blaming society for individuals' faults, Rousseau absolves the individual for responsibility for his actions. Instead, Rousseau transfers moral responsibility to the State. But no State can be a moral body. Only individuals with their senses of good and of bad and of guilt and of shame can be moral. A collective body cannot, intrinsically, be moral, since it is not sentient. It cannot *feel* responsibility, it cannot lie awake at night, worrying. As Thurlow said: 'Did you ever expect a corporation to have a conscience, when it has no soul to be damned, and no body to be kicked?'

The failure, both moral and therefore practical, of the State was never illustrated better than by the fate of Rousseau's own children. Rousseau is the father of modern education who, in *Emile*, maintained that education was so important that it had to be entrusted to the State, because only that body, incorporating the General Will, was moral enough to mould young minds. Rousseau acted out his beliefs in his own life. This is what he writes in his autobiography: 'In handing my children over for the State to educate. . . . I was acting as a citizen and a father, and looked upon myself as a member of Plato's republic. . . . I have often blessed Heaven for having thus safeguarded them. . . . My third child, therefore, was taken to the Foundling Hospital like the others, and the next two were disposed of in the same way, for I had five in all. This arrangement seemed so good and sensible and right.'

Rousseau, therefore, sent all five of his children (by Therese Lavaseur, his life-long companion) to the State's Foundling Hospital within 24 hours of their births. What happened to them there? In 1746, the *Mercure de France* performed an actuarial analysis. The death rate was so high that only a third of infants survived to the age of one, only 14 per cent to the age of seven, and only 5 per cent survived to adulthood [see L. G. Crocker's *Jean-Jacques Rousseau: The Quest, 1712–1758*, (New York, 1974, p. 263)]. The majority of those were unqualified for any professions other than that of begging, crime or prostitution. These statistics were not secret. In 1758, Rousseau himself noted that, with 5082 children, the Hospital was inundated with children it could not look after. Thus we see how the inspirer of the modern view of the state continued to sacrifice his children's lives to that view, rather than acknowledge its failure. As I argue in this book, the national-isation of science has been almost as damaging to science and to scientists, even if its proponents cannot see it.

Interestingly, Marx also abandoned one of his children, the illegit-imate one he fathered on his wife's maid.

POSTSCRIPT ON THE PUBLIC UNDERSTANDING OF SCIENCE

It is a truth universally accepted by scientists that the world is sadly ignorant of science. Thus surveys are regularly published to show that less than 50 per cent of the population knows that the earth rotates around the sun or that dolphins are mammals, not fish. These surveys are then brandished by scientific activists as arguments to improve the public understanding of science.

Why? It could be argued that the law, say, is as fundamental to the health of society as is science, but do lawyers agitate to educate the masses in the finer points of tort or chancery? Of course not. As long as the fundamental concepts of justice are widely understood – and they are – then the details can be left to the professionals. So it is with science. As long as the fundamental principles of the scientific *method* are widely disseminated – and they are, outside a handful of isolated aboriginals – then the detailed facts of science cease to hold a central cultural importance.

The scientists agitate for the public understanding of science, of course, because they want more money (which is why the lawyers are not concerned with the public understanding of law – they are already rich). Thus the first organisation in Britain specifically created to educate the public in science, the British Association for the Advancement of Science, was created as part of the campaign by Charles Babbage to persuade the Government to fund research. But, like euphemisms, or-ganisations that claim to campaign for the public understanding of science inexorably fail, and new ones are continuously being invented. We now have a Professor for the Public Understanding of Science, Imperial College's John Durrant who tours the country trying to educate the masses; while Conservative activists like Simon Mitton actually campaign for the research councils to set aside 1 per cent of their precious research funds for the public education of science.

These initiatives are misconceived. The cultural battle over the scientific method was won during the Age of Reason, and the subsequent elucidation of science facts is just that, the elucidation of facts. C. P. Snow was wrong, it does not matter if the general public remains ignorant of the Second Law of Thermodynamics. Obviously some facts are useful, but just as one can drive a car without understanding internal combustion (though the understanding can help one mend a breakdown), so one can similarly thrive in all contemporary milieux as a consumer without needing a producer's expertise (though expert knowledge is always nice).

POSTSCRIPT ON EDUCATION AND THE FREE MARKET

Education, like science, is generally believed to be the responsibility of Government, and almost all countries today enjoy 'free' universal, compulsory schooling provided by the state ('free' means, of course, paid out of taxes or of inflation rather than by fees). Yet the history of education parallels to a remarkable degree the history of science funding, and that history disproves the suggestion that government need be involved.

One of the first countries to boast a fully literate population was Britain, which by 1891 enjoyed 100 per cent literacy amongst its school-leavers (R. S. Schofield, 'Dimensions of Illiteracy', *Explorations in Economic History* (1973) vol. 10).* Yet, up to 1891, education in Britain was largely *laissez faire.*

---

* Scholars today, working retrospectively, determine literacy by cataloguing the percentage of the population that can sign their names – rather than make a mark – on their marriage registry. That may now seem a very trivial achievement, but in the nineteenth century children were taught to read first, and only to write as an advanced skill. The ability to sign one's name, therefore, was not trivial, and reflected competence at reading and as well as some competence at writing. This has been confirmed by Furet and Ozouf. During the nineteenth century, the French army subjected its recruits to very comprehensive tests of reading and writing, and Furet and Ozouf, working on those French army tests, and comparing them with the signing of marriage certificates, showed a very high degree of correlation between the two: during the nineteenth century, the ability to sign one's name reflected a competent literacy (Furet and Ozouf's study *Lire et Ecrire* (Paris, 1977) is described in Cressy's *Literacy and the Social Order* (Cambridge, 1980)].

Historically, the British Government distrusted the education of the masses – it might breed revolution. Henry VIII's *Act for the Advancement of True Religion* (1543) forbad 'women, nor artificiers, 'prentices, journeymen, servingmen of the degrees of yeoman or under, husbandmen nor labourers' from reading the Bible in English. For centuries, the wealthy classes feared that education might fuel sedition. In his *Advice to a Son* (1656) Francis Osborne stated: 'A too universally dilated learning hath been found upon Trial in all Ages no fast friend to Policy or Religion.' In an age when rioting was all too common, the view was deep-seated. In his *Free Enquiry into the Nature and Origin of Evil* (1757), Soame Jenyns credited the ignorance of the poor as 'a cordial administered by the gracious hand of providence'. Historically, therefore, education in Britain was neglected by the Government, and entrusted to the private sector. Naturally, therefore, it thrived.

By 1725 literacy was 63 per cent in men and 38 per cent in women, achieved entirely by private education (Schofield, ibid.). That is a very important statistic, because Anderson has shown that it only requires 40 per cent literacy to transform an economy out of feudalism into industrialisation [C. A. Anderson and M. J. Bowman *Education and Economic Development* (London, 1966)]. Even today, over 50 per cent of the jobs in a highly developed, post-industrial economy need only the degree of literacy that is provided by education at primary school.

The upper and middle classes educated their own children under *laissez faire*, of course, and the free market had no difficulty in meeting that need. The 1851 Education Census counted 44 836 schools in England and Wales, all private. But it was the education of the poor that, perhaps, surprises us now. The poor were desperate to educate their children. The first British education survey to be commissioned by Parliament, in 1816, the *Report on the Education of the Lower Orders in the Metropolis*, found that: 'Your Committee are happy in being able to state, that in all returns, and in all the other information laid before them, there is the most unquestionable evidence that the anxiety of the poor for education continues not only unabated, but daily increasing; that it extends to all parts of the country.'

But the education of the poor was, unsurprisingly, restricted by their lack of money. Yet private philanthropy, channelled through the churches, was stunningly generous. The Society for Promoting Christian Knowledge (SPCK) was established in 1699 to support the charity schools, and by 1719 Londoners alone were donating over £10 000 a year to the capital's schools – a huge sum (C. Rose, *History Today*, March 1990).

As Britain entered the nineteenth century, the churches set out to educate the whole nation. To build schools systematically the Church of England established the National Society in 1811, the non-conformist churches established the British and Foreign Society in 1814 and the Roman Catholics eventually established the Catholic Poor School Committee. By 1870, the churches were running 6382 primary schools. The success of private education in Britain was stunning. The 1861 Royal (Newcastle) Commission on Popular Education reported that 2 535 462 children were at school, out of a total school-age population of 2 655 767. The average length of attendance was 5.7 years. In 1869 Horace Mann, the statistician in charge of the education section of the Registrar General, wrote 'the number of children under tuition at the present time . . . is not far short of the highest proportion practicable' [quoted in E. G. West, *Education and the Industrial Revolution*, (London, 1975)].

But in 1870, the government created its first state schools. Why then, of all times? The answer is illuminating. Prussia inaugurated universal state education in 1806, making it compulsory in 1826, and France soon followed. Thereafter, the House of Commons resonated with claims that unless Britain copied, it would be left behind economically. From 1807 onwards, the Commons repeatedly passed bills to introduce universal state education, and each bill was vetoed by the Lords. The bishops sat in the House of Lords, and they always persuaded their peers to block any state competition with the church schools.

But by 1870, church – and private-school attendance had reached 99 per cent, so the bishops allowed through the 1870 Education Act to create the first 'board' or state schools. The bishops felt they had nothing to fear from them, and how right they were. By 1882 there were 11 620 church primary schools (and their numbers kept on climbing to meet the needs of the expanding population) but only 3200 board schools and they went half empty. By 1875 there were 1 193 000 excess places for children in schools – a dreadful waste of taxpayers' money – but parents preferred the church schools which were seen as offering better education in a less rough environment.

So, in 1891, Parliament passed the Free Schooling Elementary Education Act. This killed the church schools. Before 1891, all schools (church and state) charged fees of around 10*s* a year, a lot of money then [C. Birchenough, *Elementary Education* (Cambridge, 1938)]. That did not pay for all the costs of education, which were supplemented with government grants and by voluntary contributions. The voluntary contributions to church schools were huge, some £10 million between

1870 and 1891, the rough equivalent of £450 million today. But in 1891, the Government paid the board schools an extra 10*s* a year so that they could abolish fees. The extra 10*s* a year came out of increased rates and taxes, which hit the poorer classes hard. Increasingly, parents could no longer afford to pay fees to the church schools in the face of the raised taxes and the 'free' board schools, so they transferred their children. And by 1902, the church schools were compelled to apply for nationalisation. In the face of the unequal competition their rolls were falling.

The abolition of fees could not be justified in educational terms. Poor children had always received a free education in the church schools, but the bulk of parents paid. It improved education. On its foundation, the National Society had not initially charged fees – they were introduced in 1828 when teachers noted that they improved attendance. The Clerical Superintendent of the Society told the 1834 Parliamentary Committee that parents, when they paid for education, valued it more. This increased valuation was transmitted to the children, who worked harder and longer. For the same reasons, the British and Foreign Society had imposed fees in 1816, the congregational schools did so in 1848, and the Wesleyans followed in 1854.

Nor had the private education of Britain harmed its economic growth. Between 1830 and 1890, when France and Germany had universal state education and Britain did not, the three economies marched together. Britain was 1.5 times richer than Prussia or France in 1830, and it was 1.5 times richer in 1890, too.

Per capita *GNP 1830–90 (in 1960 US dollars and prices)*

Country	1830	1840	1850	1860	1870	1880	1890
Britain	346	394	458	558	628	680	785
Germany	245	267	308	354	426	443	537
France	264	302	333	365	437	464	515

SOURCE: P. Bairoch, 'Europe's Gross National Product: 1800–1975', *Journal of European Economic History* (1976) vol. 5, pp. 273–340.

Curiously, Britain's relative economic decline seems to have accelerated after the introduction of universal compulsory free education in 1891 (and of the government funding of science shortly thereafter).

Why is it so hard to persuade the general public that private education, linked to philanthropy, works? One reason is that the protagonists of state education have vested interests to defend. Thus it was the same protagonists for the state funding of science who also argued for the state funding of education. It was, therefore, Lyon Playfair (see Chapter 8) who argued in his *Industrial Instruction on the Continent* (London, 1852) that: '[the] general character of all secondary education in Germany is tending towards giving instruction in the wants of the nineteenth century, and not stopping at that considered sufficient in the thirteenth, as many of our classical schools do'. It was Matthew Arnold (see this chapter) who claimed in his *Schools and Universities on the Continent* (London, 1868) that: 'Britain is trying to meet the calls of a modern epoch . . . with a working class not educated at all, a middle class educated on the second plane, and the idea of science absent from the whole course and design of our education.'

And, today, it is the same advocates of the state funding of science who also argue for state education. Here is an extract from Corelli Barnett's *Audit of War* (London, 1986): 'In the 1860s there were still fewer children in grant-aided elementary schools of efficient standard than there were children receiving no form of education whatsoever.'

Yet these statements are deeply misleading, and they flow from an appalling deception of the House of Commons. The occasion was the introduction on 17 February 1870 of the monstrous Elementary Education Bill by William Edward Forster, the Vice-President in charge of the Education Department during Gladstone's first administration [see vol. 199 of *Hansard* (third series) cols 438–66].

For over a century, fashionable radicals have hailed the speech as a parliamentary *tour de force*. In his *Charles Dilke* (London, 1958), for example, Roy Jenkins, now Lord Jenkins, described it as 'a speech of high distinction'. To the reader coming fresh to it, however, the speech reads most oddly, starting with such statements as:

I need not detain the House with any reasons for bringing an Education Bill forward. (col. 439)

I am not going to detail the House with any long statements of facts, and still less do I intend to weary you with statistics. (col. 440)

I am not going to deal with facts at any length tonight. (col. 442)

So why did Forster introduce his Bill? His reasons will leave the reader incredulous:

I may also refer to the Report which will be speedily in the hands of Hon. Members ... concerning the educational condition of four great towns – Liverpool, Manchester, Leeds and Birmingham. That Report, I have reason to believe, will abundantly confirm my statement that we cannot depend upon the unaided and uninspected schools [i.e. private schools]. I have not myself had the opportunity of reading that Report, for I was so anxious that it should be laid before the House with the least possible delay that I did not keep it in my hands for a single hour. (col. 442)

The most important education Act in our history, then, was tabled by a Minister on the basis of a report which neither he nor the House had actually read. At least Forster had not yet descended into deceit. But then he carried on:

But I have had the privilege of corresponding with the two gentlemen who conducted the inquiries; and therefore I believe I can give pretty correctly the figures with regard, at all events, to Liverpool, and they are figures which may well alarm us. It is calculated that in Liverpool the number of children between five and thirteen who ought to receive an elementary education is 80,000; but, as far as we can ascertain, 20,000 of them attend no school whatever, while at least another 20,000 attend schools where they get an education not worth having. In Manchester – that is, in the borough of Manchester, not including Salford, there are about 65,000 children who might be at school, and of this number about 16,000 go to no school at all. (cols 441–2)

Here, then, is the origin of the belief that during the 1860s half or so of all children went uneducated. But Lord Montagu, the opposition spokesman, promptly explained to the House how those very figures proved that Liverpool enjoyed 100 per cent school attendance. The numbers of children between five and <u>thirteen</u> in Liverpool might have been 80 000, but thirteen was not the upper age of elementary education; it was eleven. On Forster's own figures, there were 60 000 children in Liverpool between five and eleven, and 60 000 children attended elementary schools. Attendance was 100 per cent. Moreover, the expression 'schools where they get an education not worth having' was, as Montagu explained, the only way Forster could bring himself to describe the private schools; which was odd because Forster himself had attended one and, moreover, private schools have consistently

TABLE 12.1   *Sources of funds for higher education research and development, OECD countries, 1991, as percentages of total*

Country	Government	Private
Austria (1989)	97.4	2.26
Belgium (1990)	83.1	16.9
Canada	**77.4**	**22.6**
Denmark	89.6	10.4
Finland (1990)	89.0	11.0
France (1990)	92.9	7.1
Italy	96.7	3.3
Japan	**49.5**	**50.5**
Holland (1990)	95.6	4.4
Portugal (1990)	94.6	5.4
Spain (1990)	89.3	10.7
Sweden (1989)	85.3	14.7
Switzerland (1989)	85.9	14.1
UK	**72.0**	**28.0**
USA	85.0	15.0

SOURCE: OECD data published in the *Times Higher Education Supplement*, 18 November 1994, p. 14.

provided a better education than those of the state.

This is not the place to discuss the sad inability of most education-alists to understand how the free market will fund education for the most obvious of reasons (parents want to educate their children, and they also want to see the children of the poor enjoy every advantage). Let us not even try to understand why most commentators 'see' Forster's claim but fail to 'see' Montagu's rebuttal of them. Let us ignore the obvious point that the British had opted for universal mass education voluntarily, and that the Prussian and French states had then imposed theirs on their peasants compulsorily to catch up. Let us not even try to understand why Britain then felt it had to copy the very people who were copying Britain. Let us simply acknowledge that public affairs are rarely decided by cool, rational analysis. In the words of David Hume: 'All human affairs are entirely governed by *opinion*.'

In science and education, sadly, opinion is almost totalitarian, and disturbers of it are treated, in Solzhenitzyn's phrase, to 'the censorship of fashion'. In this book, we have obeyed another of Solzhenitsyn's aphorisms, to 'pursue the truth regardless of the consequences'.

Notes

Notes to Chapter 1: Francis Bacon and Adam Smith

1. G. Porter, *Knowledge Itself is Power*, 1988 Dimbleby Lecture (London: BBC Publications, 1988).
2. J. B. Bury, *The Idea of Progress* (London, 1920).
3. F. Bacon, *Cogitata et Visa* (1607).
4. Ibid.
5. F. Bacon, *The Advancement of Learning* (1605) Chapter 1, Book 2.
6. Ibid., Chapter 8.
7. Ibid.
8. Ibid., Chapter 10, Book 2.
9. Ibid., Chapter 14.
10. Ibid., Chapter 13, Book 2.
11. Ibid., Chapter 9.
12. R. Gregory, *Discovery* (London, 1920).
13. A. Smith, *Wealth of Nations* (1776), Book 2, Chapter ii.
14. A. Smith, *Lectures on Jurisprudence*, ed. R. L. Meek, D. D. Raphael and P. G. Stein (Oxford: Clarendon Press, 1978).
15. Smith, *Wealth of Nations*, Book 1, Chapter i.
16. Ibid.
17. Ibid.
18. Smith, *Lectures on Jurisprudence*.
19. A. Smith, *Wealth of Nations*, ed. A. Skinner (London: Penguin, 1776), p. 272.
20. B. R. Mitchell and P. Deane, *Abstract of British Historical Statistics* (Cambridge, 1962).
21. Smith, *Wealth of Nations*.
22. Ibid., Concluding paragraphs, Book 1.
23. Ibid., Chapter 1, Section C, paragraph 1.

Notes to Chapter 2: Research and Development in Antiquity

1. O. Neugebauer, *The Exact Sciences in Antiquity* (Providence, Rhode Island: Brown University Press, 1957).
2. P. Bairoch, 'International Industrialisation Levels from 1750 to 1980', *Journal of European Economic History*, **11** (1982) pp. 269–333.
3. Arthur S. Hunt and J. G. Smyli, *The Tebtunis Papyri*, vol. III–I, no. 703 (Oxford, 1933) lines 222–34.
4. *Iliad* V, trans. E. V. Rieu (Harmondsworth, Middx: Penguin, 1950) (I 11.1258b).
5. Aristotle, *Politics*, 1.77.1258b.

6. Ibid., 111.4.1277a–1278a.
7. Xenophon, *Oeconomicas*, IV, 2–3.
8. Alexis Fragment 30, trans. T. F. Higham, *The Oxford Book of Greek Verse in Translation* (Oxford: Clarendon Press, 1953).
9. Herodotus, III. 60.
10. Philo of Byzantium, *Mechanics*, IV 3.5, trans. M. Cohen and I. E. Drabkin, *A Source Book in Greek Science* (Cambridge, Mass.: Harvard University Press, 1958).
11. Aristotle, *Airs,Waters and Places*, 16; cf 24.
12. Aristotle, *Metaphysics*, I 1.981b.
13. Euclid, *Athenaeus*, IV 162 EF.
14. Petronius, *Satyricon*, 51.
15. Suetonius, *Vespasian*, 18.
16. *De Rebus Bellicus*, preface 4, ll.2–3, trans. E. A. Thompson, *A Roman Reformer and Inventor* (Oxford, 1952).
17. Ammianus Marcellinus, XXX 9.4, Aurelius Victor *de Caesaribus*, 45.
18. R. A. Buchanan, *Technology and Social Progress* (Oxford, 1965), p. 2.

Notes to Chapter 3: The So-called Dark Ages

1. Tacitus, *Annals*, XII 45.
2. Zosimus, V 21.2.
3. Procopius, *Gothic Wars*, VIII, ll.27–31.
4. D. W. Anthony and D. R. Brown, *Antiquity*, **65** (1991) pp. 22–38.
5. Lynn White Jr, *Medieval Technology and Social Change* (Oxford, 1962).
6. Des Noettes, *L'attelage et le Cheval de Selle a Travers les Ages* (Paris, 1931).
7. A. Steensberg, *Farms and Mills in Denmark during Two Thousand Years* (Copenhagen, 1952).
8. Vitruvuis, *De Architectura*, X 5.
9. M. Bloch, 'Avenement et Conquetes du Moulin a Eau', *Annales d'Histoire Economique et Sociale*, **VII** (1936) pp. 515–22.
10. Pliny, *Natural History*, lib. XXVIII, cap. 51.
11. Galen, *Works*, ed. Kuhn, X 569 (1828).
12. F. W. Gibbs, 'The History of the Manufacture of Soap', *Annals of Science*, **iv** (1939) pp. 169–90.
13. P. Boissonade, *Life and Work in Medieval Europe* (London, 1927).
14. T. R. Holmes, *Classical Quarterly*, **iii** (1909) pp. 26–39.
15. H. H. Brindley, 'Early Pictures of Lateen Sails', *Mariner's Mirror*, **xii** (1926) pp. 9–22.
16. *Decretalis Gregorii*, IX, Lib. III, tit. 30, cap. 23.
17. Ambroise, *L'Estoire de la Guerre Sainte*, ll.3227–9, trans. M. J. Hubert (New York, 1941).
18. Lynn White Jr, *Medieval Technology and Social Change* (Oxford, 1962).
19. Lynn White Jr, 'Technology and Invention in the Middle Ages', *Speculum*, **XV** (1940) pp. 141–59.

Notes to Chapter 4: The Commercial Revolution

1. A. Maddison, 'Comparative Productivity Levels in the Developed Countries', *Banca Nazionalo del Lavoro Quarterly Review*, **20** (1967) pp. 295–315.
2. P. Laslett and K. Oosterrea, 'Long-term Trends in Bastardy in England', *Population Studies*, **27** (1973) pp. 255–86.
3. R. D. Lee and R. S. Schofield, in *Economic History of Britain Since 1700*, eds R. Floud and D. McCloskey (Cambridge Universtiy Press, 1981).
4. E. A. Wrigley, 'Family Limitation in Pre-industrial England', *Economic History Review*, 2nd Ser. **19** (1966) pp. 82–109.
5. C. de la Ronciene, in *A History of Private Life*, ed. G. Duby (Harvard, 1988) p. 215.
6. P. Mathias, *The First Industrial Nation* (London, 1983).
7. P. H. S. Hartley and H. R. Aldridge, *Johannes de Mirfield of St Bartholomew's Smithfield*, p. 133, quoted in T. Jones, *Chaucer's Knight* (London, 1980).
8. Sombart, *Enterprise and Secular Change*, ed. F. C. Lane and Riemersa (Illinois: R. D. Irwin, 1953).
9. 'Thomas Wimbledon's Sermon', ed. Nancy M. Owen, *Medieval Studies*, **XXVIII** (1966) p. 178.
10. *Rotuli Parliamentorum*, II, p. 147.
11. Froissart *Cronycle*, trans. Berners, IV 306–8.
12. Walsingham, *Historia Anglicana*, II, p. 127.
13. Gower, *Mirour de l'Omme*, 24061–72, trans. T. Jones.
14. Walsingham, *Historia Anglicana*, I, p. 371.

Notes to Chapter 5: The Agricultural Revolution

1. E. L. Jones *et al.* in *Economic History of Britain Since 1700*, ed. R. Floud and D. McClosky (Cambridge: Cambridge University Press, 1981).
2. J. H. Smith, *The Gordon's Mill Farming Club 1758–1764* (Edinburgh, 1962).
3. E. L. Jones, in *The Economic History of Britain 1700–1960*, ed. R. Floud and D. McCloskey (Cambridge: Cambridge University Press, 1981).
4. Ibid.
5. K. Hudson, *Patriotism with Profit; British Agricultural Societies in the Eighteenth and Nineteenth Centuries* (London, 1972).
6. G. Huecke, in *The Economic History of Britain 1700–1960*, ed. R. Floud and D. McCloskey (Cambridge: Cambridge University Press, 1981).
7. E. L. Jones, in *The Economic History of Britain 1700–1960*, ed. R. Floud and D. McCloskey.
8. C. Pray *et al.*, 'Private Research and Public Benefit: The Private Seed Industry for Sorghum and Pearl Millet in India', *Research Policy*, **20** (1991) pp. 315–24.
9. Ibid.

Notes to Chapter 6: The Industrial Revolution

1. Angus Maddison, *Phases of Capitalist Development* (Oxford, 1982).
2. Ibid.
3. P. Deane and W. A. Cole, *British Economic Growth 1688–1959. Trends and Structure* (Cambridge, 1967).
4. *Encyclopaedia Britannica* 15th edn (1974).
5. D. S. L. Cardwell, *Steampower in the Eighteenth Century* (London, 1963).
6. J. D. Bernal, *Science in History* (London, 1954).
7. D. L. S. Cardwell, *Technology, Science & History* (London, 1972).
8. Ibid.
9. Ibid.
10. D. S. L. Cardwell, *From Watt to Causius* (New York: Cornell University Press, 1976).
11. I. B. Cohen, *Science: Servant of Man* (Boston: Little Brown & Co., 1948).
12. J. G. Crowther, *Men of Science* (New York: W. W. Norton & Co., 1936).
13. Simon Schama, *Citizens* (London and New York, 1989).
14. A. P. Wadsworth and J. de Lin Mann, *The English Cotton Trade and Industrial Lancashire 1600–1780* (Manchester, 1931).
15. R. M. Macleod, 'The Support of Victorian Science', *Minerva.* **9** (1971) pp. 197–230.
16. Ibid.
17. G. Pickering, *Creative Malady* (London, 1974).

Notes to Chapter 7: Economic History since 1870

1. Quoted in R. H. Heindel, *The American Impact on Great Britain 1898–1914* (New York, 1968).
2. L. Playfair, *Journal of the Society of Arts,* **15** (7 June 1867) p. 477.
3. Select Committee on Scientific Instruction, 1868.
4. David Landes, *The Unbound Prometheus* (Cambridge: Cambridge University Press, 1969).
5. Correlli Barnett, *Audit of War* (London: Macmillan, 1986).
6. D. H. Aldcroft, 'The Entrepreneur and the British Economy 1870–1914', *Economic History Reviews, 2nd series,* **17** (1964) pp. 113–34.
7. T. Williams, *The Triumph of Invention* (London: MacDonald, 1987).
8. M. Abramovitz, 'Resource and Output Trends in the United States since 1870', *American Economic Review, Papers and Proceedings,* (May 1956) pp. 5–23.
9. K. W. Kendrick, *Productivity Trends: Capital and Labor* (Ann Arbor, Mich.: National Bureau of Economic Research) Occasional Paper no. 53 1956.
10. R. Solow, 'Technical Change and the Aggregate Production Function', *Review of Economics and Statistics,* **39** (1957) pp. 312–20.
11. F. Denison, *Trends in American Economic Growth 1929–1982* (Washington, DC: The Brookings Institution, 1985).
12. D. W. Jorgenson, F. M. Gollop and B. M. Fraumeni, *Productivity and*

US Economic Growth (Cambridge, Mass.: Harvard University Press, 1987).
13. M. Abramovitz, *Thinking About Growth* (Cambridge, 1989).
14. J. Frame, 'Modelling National Technological Capacity with Patent Indicators', *Scientometrics*, **22** (1991) pp. 327–39.
15. *OECD International Statistical Year for Research and Development: A Study of Resources Devoted to R&D in OECD Member Countries in 1963/64* (Paris: OECD, 1968).
16. D. E. H. Edgerton, 'British Industrial R&D, 1900–1970', in *Science and Industrial Technologies*, ed. J. Davies and P. Mathias (Oxford: Blackwell, 1993).
17. D. Landes, *The Unbound Prometheus* (Cambridge, 1969) p. 521.
18. *Industrial Training and Technical Innovation: A Comparative and Historical Study*, ed. Howard Gospel (Routledge, 1991).
19. R. Beason and D. Weinstein, *Growth, Economics of Scale and Targetting in Japan (1955–1990)* (Harvard Institute of Economic Research, 1994) Discussion Paper 1644.
20. *The Economist,* 26 February 1994, p. 91.
21. A. Maddison, *Phases of Capitalist Development* (Oxford, 1982).
22. Quoted in P. Curwen, *Understanding the UK Economy* (London, 1992), p. 146.
23. T. Burham and G. Hoskins, *Iron and Steel in Britain, 1870–1913* (London, 1943).
24. C. Barnett, *Audit of War* (London: Macmillan, 1986).
25. J. Parry, *The Rise and Fall of Liberal Government in Victorian Britain* (London: Yale University Press, 1993) p. 29.
26. Leo Amery, *Geographical Journal*, **23** (1904), p. 441.
27. P. Noble, 'The Paradox of Statistics on Science and its Funding', *The Biochemist*, **11** (1989) pp. 13–15.
28. H. W. Richardson, 'Chemicals', in *The Development of British Industry and Foreign Competition 1875–1914*, ed. D. H. Aldcroft (1968) pp. 274–306.
29. I. C. R. Byatt, 'Electrical Products', in *The Development of British Industry and Foreign Competition, 1875–1914*, ed. D. H. Aldcroft (1968) p. 238–73.
30. P. Lindert and K. Trace, *Essays on a Mature Economy*, ed. D. McCloskey (Princeton, 1971).
31. M. Gowing, *An Old and Intimate Relationship*, Spencer Lecture Oxford 1982. See also M. M. Gowing, *Contemporary Record*, **1** (1987) p. 15.
32. P. Bairoch, 'International Industrialization Levels from 1750 to 1980', *Journal of European Economic History*, **11** (1982) pp. 269–333.
33. P. Bairoch, 'Europe's Gross National Product: 1800–1975', *Journal of European Economic History*, **5** (1976) pp. 273–340.
34. P. Panayi, 'Germans in 19th Century Britain', *History Today* (January 1993) pp. 48–53.
35. Quoted in Panayi, ibid.
36. W. K. Hancock and M. M. Gowing, *The British War Economy* (London: HMSO, 1949) pp. 102–3.
37. D. H. Calhoun, *The American Civil Engineer: Origins and Conflict* (Cambridge, Mass., 1960) pp. 24–7.

38. Rudyard Kipling, *Something of Myself* (Macmillan, 1937) pp. 90–1.
39. M. Wiener, *English Culture and the Decline of the Industrial Spirit* (Cambridge, 1981).
40. T. Liesner, *One Hundred Years of Economic Statistics* (Economist Publications, 1989) p. 53.
41. D. Kleppner, 'The Mismeasure of Science', *The Sciences* (May/June 1991) pp. 18–21.
42. K. L. Sokoloff, 'Inventive Activity in Early Industrial America: Evidence from Patent Records, 1790–1846', *Journal of Economic History*, **48** (1988) pp. 813–50.
43. J. B. Rae, 'Engineers are People', *Technology and Culture*, **16** (3) (June 1975); reproduced in *The Engineer in America*, ed. T. S. Reynolds (Chicago: University of Chicago Press, 1991) pp. 27–42.
44. B. Sinclair, 'At the Turn of a Screw: William Sellers, The Franklin Institute, and a Standard American Thread', *Technology and Culture*, **10** (1) (January 1969); reproduced in *The Engineer in America*, ed. T. S. Reynolds, pp. 151–68.
45. W. Sellers, *Report to the Franklin Institute of the State of Pennsylvania, for the Promotion of the Mechanic Arts, Relative to the Metric System of Weights and Measures* (Philadelphia, 1876) p. 5, quoted in B. Sinclair, ibid.
46. *OECD Science and Technology Indicators No. 3* (Paris: OECD, 1989).
47. K. Arrow, 'The Economic Implications of Learning by Doing', *Review of Economic Studies*, **29** (1962) pp. 155–73.
48. *Reports of the DFG 1/92* (Bonn, 1992).
49. M. L. Dertouzos, R. K. Lester and R. M. Solow, *Made in America: Regaining the Productive Edge* (MIT Press, 1989).
50. T. Braun, W. Glanzel and A. Schubert, 'One More Version of the Facts and Figures on Publication Output and Relative Citation Output of 107 Countries 1978–1980', *Scientometrics* **11** (1987) pp. 9–15.

Notes to Chapter 8: Science Policies of the Twentieth Century

1. Paul Kennedy, *Rise and Fall of the Great Powers* (Random House, USA, 1988) p. 241.
2. Ibid., p. 598.
3. Nathan Reingold, 'Science in the Civil War', *Isis*, **49** (1958) pp. 307–18.
4. Quoted by Reingold, ibid.
5. Ibid.
6. Ibid.
7. Quoted by Daniel J. Kevles in George Ellery Hale, 'The First World War, and the Advancement of Science in America', *Isis*, **59** (1968) pp. 427–37.
8. Ibid.
9. Ibid.
10. Ibid.
11. Ibid.

12. Quoted in *Science Policy Study Background Report No. 1* (Washington, DC: US Government Printing Office, 1986).
13. Ibid.
14. Quoted by A. Hunter Dupree in *Science in the Federal Government* (Harvard University Press, 1957).
15. *Science*, **81** (1935) p. 46.
16. Quoted by H. Dupree, *Science in the Federal Government* (Harvard University Press, 1957).
17. H. A. Wallace, *Science*, **79** (1934) p. 2.
18. Ibid.
19. Ibid.
20. V. Bush, *Science – The Endless Frontier* (Washington, 1945) p. 12.
21. Ibid., p. 11.
22. Ibid., p. 10.
23. Ibid., p. 19.
24. H. Dupree, *Science in the Federal Government* (Harvard University Press, 1957).
25. Quoted in *Science Policy Study Background Report No. 1* (Washington, DC: US Government Printing Office, 1986).
26. Ibid.
27. *The Politics of American Science: 1939 to the Present*, ed. James L. Perrick, Carroll W. Pursell, Morgan B. Sherwood and Donald C. Swain (MIT Press, 1972).
28. Quoted by Robert D. Lapidus in 'Sputnik and its Repercussions: a Historical Catalyst', *Aerospace Historian*, **17** (1970) p. 89.
29. Quoted in *Science Policy Study Background Report No. 1* (Washington, DC: US Government Printing Office, 1986).
30. Ibid.
31. C. Macilwain, 'Clinton Budget Proposes Science Funding Freeze', *Nature*, **367** (1994) p. 497.
32. *Science Policy Study Background Report No. 1* (Washington, DC: US Government Printing Office, 1986).
33. Ibid.
34. Ibid.
35. *Technology Retrospect and Critical Events in Science* (Washington, DC: National Science Foundation, 1968).
36. R. Macleod, 'The Support of Victorian Science', Minerva **9** (1971) pp. 197–230.
37. *Nature*, **VI** (1872) p. 97.
38. *Nature*, **VIII** (1873) p. 21.
39. *Nature*, **344** (1990) p. 225.
40. *Nature*, **II** (12 May 1870) p. 25.
41. W. Crookes,'The Endowment of Scientific Research', *Quarterly Journal of Science*, **VI**, (October 1876) p. 485.
42. A. R. Wallace 'Government Aid to Science', *Nature*, **1** (13 January 1870) pp. 279–88.
43. *Minutes of the Government Fund Committee* (11 January 1877).
44. R. Macleod, 'The Support of Victorian Science', *Minerva*, **9** (1971) pp. 197–230.

45. G. Gore, 'On the Present Position of Science in Relation to the British Government', *Transactions, Social Science Association* (1873) p. 360.
46. *Nature*, **XIII** (23 December 1875), p. 141.
47. A. C. Doyle, 'A Physiologist's Wife', in *The Conan Doyle Stories* (London, 1929).
48. *Knowledge* (1 January 1886) pp. 93–5.
49. R. Proctor, *Wages and Wants of Science Workers* (London, 1876).
50. *Minutes of the Government Fund Committee* (19 February 1880).
51. M. Arnold, *Schools and Universities on the Continent* (London, 1868).
52. H. Perkin, *Origins of Modern English Society* (London, 1969) p. 381.
53. Quoted by John Campbell, *F. E. Smith* (Jonathan Cape, 1983) p. 203.
54. H. J. H. Gosden, *The Friendly Societies in England, 1815–1875* (Manchester: Manchester University Press, 1961) pp. 4–5, and P. H. J. H. Gosden, *Self-Help* (London, 1973), p. 91.
55. Lord Beveridge, *Voluntary Action* (London, 1948) p. 328.
56. Landsborough Thomson, 'Origin of the British Legislative Provision for Medical Research', *Journal of Social Policy*, **2** (1973) p. 43.
57. Linda Bryder, *Tuberculosis and the MRC in Historical Perspectives on the Role of the MRC*, ed. J. Austober and L. Bryder (Oxford: Oxford University Press, 1989).
58. Ibid.
59. Quoted in John Campbell, *F. E. Smith* (Jonathan Cape, 1983), p. 202.
60. *The Economist* (1 October 1994) p. 10 of 'A Survey of The Global Economy'.
61. T. R. Elliot, in *Obituary Notices of Fellows of the Royal Society*, **1** (1935) pp. 153–63.
62. J. Austober, 'Walter Morley Fletcher', in *Historical Perspectives on the Role of the MRC*, ed. J. Austober and L. Bryder (Oxford: Oxford University Press, 1989).
63. Ibid.
64. Ibid.
65. Ibid.
66. Ibid.
67. M. Weatherall, and H. Kamminga, *Dynamic Science: Biochemistry in Cambridge 1898–1949* (Cambridge Wellcome Unit Publication, 1992).
68. R. O. Berdahl, *British Universities and the State* (Cambridge: Cambridge University Press, 1959).
69. Quoted by P. Hinton and E. Vincent, *The University of Birmingham* (Birmingham, 1947) pp. 61, 62.
70. *Minutes of the Universities Deputation to the Treasury* (23 November 1918).
71. G. D. H. Cole, *History of the Labour Party from 1914* (London, 1948).
72. Quoted in Berdahl, op. cit.
73. See C. H. Shian *Paying the Piper: The Development of the University Grants Committee 1919–1946* (Lewes: Falmer Press, 1986) Table 2.
74. Quoted in Berdahl, op. cit.
75. M. Walker and T. Meade (ed.), *Science, Medicine and Cultural Imperialism* (Macmillan, 1991).

76. G. W. Craig, *The Germans* (Penguin, 1991).
77. Ibid.
78. Ibid.
79. Ibid.
80. J. Pelikan, *The Idea of the University: A Re-examination* (Yale University Press, 1992) pp. 15–16.
81. Quoted in Berdahl, op. cit.
82. H. J. Laski, *Reflections on the Revolution of our Time* (New York: Viking, 1943) p. 163.
83. *Universities Quarterly*, **2** (1948) p. 215.
84. *Committee on Scientific Manpower* (1946) p. 21.
85. Quoted in Berdahl, op. cit.
86. *Universities Quarterly*, **1** (1947) p. 383.
87. *Oxford University Gazette* (8 October 1948).
88. Quoted in *Technical Education* (1956) Comd 9703, p. 3.
89. Quoted in C. Barnett, *Audit of War* (London, 1986).
90. *Report on Higher Education* (London: HMSO, 1963).
91. C. P. Snow, *The Two Cultures* (Cambridge, 1993).
92. *Report on Higher Education* (HMSO, 1963) paragraph 129.
93. *The Scotsman*, 26 November 1957.
94. H. Wilson, D. Jay and H. Gaitskell, *We Accuse* (London, 1956).
95. Quoted in M. Foot, *Aneurin Bevan* (London, 1973) vol. II, pp. 646–7.
96. Quoted in Ben Pimlott, *Harold Wilson* (London, 1992), p. 236.
97. R. H. S Crossman, *Labour in the Affluent Society*, Fabian Tract 325 (London, 1960).
98. T. Balogh, *Unequal Partners,* vol. II: *Historical Episodes* (Oxford, 1963) pp. 280–1.
99. Ben Pimlott, op. cit., p. 274.
100. Richard Crossman, see diary entries for 15 February 1963 (p. 978) and 19 February 1963 (pp. 1005–6).
101. Quoted in G. Wersky, *The Visible College* (London, 1978), p. 320.
102. H. Wilson, *Purpose in Politics: Selected Speeches* (London, 1964) pp. 18–27.
103. Barbara Castle's diary, quoted in Ben Pimlott, op. cit. p. 468.
104. Quoted in D. Butler and M. Pinto-Duschinsky, *General Election of 1970* (London, 1970) pp. 123–4.
105. Quoted in G. Macdonogh, *Prussia: The Perversion of an Idea* (London, 1994).
106. A speech of Harold Wilson's quoted in V. Kitzinger, *The Second Try: Labour and the EEC* (Oxford, 1968) p. 85.
107. Quoted in J. Laughland, *The Death of Politics: France under Mitterand* (London 1994).

Notes to Chapter 9: The Economics of Research: Why the Linear Model Fails

1. *Nature*, **356** (1992) p. 273.

2. *Understanding the UK Economy*, ed. P. Curwen (Macmillan, 1992), p. 265.

3. Ibid.

4. A. Arrow, 'The Economic Implications of Learning by Doing', *Review of Economic Studies*, **29** (1962) pp. 155–73.

5. D. C. Mowery and N. Rosenberg, 'Technical Change in the Commercial Aircraft Industry 1925–1975', *Technological Forecasting and Social Change*, **20** (1981) pp. 347–58.

6. Ian McIntyre, *Dogfight* (Praeger, 1993).

7. D. Edgerton, *England and the Aeroplane* (London, 1992).

8. Edwin Mansfield, 'Academic Research and Industrial Innovation', *Research Policy*, **20** (1991) pp. 1–12.

9. *Indicators of International Trends in Technological Innovation*, Report to the National Science Foundation (Washington, DC, 1976).

10. J. Langrish *et al.*, *Wealth from Knowledge: A Study of Innovation in Industry* (Macmillan and Wiley, 1972).

11. Quoted by N. Rosenberg, *Inside the Black Box* (Cambridge: Cambridge University Press, 1982) p. 155.

12. Quoted from E. Braun and S. MacDonald, *Revolution in Miniature* (Cambridge: Cambridge University Press, 1978) pp. 126–7.

13. W. Ashworth, *The History of the British Coal Industry 1946–1982: The Nationalized Years* (Oxford, 1986) vol. 5.

14. Tim Jackson, *Turning Japanese: The Fight for Industrial Control of the New Europe* (Harper Collins, 1993).

15. *The Economist* (26 February 1994).

16. Quoted in Anthony Sampson, *The Essential Anatomy of Britain* (London: Hodder & Stoughton, 1992).

17. *Financial Times* (9 February 1993).

18. E. Leigh, *Spectator* (5 June 1993) pp. 16–19.

19. M. Porter, *Competitive Advantage of Nations* (London, 1990).

20. A. Seldon, *Capitalism* (Oxford, 1990).

21. R. Nelson, 'The Simple Economics of Basic Scientific Research', *Journal of Political Economy*, **67** (1959) pp. 297–306.

22. K. Arrow, 'Economic Welfare and the Allocation of Resources for Invention', in *The Rate and Direction of Inventive Activity* (Princeton University Press, 1962).

23. E. Mansfield, 'Basic Research and Productivity Increase in Manufacturing', *American Economic Review*, **70** (1980) pp. 863–73.

24. S. Griliches, 'Productivity, R&D and Basic Research at the Firm Level in the 1970s', *American Economic Review*, **76** (1986) pp. 141–54.

25. M. Leiberman, and D. Montgomery, 'First Mover Advantages', *Strategic Management Journal*, (1988) pp. 41–58.

26. M. Lynn, *The Billion Dollar Battle: Merck v Glaxo* (London, 1991).

27. Ibid., p. 192.

28. H. Odagiri and N. Murakimi, 'Private and Quasi-social Rates of Return on Pharmaceutical R&D in Japan', *Research Policy*, **21** (1992) pp. 335–45.

29. M. Lynn, *The Billion Dollar Battle: Merck v Glaxo* (London, 1991) p. 178.

30. N. Rosenberg, 'Why Do Firms Do Basic Research with their Own Money?', *Research Policy*, **19** (1990) pp. 165–74.
31. P. Romer, 'Endogenous Technological Change', *Journal of Political Economy*, **98** (1990) pp. S71–S102.
32. E. Mansfield, 'Academic Research and Industrial Innovation', *Research Policy*, **20** (1991) pp. 1–12.
33. R. Pool, 'The Social Return of Academic Research', *Nature*, **352** (1991) p. 661.

Notes to Chapter 10: The Real Economic of Research

1. J. D. Frame, 'Modelling National Technological Capacity with Patent Indicators', *Scientometrics*, **22** (1991) pp. 327–39.
2. *OECD Historic Statistics 1960–1985* (Paris: OECD, 1987).
3. *OECD Science and Technology Indicators No. 3* (Paris: OECD, 1989).
4. *Nature*, **364** (1993) p. 182. See also *Nature*, **376** (1995) p. 207.
5. *OECD Science and Technology Indicators No. 3* (Paris: OECD, 1989).
6. A. Galal, L. Jones, P. Tandon and I. Vogelsgang, *Welfare Consequences of Selling Public Enterprises* (World Bank, 1992); the conclusions of the report were summarised in the *Economist* (13 June 1992) pp. 83–4.
7. *New Scientist*, **136** (1992) p. 8.
8. E. Mansfield, 'Academic Research Underlying Industrial Innovations: Sources and Characteristics', presented at the American Economic Association, January 1993.
9. B. Martin, J. Irvine, F. Narin, C. Sterritt and K. A. Stevens, *Science and Public Policy* **17**, 1990, pp. 14–26.
10. L. T. C. Rolt, *Victorian Engineering* (Allen Lane, Penguin Press, 1970).
11. T. Braun, W. Glanzel and A. Schubert, 'One More Version of the Facts and Figures on Publication Output and Relative Citation Output of 107 Countries 1978–1980', *Scientometrics* **11** (1987) pp. 9–15.
12. Nigel Lawson, *The View from No. 11* (London: Bantam Press, 1992) p. 720.
13. *The Future of the Science Base* (London: The Royal Society, 1992).
14. See report in the *Times Higher Education Supplement*, (22 November 1991) p. 1.
15. S. Fortesque, *Science Policy in the Soviet Union* (Harvard, 1990).
16. L. R. Graham, *Science and the Soviet Social Order* (Routledge, 1990).
17. J. M. Buchanan and G. Tulloch, *The Calculus of Consent* (Ann Arbor, Mich., 1962).
18. R. Macleod, 'The Support of Victorian Science', *Minerva*, **9** (1971) pp. 197–230.
19. E. Mansfield, 'Basic Research and Productivity Increase in Manufacturing', *American Economic Review*, **70** (1980) pp. 863–73.
20. S. Griliches, 'Productivity, R&D and Basic Research at the Firm Level in the 1970s', *American Economic Review*, **76** (1986) pp. 141–54.
21. *Nature* **371** (1994) p. 188.
22. D. Spurgeon, 'Canadian Drug Companies Fund Universities', *Nature*, **363** (1993) p. 484.

23. A. Guillaume, *Islam* (London: Penguin, 1954).
24. Andrew Sinclair, *The Need to Give* (London: Sinclair Stevenson, 1990).
25. D. Fraser, *The Evolution of the British Welfare State* (London: Macmillan, 1973) pp. 114–5.
26. Quoted in *Nature*, **364** (1993) p. 742.
27. *Science and Engineering Indicators 1993*, Washington, quoted in *Nature*, **367** (1994) p. 583.
28. *The Economist* (4 June 1994) pp. 80–1.

Notes to Chapter 11: The So-called Decline of British and American Science

1. B. R. Martin, J. Irvine and R. Turner, 'The Writing on the Wall for British Science', *New Scientist* **104** (1984) pp. 225–9.
2. J. Irvine, B. Martin, T. Peacock, and R. Turner, *Nature* **316** (1985) pp. 587–90.
3. J. Irvine and B. R. Martin, 'Is Britain Spending Enough on Science?', *Nature* **323** (1986) pp. 591–4.
4. B. R. Martin, Irvine, F. Narin and C. Sterritt, 'The Continuing Decline of British Science', *Nature*, **330** (1987) pp. 123–6.
5. Editorial, 'Bringing Research Back to Life', *Nature*, **344** (1990) p. 275.
6. *Nature*, **323** (1986) pp. 681–4.
7. D. da S. Price, *Little Science, Big Science* (New York: Columbia University Press, 1963).
8. J. Irvine *et al., Nature*, **316** (1985) pp. 587–90.
9. M. Bradbury, *Unsent Letters* (London, 1988).
10. *University Statistics 1986–87* (UGC, 1988).
11. *Bursting at the Seams* (London: AUT, 1993). See also H. Atkinson, P. Rogers and R. Bond, *Research in the United Kingdom, France and West Germany: A Comparison* (London: SERC, 1990).
12. *Science Policy Research Unit Annual Report* 1989–1990 (University of Sussex, 1990).
13. *New Scientist*, **104** 1984 pp. 25–9.
14. *Nature*, **323** (1986) pp. 591–4.
15. J. Bray, *Science for the Citizen* (London: Labour Party, 1989).
16. *The Case for Increased Investment in our Universities* (London: Association of University Teachers, 1989).
17. *New Scientist* (20 April 1991) p. 15.
18. Ibid.
19. *Policy Study no. 1* (London: Science and Engineering Policy Studies Unit, 1987).
20. H. Atkinson, P. Rogers and R. Bond, *Research in the United Kingdom, France and West Germany: A Comparison* (London: SERC, 1990).
21. J. Langrish, *et al., Wealth from Knowledge: A Study of Innovation in Industry* (London and New York, 1972).
22. B. R. Martin, J. Irvine, F. Narin, C. Steritt and K. A. Stevens, *Science and Public Policy*, **17** (1990) pp. 14–26.

23. T. P. Stossel and S. C. Stossel, 'Declining American Representation in Leading Clinical Research Journals', *New England Journal of Medicine,* **322** (1990) pp. 739–42.
24. Ibid.
25. J. B. Wyngaarden, *New England Journal of Medicine,* **301** (1979) pp. 1254–9.
26. B. Healey, *New England Journal of Medicine,* **319** (1988) 1058–64.
27. C. Anderson, 'Science Defeats all Odds in US Budget', *Nature,* **347** (1990) p. 697.
28. D. Kleppner, 'The Mismeasure of Science', *The Sciences,* (May/June 1991) pp. 18–21.
29. Quoted in David S. Greenberg, 'Congress, Can You Spare a Grant?', *Lancet,* **337** (1991) pp. 542–3.
30. Klepper, op. cit.
31. *Developments in Biotechnology* (London: ACOST, HMSO, 1990).
32. *The Competitive Strength of US Industrial Science and Technology: Strategic Issues* (Washington, DC: National Science Board, 1992).
33. This argument of M. R. Darby was summarised in the *Economist* of 26 December 1992 (**325** (7791), p. 87).
34. *British Science: Benchmarks for the Year 2000* (Oxford: SBS, 1990).
35. D. Noble, Speech to the British Association for the Advancement of Science, Southampton, August 1992.
36. *Nature,* **361** (1993) p. 584.
37. T. Kealey, 'The Conditions of British Science', *Nature,* **344** (1990), p. 806; T. Kealey, 'The Growth of British Science', *Nature,* **350** (1991) p. 370.
38. J. Merris, 'NSF Falls Short on Shortage', *Nature,* **356** (1992) p. 553.
39. Ibid.
40. *Federally Funded Research: Decision for a Decade* (Washington, DC: Office of Technology Assessment, US Congress, 1991).
41. F. A. Hayek, *The Road to Serfdom* (Boston and London: Routledge & Kegan Paul, 1994) p. 39.

Notes to Chapter 12: Dr Pangloss was Right

1. Adam Smith, *The Wealth of Nations* (1776) vol. II, Book iv, Chapter 7, part iii.
2. Adam Smith, *The Wealth of Nations* (1776).
3. Adam Smith, *Theory of Moral Sentiments* (1759), ed. D. D. Raphael and A. L. Macfie (Oxford, 1976) p. 77.
4. Philip Sidney, *A Defence of Poesie* (1591), ed. J. A. Van Dorsten (Oxford, 1966).
5. Ibid.
6. J. Golding, *Visions of the Modern* (London, 1994).
7. Philip Sidney, *A Defence of Poesie* (1591), ed. J. A. Van Dorsten (Oxford, 1966).
8. Ibid.

9. Jean-Jacques Rousseau, *Confessions* (1782), Everyman edn (London: Everyman, 1904).
10. Quoted on p. xxiii of *The Social Contract and The Discourses* (Everyman edn, London, 1973).
11. J. J. Rousseau, *Discourse on the Arts and Sciences* (1750), Everyman edn (London: Everyman, 1973).
12. Ibid.
13. Ibid.
14. Ibid.
15. D. C. Beardslee and D. D. O'Dowd, 'The College-student Image of the Scientist', in *The Sociology of Science*, ed. B. Barber and W. Hirsch (New York, 1962).
16. P. B. Shelley to Elizabeth Hitchener, in F. L. Jones (ed.), *Letters of Percy Byshe Shelley* (Oxford, 1964) vol. I, pp. 116–7.
17. J. J. Gaskin (ed.), *The Viceregal Speeches and Addresses, Lectures and Poems of the Late Earl of Carlisle* (Dublin, 1866) p. cxxxiv.
18. E. J. Hobsbaum, *Nations and Nationalism Since 1780* (Cambridge, 1990).
19. Joseph Stalin, *Marxism and the National and Colonial Question* (London, 1936).
20. K. Marx, *Manifesto of the Communist Party* (1848). In *The Portable Karl Marx*, ed. Eugene Kamenka (London, 1983) (Kamenka denies that Engels was a significant author).
21. J. Irvine, B. R. Martin and P. A. Isaurd, *Investing in the Future* (Aldershot and Brookfield: Edward Elgar).
22. B. R. Martin, J. Irvine, F. Narin, C. Sterritt, and K. A. Stevens, *Science and Public Policy*, **17** (1990) pp. 14–26.
23. Editorial: 'Confusing Tale of Two Research Budgets', *Nature*, **367** (1994) p. 495.
24. Reported in the *Daily Telegraph* of 15 February 1994.
25. A. Yamamoto, 'Japanese Universities Feel the Chill', *Nature*, **339** (1989) pp. 575–6.
26. D. Swinbanks, 'Survey Pans University Lab', *Nature*, **350** (1991) p. 544.
27. D. Swinbanks, 'Bribery Case Underlining Japanese Bureacracy', *Nature*, **367** (1994) p. 210.
28. Quoted in *The Times*, 18 January 1994, p. 16.
29. A. M. Johnson *et al.*, *Nature*, **360** (1992) pp. 410–12.
30. National Opinion Research Centre, *Sex in America: A Definitive Survey* (Little, Brown & Co, 1994).
31. *Nature*, **371** (1994) p. 367.
32. A. Abbot, 'Italy Draws Order out of Chaos', *Nature*, **367** (1994) p. 6.
33. G. F. Bignami, 'La Dolce Vita?', *Nature*, **366** (1993) p. 642.
34. A. Abbot, 'Italy to Quit European Biology Lab in Protest over Staffing Policy', *Nature*, **367** (1994) pp. 205–6.
35. *Bursting at the Seams* (London: Association of University Teachers, 1993).
36. J. J. Rousseau, *Confessions*.
37. Ibid.
38. J. M. Buchanan and G. Tulloch, *The Calculus of Consent* (Ann Arbor, Mich., 1962).

39. Adam Smith, *Theory of Moral Sentiments* (1759), ed. D. D. Raphael and A. L. Macfie (Oxford, 1976).
40. J. Watson, *The Double Helix* (London, 1970).
41. H. Krebs, *Otto Warburg* (Oxford, 1981).
42. T. Kealey, *Spectator* (28 March 1987) pp. 15–6.
43. T. Kealey, *Science Fiction and the True Way to Save British Science* (London: Centre for Policy Studies Policy Study, 1989) No. 15.
44. Quoted in H. Krebs, *Otto Warburg* (Oxford, 1981) p. 7.
45. R. Southey, *Sir Thomas More* (London, 1829) vol. 2, pp. 424–5.
46. M. Arnold, *Culture and Anarchy* (London, 1869).
47. A. Musson and E. Robinson, *Science and Technology in the Industrial Revolution* (Manchester: Manchester University Press, 1969).
48. W. Parker, 'Economic Development in Historical Perspective', *Economic Development and Cultural Change*, **10** (1961) pp. 1–7.
49. W. Rostow, *The Stages of Economic Growth* (Cambridge, 1960).
50. N. Rosenberg, 'Science, Invention and Economic Growth', *Economic Journal*, **84** (1974) pp. 90–108.
51. J. Schmookler, *Invention and Economic Growth* (Harvard, 1966).
52. R. Zevin (1971), 'The Growth of Cotton Textile Production after 1815', in *The Reinterpretation of American Economic History*, ed. R. Fogel and S. Engerman (New York, 1971).
53. J. Schumpeter, *Capitalism, Socialism and Democracy* (New York, 1942).
54. S. Kuznets (1971), *Economic Growth of Nations* (Cambridge, Mass., 1971).
55. A. Usher, *A History of Mechanical Inventions* (Harvard, 1954).
56. S. Gilfillan, *The Sociology of Invention*, (Chicago: Follett, 1935).
57. D. Landes, *The Unbound Prometheus* (Cambridge, 1969).
58. A. Fishbow, 'Productivity and Technological Change in the Railroad Sector 1840–1910', in *Output, Employment and Productivity in the US after 1800* (New York: National Bureau of Economic Research, 1966).
59. S. Hollander, *The Sources of Increased Efficiency* (Cambridge, Mass.: MIT, 1965).
60. A. Hall, *The Historical Relations of Science and Technology* (London, 1963).
61. J. Enos, *Petroleum, Progress and Profits* (Cambridge, Mass.: MIT, 1962).
62. J. Bernal, *Science in History* (London and Cambridge, Mass., 1971).
63. C. Gillispie, 'The Natural History of Industry', *Isis*, **48** (1957) pp. 398–407.
64. W. Henderson, *Britain and Industrial Europe 1750–1870* (Liverpool, 1954).
65. R. Fogel, *Railroads and American Economic Growth* (Baltimore, Md., 1964).

Index

369

Made in the USA
Columbia, SC
05 October 2023

23986386R00217